T0223105

Lecture Notes in Computer Science 1651

Edited by G. Goos, J. Hartmanis and J. van Leeuwen

Springer
Berlin
Heidelberg
New York
Barcelona
Hong Kong
London
Milan
Paris
Singapore
Tokyo

Ralf Hartmut Güting   Dimitris Papadias
Fred Lochovsky (Eds.)

Advances in
Spatial Databases

6th International Symposium, SSD '99
Hong Kong, China, July 20-23, 1999
Proceedings

 Springer

Series Editors

Gerhard Goos, Karlsruhe University, Germany
Juris Hartmanis, Cornell University, NY, USA
Jan van Leeuwen, Utrecht University, The Netherlands

Volume Editors

Ralf Hartmut Güting
Praktische Informatik IV, Fernuniversität Hagen
D-58084 Hagen, Germany
E-mail: gueting@fernuni-hagen.de

Dimitris Papadias
Fred Lochovsky
Department of Computer Science
Hong Kong University of Science and Technology
Clear Water Bay, Hong Kong, China
E-mail: {dimitris,fred}@cs.ust.hk

Cataloging-in-Publication data applied for

Die Deutsche Bibliothek - CIP-Einheitsaufnahme

Advances in spatial databases : 6th international symposium ;
proceedings / SSD '99, Hong Kong, China, July 20 - 23, 1999.
Hartmut Güting ... (ed.). - Berlin ; Heidelberg ; New York ; Barcelona
; Hong Kong ; London ; Milan ; Paris ; Singapore ; Tokyo : Springer,
1999
  (Lecture notes in computer science ; Vol. 1651)
  ISBN 3-540-66247-2

CR Subject Classification (1998): H.2-3, H.5, I.4, I.5, J.2, F.2

ISSN 0302-9743
ISBN 3-540-66247-2 Springer-Verlag Berlin Heidelberg New York

© Springer-Verlag Berlin Heidelberg 1999
Printed in Germany

Typesetting: Camera-ready by author
SPIN: 10703951   06/3142 – 5 4 3 2 1 0   Printed on acid-free paper

Preface

The Sixth International Symposium on Spatial Databases (SSD'99) was held in Hong Kong, China, July 20-23, 1999. This is a "ten-year anniversary" edition. The series of conferences started in Santa Barbara in 1989 and continued bi-annually in Zürich (1991), Singapore (1993), Portland, Maine (1995), and Berlin (1997). We are very pleased that on this occasion Oliver Günther, one of the initiators of the conference in 1989, agreed to give an "anniversary talk" in which he presented his view of SSD in the past as well as in the future ten years.

SSD is well established as the premier international conference devoted to the handling of spatial data in databases. The number of submissions has been stable during the last years; in 1999 there were 55 research submissions, which is exactly the same number as for the last conference in Berlin. Out of these, the program committee accepted 17 excellent papers for presentation. In addition to the "anniversary talk", the technical program contained two keynote presentations (Christos Papadimitriou, Timos Sellis), four tutorials (Markus Schneider, Leila De Floriani and Enrico Puppo, Jayant Sharma, and Mike Freeston), and one panel.

The papers included in these proceedings reflect some of the current trends in spatial databases. Classical topics such as spatial indexing or spatial join continue to be studied, as well as interesting new directions such as including generalization/scale in indexing or treating multiway instead of binary joins. Some topics such as generalization, spatial data mining, or treatment of uncertainty have been around for a while and continue to be in the focus of interest. A strong newcomer in this volume is the topic of spatio-temporal databases, especially with a "moving objects" flavor. One reason for increased interest in this area is the existence of the related European Research Networks; this was also addressed in the keynote speech by Timos Sellis.

Furthermore, for the first time SSD offered a track with presentations of "industrial and visionary applications" papers. We hope that this will further promote the exchange of ideas between practitioners and researchers, especially by pointing out research problems arising in practice, and giving researchers feedback on how suitable their solutions are within a system context. These contributions were selected by a small industrial subcommittee of the program committee.

Numerous people helped to make this conference a success. First, the members of the program committee as well as the external referees did an excellent job in providing careful reviews, and in conducting engaged conflict resolution discussions via email before the program committee meeting. The program committee members who attended the meeting in Hagen deserve special credit. We also thank Agnès Voisard for her help in promoting this event, and for a lot of good advice based on her experience with the last conference. Thanks to Dave Abel for organizing a careful selection process for the industrial contributions.

Thanks to Anne Jahn at FernUniversität Hagen for a lot of technical help in organizing the reviewing process and in producing the proceedings. We also thank several graduate students at HKUST, including Nikos Mamoulis, Panos Kalnis, and Marios Mantzourogiannis for assisting in all kinds of small problems (WWW design, bookings, etc.). Then, several offices of the University (HKUST) were quite helpful, in particular the Finance Office (registrations), the Purchasing Office (hotel accommodations) and the Educational Technology Center (poster preparation).

We are grateful to our industrial sponsors, Environmental Systems Research Institute (ESRI) and Oracle Corporation, for their generous support. Furthermore, we would like to thank IEEE HK Chapter, the Sino Software Research Institute, and the Department of Computer Science, HKUST, for providing us with initial funding. Co-operation with ACM-SIGMOD and the U.S. National Center for Geographic Information and Analysis is also gratefully acknowledged.

May 1999   Ralf Hartmut Güting
                                                        Dimitris Papadias
                                                        Fred Lochovsky

Conference Organization

General Chairs

Dimitris Papadias, Hong Kong University of Science and Technology, China
Fred Lochovsky, Hong Kong University of Science and Technology, China

Program Chair

Ralf Hartmut Güting, FernUniversität Hagen, Germany

Publicity

Agnès Voisard, Freie Universität Berlin, Germany

Panel Chair

Max Egenhofer, University of Maine, USA

Industrial Sessions Chair

Dave Abel, CSIRO, Canberra, Australia

Program Committee

Dave Abel, CSIRO, Australia
Walid Aref, Panasonic Technologies, USA
Peter Baumann, FORWISS, Germany
Claudia Bauzer Medeiros, UNICAMP, Brazil
Stefan Berchtold, stb software technologie beratung gmbh, Germany
Elisa Bertino, University of Milano, Italy
Wesley W. Chu, UCLA, USA
Max Egenhofer, University of Maine, USA
Leila De Floriani, University of Genova, Italy
Andrew Frank, Technical University Vienna, Austria
Michael Freeston, University of Aberdeen, Scotland & UCSB, USA
Volker Gaede, Oracle, Germany
Stéphane Grumbach, INRIA, France
Oliver Günther, Humboldt Univ. Berlin, Germany
Jiawei Han, Simon Fraser University, Canada
John Herring, Oracle, USA
Klaus Hinrichs, University of Münster, Germany

Martin Kersten, CWI, The Netherlands
Hans-Peter Kriegel, University of Munich, Germany
Ming-Ling Lo, IBM T.J. Watson Research Center, USA
Yannis Manolopoulos, Arist. Univ. Thessaloniki, Greece
Scott Morehouse, ESRI, USA
Beng Chin Ooi, National University, Singapore
Bernd-Uwe Pagel, SAP, Germany
Dimitris Papadias, Hong Kong Univ. (HKUST), China
Jan Paredaens, University of Antwerp, Belgium
Jignesh M. Patel, NCR, USA
Norman Paton, Univ. of Manchester, United Kingdom
Peter Revesz, University of Nebraska-Lincoln, USA
Philippe Rigaux, CNAM Paris, France
Elke A. Rundensteiner, Worcester Polytechnic Institute, USA
Hans-Jörg Schek, ETH Zürich, Switzerland
Michel Scholl, CNAM Paris & INRIA, France
Timos Sellis, NTUA, Greece
Shashi Shekhar, University of Minnesota, USA
Terence Smith, UCSB, USA
Per Svensson, FOA & KTH, Sweden
Yannis Theodoridis, Computer Technology Institute, Greece
Agnès Voisard, FU Berlin, Germany
Kyu-Young Whang, KAIST, Korea
Ouri Wolfson, University of Illinois, USA
Michael Worboys, Keele University, United Kingdom

Industrial Sessions Subcommittee

Dave Abel, CSIRO, Australia
Oliver Günther, Humboldt Univ. Berlin, Germany
John Herring, Oracle, USA
Scott Morehouse, ESRI, USA

Additional Referees

Christian Böhm	Nikos Mamoulis	Jörg Sander
Stéphane Bressan	Apostolos Papadopoulos	Markus Schneider
Li Chen	Sanghyun Park	Thomas Seidl
Lyman Do	Dragan Petrovic	Rod Son
Martin Ester	Enrico Puppo	John Stell
Kathleen Hornsby	Wilko Quak	Kian-Lee Tan
H.-Arno Jacobsen	Prabhu Ram	Anthony Tung
David Johnson	Lukas Relly	Bo Xu
Krzysztof Koperski	Uwe Röhm	Weining Zhang
Paola Magillo		

Table of Contents

Spatial Data Mining and Classification

Spatial Join

Uncertainty and Geologic Hypermaps

Industrial and Visionary Applications Track

Invited Talks

Topological Queries

Christos H. Papadimitriou*

University of California, Berkeley, USA
christos@cs.berkeley.edu

A spatial database contains information about several two-dimensional regions. Of all queries one may ask about such a database, the *topological* queries are those that are invariant under continuous invertible deformations and reflections of the database. For example, of these three queries *'region A has larger area than region B'*, *'region A is above region B'* and *'region A and region B have a connected intersection,'* only the third one is topological. Topological queries form a robust and practically interesting subclass of all possible queries one may want to ask of a spatial database. Furthermore, they give rise to an unexpectedly complex and rich theory.

The most elementary query language imaginable, the Boolean closure of the eight basic topological relations [2] between two regions —such as *disjoint*(A, B) and *includes*(B, C)— is already surprisingly complicated (it is open whether its satisfiability problem is decidable [1]), and has given rise to some novel problems in the theory of planar graphs [1]. Allowing quantifiers that range over all regions yields a much more powerful language, which ultimately can express all topological queries, but is of formidable complexity [4] (this was to be expected, as any language expressing all topological queries must necessarily be very complex). However, it turns out that all topological queries can be expressed as *relational* queries on the *topological invariant* of the spatial database, a graph structure that captures the planar embedding of the planar graph defined by the boundaries of the regions [4]; there is much known about the expressiveness and complexity of queries on the topological invariant [4,5]. The class of topological queries expressible in another possible query language, the first-order theory of reals (where the quantifiers range over real coordinates of the plane), has been delimited severely [3,4].

The major open question in the area of topological queries is the design of a natural and efficiently interpretable query language that captures all *efficiently computable* topological queries of a spatial database —that is to say, of a natural and expressive topological query language with reasonable implementation cost.

References

1. Z. Chen, M. Grigni, C. H. Papadimitriou "Planarity, revisited," *Proc. 1998 STOC.*
2. M. J. Egenhofer, J. Sharma "Assessing the consistency of complete and incomplete topological information," *Geographical Systems*, 1:47–68, 1993.

---

* Research partially supported by the National Science Foundation.

3. B. Kuijpers, J. Van Den Bussche "Capturing first-order topological properties of planar spatial databases," *Proc. 1999 ICDT,* Springer.
4. C. H. Papadimitriou, D. Suciu, V. Vianu "Topological queries in spatial databases", *Proc. 1996 PODS.* Also, special *JCSS* issue for the 1996 PODS, 1998.
5. L. Segoufin, V. Vianu "Querying spatial databases via topological invariants", *Proc. 1998 PODS.*

Research Issues in Spatio-temporal Database Systems

Timos Sellis

National Technical University of Athens, Dept. of Electrical and Comp. Engineering,
Zografou 15773, Athens, Greece
timos@dbnet.ece.ntua.gr

Abstract. Spatiotemporal database management systems can become an enabling technology for important applications such as Geographic Information
Systems (GIS), environmental information systems, and multimedia. In this paper we address research issues in spatio-temporal databases, by providing an
analysis of the challenges set, the problems encountered, as well as the proposed solutions and the envisioned research areas open to investigation.

1   Introduction

Temporal databases and spatial databases have long been separate, important areas of
database research, and researchers in both areas have felt that there are important
connections in the problems addressed by each area, and the techniques and tools
utilized for their solution. So far, relatively little systematic interaction and synergy
among these two areas have occurred. Current research aims to achieve exactly this
kind of interaction and synergy, and aims also to address the many real-life problems
that require spatio-temporal concepts that go beyond traditional research in spatial
and temporal databases. *Spatio-temporal database management systems* (STDBMSs)
can become an enabling technology for important applications such as Geographic
Information Systems (GIS), environmental information systems, and multimedia. In
this paper we address research issues in spatio-temporal databases, by providing an
analysis of the challenges set, the problems encountered, as well as the proposed
solutions and the envisioned research areas open to investigation.

Most of the ideas presented in this paper are based on the experience with a research project that has been going on since 1996, CHOROCHRONOS. CHOROCHRONOS
was established as a Research Network with the objective of studying the design,
implementation, and application of STDBMSs. The participants of the network are
the Institute of Computer and Communication Systems of the National Technical
University of Athens, Aalborg University, FernUniversität Hagen, Universita Degli
Studi di L'Aquila, UMIST, Politecnico di Milano, INRIA, Aristotle University of
Thessaloniki, Agricultural University of Athens, Technical University of Vienna, and
ETH. All these are established research groups in spatial and temporal database sys-
tems, most of which have so far been working exclusively on spatial or temporal

R.H. Güting, D. Papadias, F. Lochovsky (Eds.): SSD'99, LNCS 1651, pp. 5-11, 1999.

databases. CHOROCHRONOS enables them to collaborate closely and to integrate their findings in their respective areas.

To achieve the objective of designing spatio-temporal databases, several issues need to be addressed; these are related to (a) the ontology, structure, and representation of space and time, (b) the data models and query languages, (c) graphical user interfaces, (d) query processing algorithms, storage structures and indexing techniques, and (e) architectures for STDBMSs.

2   Overview of Research Issues

Put briefly, a spatio-temporal database is a database that embodies spatial, temporal, and spatio-temporal database concepts, and captures simultaneously spatial and temporal aspects of data.

All the individual spatial and temporal concepts (e.g., rectangle or time interval) must be considered. However, attention focuses on the area of the intersection between the two classes of concepts, which is challenging, as it represents inherently *spatio-temporal concepts* (e.g., velocity and acceleration). In spatio-temporal data management, the simple aggregation of space and time is inadequate. Simply connecting a spatial data model to a temporal data model will result in a temporal data model that may capture spatial data, or in a spatial data model that may capture time-referenced sequences of spatial data.

Rather, the temporal characteristics of spatial objects (i.e., how entities evolve in space) must be investigated in order to produce inherently spatio-temporal concepts such as unified spatio-temporal data structures, spatio-temporal operators (e.g., approach, shrink), and spatio-temporal user-interfaces.

The main topics of interest when studying the issues involved in spatio-temporal databases are:

- *Ontology, Structure, and Representation of Space and Time.* This involves the study of temporal and spatial ontologies, including their interrelations and their utility in STDBMSs. In addition, structural and representational issues as they have been articulated in spatial and temporal database research should be considered in order to obtain a common framework for spatio-temporal analysis.
- *Models and Languages for STDBMSs.* The focus here is on three topics: (i) the study of languages for spatio-temporal relations, (ii) the development of models and query languages for spatio-temporal databases, and (iii) the provision of design techniques for spatio-temporal databases. This work builds on previous proposals and covers relational and object-oriented databases.
- *Graphical User Interfaces for Spatio-temporal Information.* Research in this area has two goals: (i) the extension of graphical interfaces for temporal and spatial databases, and (ii) the development of better visual interfaces for specific applications (e.g. VRML for time-evolving spaces).
- *Query Processing in Spatio-temporal Databases.* Techniques for the efficient evaluation of queries are the focus of this area. These studies cover a variety of

optimization techniques, ranging from algebraic transformations to efficient page/object management.

- *Storage Structures and Indexing Techniques for Spatio-temporal Databases.* Research in this area involves the integration or mixing of previously proposed storage and access structures for spatial and/or temporal data.
- *The Architecture of an STDBMS.* Finally, care must be taken in developing real systems, and therefore the architecture of a STDBMS is of high interest.

After this brief outline of the research areas, we proceed to give more detailed descriptions of some of these areas.

3   Spatio-temporal Data Ontologies and Modeling

In this section we address some issues involved in the ontology of spatial entities as well as the ontology of space itself, and issues corresponding to the development of conceptual and logical models, along with respective languages, for spatio-temporal data.

3.1   Ontological Issues

Regarding the *ontology of spatial entities*, in order to model change in geographic space, a distinction is made between *life* (the appearance and disappearance, and merging and splitting of objects) and *motion* (the change of location over time) of objects. At its outset, this research must identify and investigate prototypical situations in the life and movement of objects in geographic space.

Regarding the *ontology of space,* one should observe that spatial objects are located at regions in space. The concept of *exact location* is a relation between an object and the region of space it occupies. Spatial objects and spatial regions have a composite structure, i.e., are made up of parts. The ways in which parts of objects are located at parts of regions of space are captured by the notion of *part location.* Since there are multiple ways for parts of spatial objects to be located at parts of regions of space, multiple part location relations are identified, and a classification of part location relations is needed. An example of such work is the work on *rough locations* [2]. Rough locations are characterized by sets of part location relations that relate parts of objects to parts of partition regions [1].

3.2   Models and Languages

Models and languages for spatio-temporal database systems are a central activity, as it serves as the basis for several other tasks (for example, query processing and optimization). This research may be divided into two categories: a) research that focuses on tightly integrated spatio-temporal support, and b) previously initiated efforts that have dealt mainly with temporal aspects, extended to also cover spatial aspects. We consider in turn research in each category.

An important effort focuses on identifying the main requirements for a spatio-temporal data model and a spatio-temporal DBMS. Based on a *data type approach* to data modeling, the concept of *moving objects* has been studied in [4, 5]. This has led to a series of several results leading from data model specifications (at two different, abstract levels) to implementation aspects. Having specified mappings between different data type models and their relational DBMS embeddings, a precise landscape of the models' relative expressive powers has been drawn in [6]. Finally, concrete spatio-temporal data models for moving objects have also been provided. Here the focus has been on a systematic classification of operations on relevant data types that facilitates a highly generic specification and explanation of all types and, in particular, of the operations on them. A detailed description of this model, including formal specifications of the types and operations is presented in [10] along with examples demonstrating the data model in action.

A significant effort in the area of models and languages also deals with *constraint database models*. These models constitute a separate direction that researchers are exploring in modeling spatio-temporal information. As an example, DEDALE is a prototype of a constraint database system for spatio-temporal information [7,8]. DEDALE is implemented on top of the O_2 DBMS and features graphical querying. It offers a linear-constraint abstraction of geometric data, allowing the development of high-level, extensible query languages with a potential for optimization, while allowing the use of optimal computation techniques for spatial queries. Also in the context of constraint database models, other researchers have studied the role of spatial and temporal constraints in STDBMS [14]. The efforts here concentrated on the development of a new spatio-temporal constraint-based database model. This model is based on the linear constraint database model of DEDALE and the indefinite temporal constraint database model (ITCDB) previously proposed by Koubarakis.

Other research teams pursue research in extending relational models and languages. For example, the core of an SQL-based language, STSQL, has been proposed [3]. This language generalizes previous proposals by permitting relations to include multiple temporal as well as spatial attributes, and it generalizes temporal query language constructs, to apply to both the spatial and temporal attributes of relations. Because space and time are captured by separate attributes, STSQL is intended for applications that do not involve storing the movement of continuously moving objects.

Spatial and temporal conceptual modeling extends previous work on temporal and spatial data modeling. Spatial modeling aspects, e.g., the representation of objects' "position" in space, as well as temporal modeling aspects, e.g., the capture of the valid time of objects' properties, have been studied, and resulting new modeling constructs have been applied to existing conceptual models such as the ER model. Furthermore, the structure and behavior of so-called spatio-temporal phenomena (e.g., a "storm") have been investigated, and a formal framework with a small set of new modeling constructs for capturing these during conceptual design, has been defined [19, 20, 21, 22].

Finally, modeling issues related to uncertain spatio-temporal data need to be examined. By adopting fuzzy set methodologies, for example, a general spatial data model can be extended to incorporate the temporal dimension of geographic entities and their uncertainty. In addition, the basic data interpretation operations for handling the spatial dimension of geographic data have been extended to also support spatio-temporal reasoning and fuzzy reasoning. Some work has already been initiated in this direction [13].

4   Storage Structures, Indexing Techniques, and Query Processing

Having given a brief overview of the data modeling efforts undertaken, this section concentrates on efforts to develop techniques for the efficient implementation of the proposed data models and languages.

Substantial efforts have been devoted to the study of *storage structures and indexing*. In particular, (a) efficient extensions of spatial storage structures to support motion have been proposed, and (b) benchmarking issues have been studied.

Modern DBMSs should be able to efficiently support the retrieval of data based on the spatio-temporal extents of the data. To achieve this, existing multidimensional access methods need to be extended. Work has already initiated in this area. For example, approaches that extend R-trees and quadtrees were reported in [18] and [23], respectively, along with extensive experiments on a variety of synthetic data sets.

Work on benchmarking issues for spatio-temporal data has also started and is reported in [16]. This work introduced basic specifications that a spatio-temporal index structure should satisfy, evaluated existing proposals with respect to the specifications, and illustrated issues of interest involving object representation, query processing, and index maintenance. As a second step, a benchmarking environment that integrates access methods, data generation, query processing, and result analysis should be developed. The objective is to obtain a common platform for evaluating spatio-temporal data structures and operations that are connected to a data repository and a synthetic data set generator. A platform for evaluating spatiotemporal query processing strategies has been designed andimplemented and has been already used for evaluating spatial join strategies [11]. The "*A La Carte*" environment also provides benchmarking features for spatial join operations [9]. Finally, a very important step in this direction has been the work on generating spatio-temporal data in a controlled way so that benchmarks can be run [17].

Work on *query processing and optimization* has focused thus far on (a) the development of efficient strategies for processing spatial, temporal, and inherently spatio-temporal operations, (b) the development of efficient cost models for query optimization purposes, and (c) the study of temporal and spatial constraint databases.

In [15] it was argued that expressing spatial operations, required by different application domains, is possible through a set of window searches, so that their execution could be supported by the available spatial indexing techniques. When the availability of index structures is not guaranteed, incremental algorithms have been proposed to support join operations for time-oriented data [12]. Regarding the execution of inherently spatio-temporal operations, the basic classes of spatio-temporal operations required by different application domains involving the representation and reasoning on a dynamic world should be defined [24].

5 Conclusions

In this paper we presented some issues related to the research undergone in spatio-temporal databases. Clearly, significant progress has been achieved in several areas; these include the understanding of the requirements of spatio-temporal applications, data models, indexing structures, and query evaluation.

Although the research community has made significant progress, much work remains to be done before an STDBMS may become a reality. Open areas include

- devising data models and operators with clean and complete semantics,
- efficient implementations of these models and operators,
- work on indexing and query optimization,
- experimentation with alternative architectures for building STDBMSs (e.g. layered, extensible, etc).

One can observe that this is an exciting new area and the spatial database community will have a word in it!

Acknowledgement

The research presented has been influenced by the work done in the European Commission funded Training and Mobility of Researchers project, "CHOROCHRONOS: *A Research Network for Spatio-temporal Database Systems*", contract number ERBFMRX-CT96-0056 (http://www.dbnet.ece.ntua.gr/~choros/). The input from all participating institutions and the partial sponsoring of this work by that program is gratefully acknowledged.

References

1. Bittner, T.: A Qualitative Coordinate Language of Location of Figures within the Ground. COSIT'97, 1997.
2. Bittner, T.: Rough Location, Tech. Report, Department of Geoinformation. Technical University of Vienna, 1999.
3. Böhlen, M.H., Jensen, C. S., Skjellaug, B.: Spatio-Temporal Database Support for Legacy, 1998 ACM Symposium on Applied Computing, Atlanta, Georgia, 1998.
4. Erwig, M., Güting, R.H., Schneider, M., Vazirgiannis, M.: Spatio-Temporal Data Types: An Approach to Modeling and Querying Moving Objects in Databases. FernUniversität Hagen, Informatik-Report 224, 1997, to appear in GeoInformatica.
5. Erwig, M., Güting, R.H., Schneider, M., Vazirgiannis, M.: Abstract and Discrete Modeling of Spatio-Temporal Data Types. 6th ACM Symp. on Geographic Inf. Systems, 1998.

6. Erwig, M., Schneider, M., Güting, R.H.: Temporal Objects for Spatio-Temporal Data Models and a Comparison of Their Representations. Intl. Workshop on New Database Technologies for Collaborative Work Support and Spatio-Temporal Data Management, 1998.
7. Grumbach, S., Rigaux, P., Segoufin, L.: The DEDALE System for Complex Spatial Queries, Fifteenth ACM SIGACT-SIGMOD Symp. on Principles of Database Systems, Seattle, Washington, 1998.
8. Grumbach, S., Rigaux, P., Scholl, M., Segoufin, L.: Dedale: A spatial constraint database, Workshop on Database Programming Languages, Boulder, 1997.
9. Günther, O., Oria, V., Picouet, P., Saglio, J.M., Scholl, M.: Benchmarking Spatial Joins A La Carte, 10th Int. Conf. on Scientific and Statistical Database Management, Capri, Italy, 1998.
10. Güting, R.H., Böhlen, M.H., Erwig, M., Jensen, C.S., Lorentzos, N.A., Schneider, M., Vazirgiannis, M.: A Foundation for Representing and Querying Moving Objects. FernUniversität Hagen, Informatik-Report 238, 1998, submitted for publication.
11. Papadopoulos, A., Rigaux, P., Scholl, M.: A Performance Evaluation of Spatial Join Processing Strategies. 6th Intern. Symp. on Spatial Databases (SSD'99), Hong Kong, 1999.
12. Pfoser, D., Jensen, C.S.: Incremental Join of Time-Oriented Data, TimeCenter Tech. Report-34, Aalborg University, 1998.
13. Pfoser, D., Jensen, C.S.: Capturing the Uncertainty of Moving-Object Representations. 6th Intern. Symp. on Spatial Databases (SSD'99), July 20-23, Hong Kong, 1999.
14. Skiadopoulos, S., Koubarakis, M..: Tractable Query Answering in Indefinite Linear Constraint Databases, CHOROCHRONOS Tech. Report CH-98-08.
15. Theodoridis, Y., Papadias, D., Stefanakis, E., Sellis, T.: Direction Relations and Two-Dimensional Range Queries: Optimisation Techniques, Data and Knowledge Engineering, 27(3), 1998.
16. Theodoridis, Y., Sellis, T., Papadopoulos, A., Manolopoulos, Y.: Specifications for Efficient Indexing in Spatiotemporal Databases, 10th Int. Conf. on Scientific and Statistical Database Management, Capri, Italy, 1998.
17. Theodoridis, Y., Silva, J.R.O., Nascimento, M.A.: On the Generation of Spatiotemporal Datasets. 6th Intern. Symp. on Spatial Databases (SSD'99), Hong Kong, 1999.
18. Theodoridis, Y., Vazirgiannis, M., Sellis, T.: Spatio-Temporal Indexing for Large Multimedia Applications, 3rd IEEE Conf. on Multimedia Computing and Systems, ICMCS'96, Hiroshima, Japan, 1996.
19. Tryfona, N.: Modeling Phenomena in Spatiotemporal Applications: Desiderata and Solutions, 9th Int. Conf. on Database and Expert Systems Applications, LNCS, 1998.
20. Tryfona, N., Hadzilacos, Th.: Logical Data Modeling of SpatioTemporal Applications: Definitions and a Model, Int'l. Database Engineering and Applications Symp.m 1997.
21. Tryfona, N., Hadzilacos, Th.: Evaluation of Database Modeling Methods for Geographic Information Systems, CHOROCHRONOS Tech. Report CH-97-05.
22. Tryfona, N., Jensen, C. S.: Conceptual Data Modeling for Spatiotemporal Applications, CHOROCHRONOS Tech. Report CH-98-08.
23. Tzouramanis, T., Vassilakopoulos, M., Manolopoulos, Y.: Overlapping Linear Quadtrees: a Spatiotemporal Access Method, 6th ACM Int. Workshop on Geographical Information Systems, 1998.
24. Vazirgiannis, M., Theodoridis, Y., Sellis, T.: Spatio-Temporal Composition and Indexing for Large Multimedia Applications, ACM Multimedia Systems, 6(5), 1998.

Looking Both Ways: SSD 1999 ± 10

Oliver Günther

Humboldt University, Berlin and
Pôle Universitaire Leonard de Vinci, Paris
guenther@wiwi.hu-berlin.de

Since the 1960s, researchers and practitioners in the geosciences have been working on computer solutions to match their specific needs. Commercial geographic information systems (GIS) are among the most important outcomes of these efforts. In those early years, computer scientists have been involved only marginally in the development of such systems. More generally speaking, cooperations between computer scientists and the geoscientific communities have traditionally been rare.

In the 1980s, however, this has started to change, and the increasing number of contacts is now bearing fruit. The design and implementation of data management tools for spatial applications is pursued by an interdisciplinary community of people from academia, government, and industry. Spatial databases are now considered an important enabling technology for a variety of application software (such as CAD systems or GIS), and they start to find their way into mainstream business software (such as ERP systems). Many commercial database management systems have special toolboxes for the management of spatial data. There have been several interdisciplinary research projects of high visibility, including the U.S. *National Center for Geographic Information and Analysis (NCGIA)* [6] and the *Sequoia 2000* project [10,4]. U.C. Berkeley's *Digital Environmental Library (ELIB)* project [11,12], and U.C. Santa Barbara's *Alexandria* project [9] are pursuing related goals; both projects are funded through the NSF/ARPA/NASA Digital Library Initiative.

It was also in the 1980s that a number of interdisciplinary conference series were launched. In 1984, GIS researchers initiated the first *Symposium on Spatial Data Handling (SDH)*. Five years later, an NCGIA research initiative led to the first *Symposium on Spatial Databases (SSD)*. Compared to SDH, SSD has a stronger focus on computer technology. 1993 was the year of the first *Conference on Spatial Information Theory (COSIT)* and the first *ACM Workshop on Geographic Information Systems (ACM-GIS)*. All of these conference series continue to be held annually (ACM-GIS) or biannually (COSIT, SDH, SSD). Their proceedings often appear as books with major publishers (e.g. SSD: [2,5,1,3,8]).

These and related activities have resulted in numerous interdisciplinary publications, as well as system solutions and commercial products. They also led to a certain refocus of one of the premier journals in the area, the *International Journal on Geographic Information Science* (formerly the *International Journal on Geographic Information Systems*), published by Taylor & Francis. In 1996, Kluwer Academic Publishers started another journal in this area, called *GeoInformatica*.

R.H. Güting, D. Papadias, F. Lochovsky (Eds.): SSD'99, LNCS 1651, pp. 12–15, 1999.

At the organizational level, an important event happened in 1994, when an international group of GIS users and vendors founded the *Open GIS Consortium (OGC)*. OGC has quickly become a powerful interest group to promote open systems approaches to geoprocessing [7]. It defines itself as a "membership organization dedicated to open systems approaches to geoprocessing." It promotes an *Open Geodata Interoperability Specification (OGIS)*, which is a computing framework and software specification to support interoperability in the distributed management of geographic data. OGC seeks to make geographic data and geoprocessing an integral part of enterprise information systems. More information about OGC including all of their technical documents are available at the consortium's Web site, http://www.opengis.org.

This short historical overview shows that spatial data management has come a long way, both as an academic discipline and as an application of advanced computing technology. The SSD conference series had an important role in these developments. Since SSD was first held 10 years ago in Santa Barbara, California, the conference series has established itself as an important cornerstone of the community. It continues to be an important opportunity to meet one's peers and to "talk shop," i.e. to discuss highly specific technical issues. Interdisciplinary communication between geoscientists and computer scientists is the norm, as is technology transfer between researchers and practitioners. In all these respects, SSD is a good example of the smaller, more focused conferences that complement the larger venues, such as SIGMOD or VLDB.

If one compares the programs of SSD 1989 and SSD 1999, one observes that, on the one hand, the 1999 program reflects some of the major paradigm changes in data management. There are several papers on Internet-related topics and on data mining – two topics that arguably had the most fundamental impact on data management during the past 10 years.

On the other hand, many of the principal topics do *not* seem to have changed in a major way. Most papers presented at SSD 1999 can still be grouped into the following five major categories that were already present in 1989:

- physical data management: access methods, query optimization;
- logical data management: topology, modeling, algebras;
- distributed data management: Internet-related aspects, fault tolerance;
- spatial reasoning and cognition;
- GIS applications.

The work performed under these headings is mostly a natural continuation of the work conducted 10 years ago. Some of these categories are represented stronger than others, partly to complement other conference series on spatial data management. COSIT, for example, is traditionally strong on spatial reasoning and cognition issues. SDH serves as an important outlet for application-oriented work by the geoscientific community. Both of these areas take a somewhat weaker role at SSD. Other categories are conspicuously missing at all three conference series: papers on human-computing interfaces, for example, are rare.

I see two major reasons for this relative constancy. First, the essential technical requirements regarding spatial databases and geographic information systems have not changed since ten years ago, except for certain technical issues raised by the increased importance of the Internet. In particular, the much-heralded switch to main memory data management has not happened (yet). Second, researchers do not always just do what the marketplace demands. Many of us choose their focus based on the intellectual stimulus provided by certain classes of problems, as well as their apparent solvability in principle. This leads to a longer-term orientation of one's research focus. Let me give two examples.

The definition of a consistent and complete spatial algebra, for example, may seem of secondary relevance to most commercial GIS vendors and users. On the other hand, work on algebras can be great fun for researchers who are mathematically inclined and who like the idea of bringing structure to what used to be an unorganized collection of data types and operators. Even though a number of researchers has worked on related problems, the issue has not been completely solved, and I certainly expect related papers on the program of SSD 2009. This includes extensions to the classical approach, such as spatio-temporal algebras.

Spatial access methods, my own area, has matured greatly during the past 10 years. Great papers were written – some of them even received awards for their long-term impact. Of course, there were also many papers that did not really help to move the field forward: yet another structure was proposed, yet another set of very narrow performance evaluations were conducted, and the reader was left puzzled how and why the structure would be better than any of the 100 methods already known. Technology transfer into modern GIS has somewhat slowed down, after some initial successes. Nevertheless, the field still attracts researchers young and old. Why? Because the related problems are well-defined, solving them is intellectually stimulating, and even if one does not have an earthshattering new idea, the chances of obtaining a publishable paper after a few months of relatively straightforward work are not bad. In addition, there are many areas of possible extensions of high practical relevance, including high-dimensional applications (such as data mining) or spatio-temporal modeling. Once again, I am certain to find related papers on the agenda of SSD 2009.

Obviously there is a certain discrepancy between short- and mid-term practical needs on the one hand and researcher's interests on the other hand. Personally speaking, I do not find this problematic, even though we are working in an application-oriented field. On the contrary, I find this discrepancy essential for the research paradigm we have adopted. Like most of my colleagues in research, I strongly believe that a research community can deliver "useful" results only if researchers have the right and the means to *play*, i.e., to work on issues that they just enjoy working on. It is our duty as a research community, however, that we help policy makers in defining about how many people should have this privilege, and that we identify the right people among our graduate students and junior colleagues to enjoy it. Having too many journals and conferences with too many papers that nobody is really interested in can be as detrimental to a research community as having not enough outlets.

With regard to SSD, this means that we should continue to be selective, and that we should continue to honor work that balances intellectual stimulus with practical relevance. More concretely: we have to continue to honor work that solves difficult intellectual problems *without* regarding their immediate practical relevance. On the other hand, we also have to continue to honor good practical solutions that reflect the technical state-of-the-art, even though the underlying concepts may seem somewhat obvious to an academic researcher. The challenge for current and future SSD program committees has always been to balance these two approaches to research. So far I believe we have succeeded in finding the right balance, and I am looking forward to a continuation of this tradition. SSD – ad multos annos!

References

1. D. Abel and B. C. Ooi, editors. *Advances in Spatial Databases*. LNCS 692. Springer-Verlag, Berlin/Heidelberg/New York, 1993.
2. A. Buchmann, O. Günther, T. R. Smith, and Y.-F. Wang. *Design and Implementation of Large Spatial Databases*. LNCS 409. Springer-Verlag, Berlin/Heidelberg/New York, 1990.
3. M. J. Egenhofer and J. R. Herring, editors. *Advances in Spatial Databases*. LNCS 951. Springer-Verlag, Berlin/Heidelberg/New York, 1995.
4. J. Frew. Bigfoot: An earth science computing environment for the Sequoia 2000 project. In W. K. Michener, J. W. Brunt, and S. G. Stafford, editors, *Environmental Information Management and Analysis: Ecosystem to Global Scales*. Taylor & Francis, London, 1994.
5. O. Günther and H.-J. Schek, editors. *Advances in Spatial Databases*. LNCS 525. Springer-Verlag, Berlin/Heidelberg/New York, 1991.
6. National Center for Geographic Information and Analysis (NCGIA), 1998. URL http://www.ncgia.ucsb.edu.
7. The Open GIS Consortium and the OGIS Project, 1997. URL http://www.opengis.org.
8. M. Scholl and A. Voisard, editors. *Advances in Spatial Databases*. LNCS 1262. Springer-Verlag, Berlin/Heidelberg/New York, 1997.
9. T. R. Smith. A digital library for geographically referenced materials. *IEEE Computer*, 29(5), 1996.
10. M. Stonebraker. The Sequoia 2000 project. In D. Abel and B. C. Ooi, editors, *Advances in Spatial Databases*, LNCS 692, Berlin/Heidelberg/New York, 1993. Springer-Verlag.
11. University of California at Berkeley, Digital Library Project, 1998. URL http://elib.cs.berkeley.edu.
12. R. Wilensky. Toward work-centered digital information services. *IEEE Computer*, 29(5), 1996.

Multi-resolution and Scale

Generalizing Graphs
Using Amalgamation and Selection

John G. Stell and Michael F. Worboys

Department of Computer Science, Keele University
Keele, Staffs, ST5 5BG, U. K.
{john,michael}@cs.keele.ac.uk

Abstract. This work is a contribution to the developing literature on multi-resolution data models. It considers operations for model-oriented generalization in the case where the underlying data is structured as a graph. The paper presents a new approach in that a distinction is made between generalizations that amalgamate data objects and those that select data objects. We show that these two types of generalization are conceptually distinct, and provide a formal framework in which both can be understood. Generalizations that are combinations of amalgamation and selection are termed simplifications, and the paper provides a formal framework in which simplifications can be computed (for example, as compositions of other simplifications). A detailed case study is presented to illustrate the techniques developed, and directions for further work are discussed.

1   Introduction

Specialist spatial information systems (SIS) play an increasingly important role within the Information Technology industry [Abe97]. For the potential of SIS to be fully realised, spatial database functionality needs to be integrated with other more generic aspects of database technology. Spatial data comprise a valuable subset of the totality of data holdings of an enterprise and their utility is optimized when they are flexible enough to be capable of integration with other data sets in a variety of ways. The focus of this paper is a contribution towards the provision of flexibility with regard to the scale or resolution at which data are handled. Resolution is concerned with the level of discernibility between elements of a phenomenon that is being represented by the data, and higher resolutions allow more detail to be observed in the components of the phenomenon. Flexibility in handling resolution is advanced by provision of multi-resolution data models, where data are managed in the SIS at a variety of levels of detail.

The issue of multi-resolution spatial datasets has been taken up by several authors (e.g. [PD95, RS95]). In our own earlier work [SW98, Wor98a, Wor98b] we proposed a general model that helps to provide a formal basis for processing and reasoning with spatial data that are heterogeneous with regard to semantic and geometric precision. For multi-resolution data models to be effective, there must be the means to make appropriate transitions between levels of detail

R.H. Güting, D. Papadias, F. Lochovsky (Eds.): SSD'99, LNCS 1651, pp. 19–32, 1999.
© Springer-Verlag Berlin Heidelberg 1999

in the data. Transition from higher to lower resolutions is often referred to as *generalization*. Cartographic generalization has been the subject of a great deal of research by the cartographic and GIS communities, particularly on the geometric aspects of the generalization process (see for example [BM91, MLW95]). When the word 'generalization' is used in this paper it usually refers to model-oriented generalization, in the sense of [M$^+$95], as we are not concerned here with the particular form of the cartographic representation of the data.

The focus of this paper is on the geometric components of geospatial data. In particular, the emphasis here is on network data structures, as they provide a simpler case than fully two-dimensional data structures, and yet have many applications to real world systems. The paper seeks to make a clear formal distinction between model-oriented generalizations that are based on selection of data and those that are based on data amalgamation. This is a distinction that is somewhat blurred in some of the earlier multi-resolution spatial data models.

In the next section, the background to this research is outlined, particularly in the context of multi-resolution data models, generalization and functionality in databases for handling graphs. The following section makes precise the distinction between selection and amalgamation transformations from higher to lower levels of detail. The remainder of the paper is devoted to working out in detail the formal properties of selection and amalgamation operations on graphs, and includes consideration of a detailed case study.

2   Background

2.1   Multi-resolution Data Models and Generalization

Generalization is the process of transforming a representation of a geographic space to a less detailed one. The representation may be in terms of a data/process model, in which case the transformation is called *model generalization*, or involve visualization of the space on an output device or hard copy, in which case the transformation is called *cartographic generalization*. Cartographic generalization has been the subject of a great deal of research by the cartographic and GIS communities, particularly on the geometric aspects of the generalization process (see for example [BM91, MLW95]). Model-oriented generalization was introduced by Müller *et al.* [M$^+$95]. Rigaux and Scholl [RS95] discuss the impact of scale and resolution on spatial data modelling and querying. They develop a theory with spatial and semantic components and apply the ideas to a partial implementation in the object-oriented DBMS O_2.

A *multi-resolution model* of a geographic space affords representations at a variety of levels of detail as well as providing a structure in which these representations are located. In such models, generalization operators are required in order that transitions between different locations in the structure can be made. Puppo and Dettori [PD95] provide a formal model of some of the topological and metric aspects of multi-resolution using abstract cell complexes and homotopy theory - both topics within algebraic topology. These ideas are further developed by Bertolotto [Ber98], who proposes a definition of a base set of transformations, from which set a significant class of generalization operators can be obtained. This body of work provides one of the motivations for the current work, in that the earlier work does not seek to make a distinction between *selection* of features, where certain features of the phenomena are chosen and others omitted, and *amalgamation* of features, where certain features originally considered distinguishable are made indistinguishable. In current multi-resolution models, selection and amalgamation are not distinguished, yet they are conceptually quite distinct.

We use the term *simplification* for a generalization which can be described as a selection followed by an amalgamation. We are aware that 'simplification' does refer to a very specific operation in the literature on cartographic generalization, and that we are using the word in a more general sense here. However, this word has been used by Puppo and Dettori [PD95, p161] in a way that fits very closely with the present paper. Puppo and Dettori define 'simplification mappings' which are certain mappings between cell complexes. A particular simplification mapping $F : \Gamma \to \Gamma'$ provides a reason why the cell complex Γ' is a simplified version of Γ. This leads to a category [BW95], where the objects are cell complexes and the morphisms are the simplification mappings. In our work we have a category where the objects are graphs, and the morphisms are simplifications in our sense.

Further work on the amalgamation properties of multi-resolution data models is discussed by Worboys [Wor98b], where a lattice of resolution is constructed and properties of entities represented at differing degrees of granularity considered. This theme is pursued further in [Wor98a] by showing that the resolution lattice can be applied to both geometric and semantic resolutions. Stell and Worboys [SW98] develop the formal properties of the resolution lattice, showing how each resolution in the lattice gives rise to a space of spatial data representations all with respect to that resolution, and the totality of spatial data representations being stratified by the resolution lattice. Generalization operators and their inverses can be considered as transitions between layers in the stratified spatial data space. Stell has also recently provided [Ste99] an analysis of different notions of granularity for graphs.

2.2   Handling Graphs in Databases

In this paper, the techniques developed with regard to selection and amalgamation operators are applied as transitions between resolutions in a multi-resolution data model in the particular context of graphs. This is done for three reasons:

1. Graphs provide an intermediate level between non-spatial data and full planar spatial data, and are sufficiently rich to illustrate the application of the approach.
2. Graphs have many applications in spatial information systems, for example road and rail networks, and cable and other utility networks.

3. The dual graph of an areal partition of the plane (where nodes of the dual graph are associated with areas in the partition, and nodes are connected by an edge if and only if the areas are adjacent in the partition) is an important indicator of the topological relationships between the areas in the partition.

The database community has put some effort into considering how generic database technology can be used to provide functionality for handling graph data structures. Mannino and Shapiro [MS90] survey work on extending the relational database model to incorporate functionality for handling graphs, including extensions to query languages for graph traversal. Güting [Güt92], presented an approach that extended the relational data model with data types for planar-embedded graphs. Güting's 1992 paper was followed by a series of papers in which he and colleagues developed the theme of incorporating graph handling capabilities in database systems [Güt94, EG94, BG95]. Erwig and Schneider [ES97] pose the question of the meaning of vagueness with reference to a graph. Stell and Worboys [SW97] have discussed the algebraic structure of the set of subgraphs of a graph.

3   Selection and Amalgamation

A major motivation of this work is to clarify the distinction between selection and amalgamation generalization operations. In this section we explore the foundations of the concept *"less detailed than"*, based on the notions of selection and amalgamation. At the most abstract level, there are two ways in which data represented by a structure X can be less detailed than data represented by a structure Y.

selection: X can be derived from Y by selecting certain features, and possibly leaving out others.

amalgamation: X can be derived from Y by amalgamating some features of Y so that some distinct things in Y are regarded as indistinguishable and become just one thing in X.

3.1   Amalgamation and Selection for Sets

The two operations are illustrated in the case of sets X and Y, the simplest formal structures, by the following concrete examples.

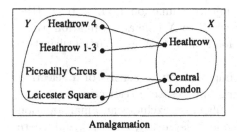

Amalgamation

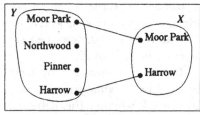

Selection

In the amalgamation example, the set Y consists of four stations on the London Underground. For some application it may be inappropriate to distinguish between the two individual stations at Heathrow Airport. Similarly, the stations Piccadilly Circus and Leicester Square are physically close together, and at a lower level of detail, the distinction between them may not be important. By avoiding the distinctions between these pairs of stations, we arrive at the set X as a less detailed representation of the data in Y.

The example of selection is also derived from actual data about the London Underground. Here again Y is a set representing four individual stations which is represented at a lower level of detail by a set X containing only two elements. However, in this case the operation performed on Y to produce X is quite different. The stations present in X are selected from those in Y because of their relative importance. Northwood and Pinner are minor stations, and many trains which do stop at Moor Park and Harrow do not stop at the two smaller stations.

When X and Y are sets it is straightforward to formalize the notions of amalgamation and selection. If the relationship of X to Y is one of selection, then there is an injective (or one-to-one) function from X to Y. If the relationship is one of amalgamation, then there is a surjective (or onto) function from Y to X.

3.2   Combining Amalgamation and Selection for Sets

The above examples deal with two ways in which X may be a less detailed representation of Y. In more complicated examples the relationship need not be solely one of amalgamation or selection. In general, a loss of detail relationship between X and Y will involve both selection and amalgamation. This entails a set Z which is obtained from Y by selection, and which is amalgamated to produce X. A simple example appears in the the following diagram.

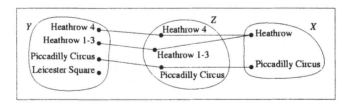

A pair consisting of a selection followed by an amalgamation will be called a *simplification* from Y to X. Formally, a simplification from a set Y to a set X consists of three things: a set Z, an injective function from Z to Y (the selection part) and a surjective function from Z to X (the amalgamation part). Alternatively we can describe a simplification from Y to X as a partial surjective function from Y to X.

It might appear that by defining a simplification to consist of a selection followed by an amalgamation, we are being unnecessarily restrictive. It is natural to ask whether this definition of simplification excludes an amalgamation followed by a selection, or a sequence of the form $[s_1, a_1, s_2, a_2, \cdots, s_n, a_n]$ where each a_i is an amalgamation, and each s_i is a selection. In fact, provided we are dealing with simplifications of graphs, or of sets, every sequence of the above form can be expressed as a single selection followed by a single amalgamation. The justification for this lies in the fact that simplifications can be composed. For simplifications of sets, this is discussed in the following paragraph. For simplifications of graphs, composition is illustrated by an example in section 4, and defined formally in section 5.4. It is worth noting that a single selection on its own is still a simplification. This is because it can be expressed as a selection followed by the trivial amalgamation in which no distinct entities are amalgamated. Similarly, a single amalgamation on its own is a simplification, since it is equal to the trivial selection, which selects everything, followed by the amalgamation.

A simplification from Y to X gives a way of modelling a *reason why* X is less detailed than Y. As the earlier examples showed, X can be a simplification of Y for many different reasons, thus it is necessary to keep track of the specific amalgamations and selections involved. It is also important to be able to compose simplifications. If σ_1 is a simplification from Y to X, and σ_2 a simplification from X to W, we need to be able to construct a simplification $\sigma_1; \sigma_2$ from Y to W which represents σ_1 followed by σ_2. The usual notion of composition for partial functions provides the appropriate construction in the current context.

3.3   Amalgamation and Selection for Graphs

Simplifications between sets are a useful way of illustrating the concepts of selection and amalgamation, but to handle more complex kinds of data we need more elaborate structures than sets. In this section simple examples of amalgamation and selection for graphs are introduced. Further examples appear in the detailed case study in section 4.

The following two examples develop the previous treatment of amalgamation and selection for sets by adding edges between the elements of the sets to represent how the stations are joined by railway lines.

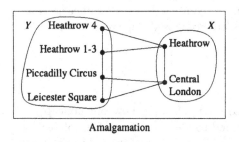
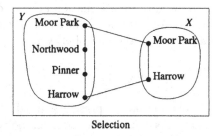

           Amalgamation   Selection

In the amalgamation example, the two stations at Heathrow airport collapse into a single entity, as before, but note that the edge between them in Y is not present in X. This disappearance of an edge can be understood in terms of amalgamations of paths. Roughly speaking, a path is a sequence of zero, one, or more edges in which each edge in the sequence shares one end with the next edge in the sequence and one end with the previous edge in the sequence. However, because we are using undirected edges, a more careful formal treatment is needed, and appears in section 5 below. In the concrete example being discussed here, four paths in Y become amalgamated into a single edge in X. The four paths distinguished in Y which are amalgamated in X are as follows.

Heathrow 4 ——	Heathrow 1-3 ——	Piccadilly Circus ——	Leicester Square
	Heathrow 1-3 ——	Piccadilly Circus ——	Leicester Square
Heathrow 4 ——	Heathrow 1-3 ——	Piccadilly Circus	
	Heathrow 1-3 ——	Piccadilly Circus	

In the selection example, the edge in the graph X is not selected from the edges present in Y, but is selected from the paths in Y. The use of paths in both the amalgamation and selection operations is an important feature of our work. Formally we treat amalgamations and selections as particular kinds of morphisms between graphs. These morphisms are mappings taking nodes to nodes, but which may map edges to paths, and not merely to edges.

A significant distinction between our work and that of both Puppo and Dettori [PD95] and Bertolotto [Ber98] is illustrated in the selection example. As a graph is a particular kind of 1-dimensional abstract cell complex, the technique of using continuous mappings between abstract cell complexes to model simplifications, which these authors use, can be applied to graphs. However, this technique would force us to use a mapping sending the two intermediate stations as well as the three edges in the graph Y to the single edge in the graph X. Conceptually this act of amalgamating Northwood and Pinner stations with three railway lines appears inappropriate if we want to model the simple idea that our graph X is obtained from Y by *omitting* certain features altogether. While continuous mappings between abstract cell complexes may be suitable for some kinds of simplification, they do not seem adequate to model the concept of selection.

These two examples of amalgamation and selection for graphs illustrate only a few of the features of our approach. As with sets, general loss of detail relationships between graphs involve both amalgamation and selection. Examples showing how amalgamation and selection are combined into simplifications for graphs, and how simplifications of graphs are composed appear in the detailed case study in section 4 below.

4   Case Study

In this section we present a detailed case study showing how our concepts of amalgamation and selection can be combined to yield a notion of simplification for graphs. The case study is drawn from genuine examples of the railway network in Britain.

The following diagram illustrates a simplification:

At the most detailed level, two stations in Birmingham are shown: Birmingham New Street and Birmingham International. These are amalgamated at the lower level. The line from Swansea to Birmingham New Street, as well as Swansea station itself are omitted at the lower level, as are the two stations intermediate between Reading and Birmingham. The route from Reading to Birmingham New Street is amalgamated with that from Reading to Birmingham International.

The simplification is made up of a selection and an amalgamation as in the following diagram:

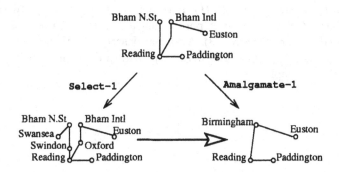

Note that a selection from a graph G need not be obtained by selecting some of the nodes and some of the edges from G. We allow a selection to take paths and not just edges from G. This technique allows selections to omit intermediate stations, such as Swindon, without being forced to omit railway lines passing through such stations. This means that we can model the fact that a line joins Reading to Birmingham New Street, even though no single edge represents this at the highest level. This use of paths is an important aspect of our work, formally it amounts to working with morphisms between graphs which take edges to paths, and is detailed in section 5 below.

The graph which appeared as the end result of the above simplification can be simplified further as in the following diagram. Here the two stations in London: Euston and Paddington have been amalgamated, as have the two routes from London to Birmingham. Reading station has also been omitted.

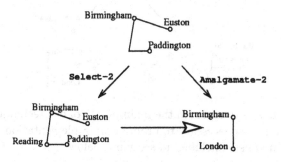

We now have two successive simplifications involving five graphs altogether as follows:

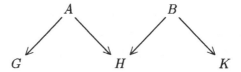

and we wish to express this as a single simplification.

The basic idea is to note that from A to B we have an amalgamation (to H) followed by a selection (from H). It is possible to interchange these, so that we can obtain B from A by first performing a selection (**Select-3**) and then an amalgamation (**Amalgamate-3**). A formal description of this construction is provided in section 5.4 below, but here we provide a diagram showing the result for our specific example.

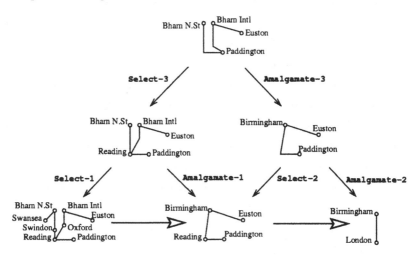

By composing **Select-3** with **Select-1**, and by composing **Amalgamate-3** with **Amalgamate-2** we obtain a single simplification:

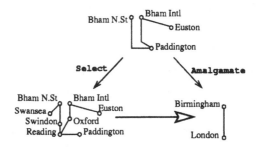

This case study has demonstrated the main points of our technique, but has not included the full details necessary to produce an implementation. A full account of the technical details is included in section 5 below.

5   Technical Details

To model structures and simplifications between them, including composition of successive simplifications, it is appropriate to use the mathematical structure known as a category [BW95]. Categories have already been used in the context of multi-resolution spatial data models by Bertolotto [Ber98]. However, unlike Bertolotto, our treatment is based on the category **Graph***, which is described in section 5.2 below. In order to present this material, some basic facts about graphs are needed first.

5.1   The Category Graph

The graphs used in this paper are undirected, are permitted to have loops, and may have multiple edges between the same pair of nodes. Much of the work in the paper can be carried out for directed graphs, but some aspects become slightly more complicated, while other aspects are easier to deal with. Limitations on space prevent us from giving details of both the undirected and directed cases.

The set of all subsets having either one or two elements, of a set, N, is denoted by $\mathbb{P}_2 N = \{\{x, y\} \mid x, y \in N\}$. A graph is then described formally as a pair of sets E and N (of edges and nodes respectively), together with an incidence function $i : E \to \mathbb{P}_2 N$.

A graph morphism $f : \langle E_1, i_1, N_1 \rangle \to \langle E_2, i_2, N_2 \rangle$ is a pair of functions $f_E : E_1 \to E_2$ and $f_N : N_1 \to N_2$, such that if the ends of edge $e \in E_1$ are x and y, then the ends of $f_E e \in E_2$ are $f_N x$ and $f_N y$. These morphisms take edges to edges in a way which preserves the incidence function. Given a graph morphism, f, the two functions f_E and f_N are referred to as the edge part and the node part of the morphism respectively. The category **Graph** has graphs as objects, and graph morphisms as its morphisms.

5.2   The Category Graph*

To define simplifications of graphs, we need another category which has the same objects as **Graph**, but more general morphisms. Given a graph G define the graph G^* to have the same nodes as G, and as edges, the set of all paths in G. A path in G can be described by a sequence, of nodes and edges of the form

$$[x_0, e_1, x_1, e_2, \ldots, e_\ell, x_\ell] \tag{1}$$

where edge e_k has ends x_{k-1} and x_k. The case of $\ell = 0$ is allowed, and gives zero length paths which are loops on each node. Two different sequences represent the same path iff each is the reverse of the other. The ends of the path are x_0 and x_ℓ.

Sometimes it is appropriate to write a path simply as $[e_1, e_2, \ldots, e_\ell]$, but in general this can be ambiguous. For example, consider the following graph with two nodes, m and n, and two edges, a and b.

$$m \overset{a}{\underset{b}{\rule{0pt}{0pt}\rule{2cm}{0.4pt}}} n$$

The paths $[m, a, n, b, m]$ and $[n, a, m, b, n]$ are quite distinct, and simply using the sequence of edges $[a, b]$ in this context would be ambiguous.

The $*$ construction is applicable not only to graphs, but also to morphisms. Given any graph morphism $f : G \to H$ we can construct a graph morphism $f^* : G^* \to H^*$. The morphism f^* has the same effect as f on nodes, and takes the edge (1) above to $[fx_0, fe_1, fx_1, fe_2, \ldots, fe_\ell, fx_\ell]$.

The graph G can always be embedded in G^* by the **Graph** morphism $\eta_G : G \to G^*$ which takes each node to itself, and an edge e to the path $[e]$ of length 1. Repeating the $*$ construction leads to a graph $(G^*)^*$. This has the same nodes as G and G^*, but the edges are paths of paths of G, which have the form

$$[x_0, [x_0, \sigma_1, x_1], x_1, [x_1, \sigma_2, x_2], \ldots, x_{n-1}, [x_{n-1}, \sigma_n, x_n], x_n] \tag{2}$$

where each σ_k is a sequence of edges and nodes of G, of the form

$$e_1, y_1, e_2, \ldots, e_{m-1}, y_{m-1}, e_m.$$

It is possible to reduce, or 'flatten', an edge in $(G^*)^*$ to one in G^* by mapping the edge (2) above to $[x_0, \sigma_1, x_1, \sigma_2, x_2, \ldots, x_{n-1}, \sigma_n, x_n]$. This assignment gives the edge part of a morphism $\mathsf{flat}_G : (G^*)^* \to G^*$ which is the identity on nodes.

The $*$ construction allows us to define the category **Graph***. This has the same objects as **Graph**, but morphisms from G to H in **Graph*** are ordinary graph morphisms from G to H^*. Given morphisms $f : G \to H$ and $g : H \to K$ in **Graph***, their composition is given by

$$G \xrightarrow{\ f\ } H^* \xrightarrow{\ g^*\ } (K^*)^* \xrightarrow{\ \mathsf{flat}_K\ } K^*$$

5.3   Selection and Amalgamation for Graphs

Definition 1 *A* selection *from a graph G is a subgraph A of G^* such that for each path π in G, there is at most one path ψ in A where flattening ψ yields π.*

The following example should help to clarify this definition.

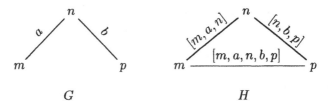

$$G \qquad\qquad\qquad\qquad H$$

In the above diagram we have a graph, G, and a graph H. The graph H is a subgraph of G^*, but is not a selection from G. This is because the two paths in H: $[m, [m, a, n], n, [n, b, p], p]$ and $[m, [m, a, n, b, p], p]$ both flatten to the same path $[m, a, n, b, p]$ in G.

If A is a selection from G and C is a selection from A, one might hope that C would be a selection from G. However, this does not happen, as can be seen from a simple example. Let G be the graph with three edges and four nodes $m \overset{a}{\rule{1.5em}{0.4pt}} n \overset{b}{\rule{1.5em}{0.4pt}} p \overset{c}{\rule{1.5em}{0.4pt}} q$, and let A be the selection from G: $m \overset{[m,a,n,b,p]}{\rule{3em}{0.4pt}} p \overset{[p,c,q]}{\rule{2em}{0.4pt}} q$. The graph C: $m \overset{[m,[m,a,n,b,p],p,[p,c,q],q]}{\rule{6em}{0.4pt}} q$ is a selection from A, but not a selection from G, since it is a selection from G^*. This failure of selections to compose is overcome by noting that applying flat_G to C yields a selection from G which is isomorphic to C itself. For our specific example, $\mathsf{flat}_G\, C$ is the graph $m \overset{[m,a,n,b,p,c,q]}{\rule{5em}{0.4pt}} q$

Definition 2 *An* amalgamation *for a graph G is a **Graph** morphism $\alpha : G \to H^*$, such that the node part of the morphism, α_N, is surjective, and for every edge e of H there is some edge e' of G for which $\alpha_E e' = [e]$, where α_E is the edge part of the morphism.*

Definition 3 *A* simplification *from a graph G to a graph H, is a pair (A, α) where A is a selection from G, and $\alpha : A \to H^*$ is an amalgamation of A.*

5.4 A Construction for Composing Simplifications

If we have two successive simplifications of graphs $G \overset{(A,\alpha)}{\longrightarrow} H \overset{(B,\beta)}{\longrightarrow} K$ we need to be able to compose them to give a simplification $G \overset{(A,\alpha);(B,\beta)}{\longrightarrow} K$. This is done by first constructing the graph C, which is the largest subgraph, X, of A^* such that $\alpha^* X = B$. By restricting α^* to C, we obtain a **Graph** morphism $\delta : C \to B^*$, so that B is an amalgamation of C.

Now C is a subgraph of A^*, and hence of $(G^*)^*$, whereas for a simplification from G to K, we need a subgraph of G^*. By applying the construction for selections of selections above, we get $\mathsf{flat}_G C$ as a selection from G, and an isomorphism $\varphi : \mathsf{flat}_G C \to C$. Finally we get the definition of the composite simplification

$$(A, \alpha); (B, \beta) = (\mathsf{flat}_G C, \varphi; \delta; \beta),$$

where $\varphi; \delta; \beta$ denotes the composite of these three morphisms in **Graph***. The overall picture is seen in this diagram in the category **Graph***:

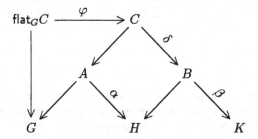

With this method of composing simplifications, we have a category where the objects are graphs, and the morphisms are simplifications.

6   Conclusions and Further Work

The work described in this paper has concerned explication of and distinction between the generalization operations that select and amalgamate data objects in a data model of a spatial phenomenon. In previous work, the functions of these two conceptually quite distinct operations have been conflated. We have termed such operations simplifications, and have provided a formal treatment of simplification in the case where the underlying data structure is a graph. In particular, we have shown that the appropriate formal home for these structures is the category **Graph***, and in that context it is possible to provide a construction for composition of simplification operations. We justified concentration on graph data structures because they were simple enough to show up clearly the main structural features of our approach, while at the same time being useful in spatial information handling with a history of treatment by SIS researchers.

Our approach is limited in several ways. Firstly, the simplification operators considered are in no way claimed to be a complete set of generalization operations, and further work is required to incorporate into this framework a richer collection of generalization operations. Secondly, the graph data structures are one-dimensional. The next step in the work is to consider simplification operators in fully two-dimensional data structures, in particular 2-complexes. The technical details for this case are more difficult. For example, the notion of a path of edges in a graph must be generalized to a gluing together of faces in a 2-complex. Work in this direction will be reported in a future publication.

There is one particular construction in which graph data structures have immediate application to fully two-dimensional data, and that is to areal decompositions of the plane. The *dual graph* of such a decomposition is a graph, where areas in the decomposition are nodes of the graph and two nodes are connected by an edge in the graph if the corresponding areas are adjacent in the decomposition. Some forms of generalization of an areal decomposition can be viewed as simplifications of its dual graph. For example, merging of two adjacent areas in an areal decomposition is equivalent in the dual graph to amalgamation of two nodes and elimination of the edge between them. One direction for further work using dual graphs would be to develop the boundary sensitive approach to qualitative location [BS98, Ste99] in a multi-resolution context. A detailed exploration of generalizations of areal decompositions of the plane in terms of simplification of the dual graph will be reported in later publications.

References

[Abe97]   D. J. Abel. Spatial internet marketplaces: A grand challenge. In *Proceedings 5th International Symposium SSD'97*, volume 1262 of *Lecture Notes in Computer Science*, pages 3–8. Springer-Verlag, 1997.

[Ber98]   M. Bertolotto. *Geometric modeling of spatial entities at multiple levels of resolution*. PhD thesis, Dipartmento di Informatica e Scienze dell Informazione, Universita di Genova, 1998.

[BG95]   L. Becker and R. H. Güting. The GraphDB Algebra: Specification of Advanced Data Models with Second-Order Signature. Informatik Berichte 183, FernUniversität, Hagen, May 1995.

[BM91]   B. Buttenfield and R. McMaster. *Map Generalisation: Making Rules for Knowledge Representation.* Longman, London, 1991.

[BS98]   T. Bittner and J. G. Stell. A boundary-sensitive approach to qualitative location. *Annals of Mathematics and Artificial Intelligence*, 24:93–114, 1998.

[BW95]   M. Barr and C. Wells. *Category Theory for Computing Science.* Prentice Hall, second edition, 1995.

[EG94]   M. Erwig and R. H. Güting. Explicit graphs in a functional model for spatial databases. *IEEE Transactions on Knowledge and Data Engineering*, 5:787–804, 1994.

[ES97]   M. Erwig and M. Schneider. Partition and conquer. In Hirtle and Frank [HF97], pages 389–407.

[Güt92]   R. H. Güting. Extending a spatial database system by graphs and object class hierarchies. In G. Gambosi, M. Scholl, and H. Six, editors, *Proc. Int. Workshop on Geographic Database Management Systems*, pages 34–55. Springer-Verlag, 1992.

[Güt94]   R. H. Güting. GraphDB: A Data Model and Query Language for Graphs in Databases. Informatik Berichte 155-2/1994, FernUniversität, Hagen, 1994.

[HF97]   S. C. Hirtle and A. U. Frank, editors. *Spatial Information Theory, International Conference COSIT'97, Proceedings*, volume 1329 of *Lecture Notes in Computer Science*. Springer-Verlag, 1997.

[M+95]   J. C. Müller et al. Generalization - state of the art and issues. In J. C. Müller, J. P. Lagrange, and R. Weibel, editors, *GIS and Generalisation: Methodology and Practice*, pages 3–17. Taylor and Francis, London, 1995.

[MLW95]   J. C. Müller, J. P. Lagrange, and R. Weibel, editors. *GIS and Generalisation: Methodology and Practice.* Taylor and Francis, London, 1995.

[MS90]   M. Mannino and L. Shapiro. Extensions to query languages for graph traversal problems. *IEEE Transactions on Knowledge and Data Engineering*, 2:353–363, 1990.

[PD95]   E. Puppo and G. Dettori. Towards a formal model for multiresolution spatial maps. In *Advances in Spatial Databases SSD'95*, volume 951 of *Lecture Notes in Computer Science*, pages 152–169. Springer-Verlag, 1995.

[RS95]   P. Rigaux and P. Scholl. Multi-scale partitions: Applications to spatial and statistical databases. In *Advances in Spatial Databases SSD'95*, volume 951 of *Lecture Notes in Computer Science*, pages 170–183. Springer-Verlag, 1995.

[Ste99]   J. G. Stell. Granularity for graphs. In *Proceedings of Conference on Spatial Information Theory, COSIT'99*, Lecture Notes in Computer Science. Springer-Verlag, to appear 1999.

[SW97]   J. G. Stell and M. F. Worboys. The algebraic structure of sets of regions. In Hirtle and Frank [HF97], pages 163–174.

[SW98]   J. G. Stell and M. F. Worboys. Stratified map spaces: A formal basis for multi-resolution spatial databases. In T. K. Poiker and N. Chrisman, editors, *SDH'98 Proceedings 8th International Symposium on Spatial Data Handling*, pages 180–189. International Geographical Union, 1998.

[Wor98a]   M. F. Worboys. Computation with imprecise geospatial data. *Computers, Environment and Urban Systems*, 22:85–106, 1998.

[Wor98b]   M. F. Worboys. Imprecision in finite resolution spatial data. *GeoInformatica*, 2:257–279, 1998.

Data Structures for Simplicial Multi-complexes

Leila De Floriani, Paola Magillo, and Enrico Puppo

Dipartimento di Informatica e Scienze dell'Informazione – Università di Genova
Via Dodecaneso, 35, 16146 Genova, ITALY
{deflo,magillo,puppo}@disi.unige.it

Abstract. The *Simplicial Multi-Complex* (SMC) is a general multiresolution model for representing k-dimensional spatial objects through simplicial complexes. An SMC integrates several alternative representations of an object and offers simple methods for handling representations at variable resolution efficiently, thus providing a basis for the development of applications that need to manage the level-of-detail of complex objects. In this paper, we present general query operations on such models, we describe and classify alternative data structures for encoding an SMC, and we discuss the cost and performance of such structures.

1 Introduction

Geometric cell complexes (meshes) have a well-established role as discrete models of continuous domains and spatial objects in a variety of application fields, including Geographic Information Systems (GISs), Computer Aided Design, virtual reality, scientific visualization, etc. In particular, simplicial complexes (e.g., triangle and tetrahedra meshes) offer advantageous features such as adaptivity to the shape of the entity, and ease of manipulation.

The accuracy of the representation achieved by a discrete geometric model is somehow related to its resolution, i.e., to the relative size and number of its cells. At the state-of-the-art, while the availability of data sets of larger and larger size allows building models at higher and higher resolution, the computing power and transmission bandwidth of networks are still insufficient to manage such models at their full resolution. The need to trade-off between accuracy of representation, and time and space constraints imposed by the applications has motivated a burst of research on *Level-of-Detail (LOD)*. The general idea behind LOD can be summarized as: *always use the best resolution you need – or you can afford – and never use more than that*. In order to apply this principle, a mechanism is necessary, which can "administrate" resolution, by adapting a mesh to the needs of an application, possibly varying its resolution over different areas of the entity represented.

A number of different LOD models have been proposed in the literature. Most of them have been developed for applications to terrain modeling in GISs (see, for instance, [1, 4, 9]) and to surface representation in computer graphics and virtual reality applications (see, for instance, [10, 8, 15, 7]), and they are strongly characterized by the data structures and optimization techniques they

R.H. Güting, D. Papadias, F. Lochovsky (Eds.): SSD'99, LNCS 1651, pp. 33–51, 1999.

adopt as well as custom tailored to perform specific operations, and to work on specific architectures. In this scenario, developers who would like to include LOD features in their applications are forced to implement their own models and mechanisms. On the other hand, a wide range of potential applications for LOD have been devised, which require a common basis of operations (see, e.g., [3]). Therefore, it seems desirable that the LOD technology is brought to a more mature stage, which allows developers to use it through a common interface, without the need to care about many details.

In our previous work, we have developed a general model, called a *Simplicial Multi-Complex (SMC)*, that can capture all LOD models based on simplicial complexes as special cases [13, 5, 14]. Based on such model, we have built systems for managing the level of detail in terrains [2], and in free-form surfaces [3], and we are currently developing an application in volume visualization.

In this paper, we consider general operations that can be performed on LOD models and propose an analysis of cost and performance of their encoding data structures. Trade-off between cost and performance is a key issue to make the LOD technology suitable to a wide spectrum of applications and architectures in order to achieve a more homogeneous and user-transparent use of LOD.

The Simplicial Multi-Complex is briefly described in Section 2, and general query techniques on such model are outlined in Section 3. In Section 4, we analyze the spatial relations among entities in the SMC, which are fundamental to support queries and traversal algorithms. In Section 5, we analyze different data structures to encode SMCs in the general case, as well as in special cases, and we discuss both the cost of such data structures, and their performance in supporting the extraction of spatial relations. In Section 6, we present some concluding remarks.

2   Simplicial Multi-complexes

In this section, we briefly review the main concepts about the Simplicial Multi-Complex, a dimension-independent multiresolution simplicial model which extends the Multi-Triangulation presented in [13, 5, 14]. For the sake of brevity, this subject is treated informally here. For a formal treatment and details see [11].

In the remainder of the paper, we denote with k and d two integer numbers such that $0 < k \leq d$. A k-*dimensional simplex* σ is the locus of points that can be expressed as the convex combination of $k + 1$ affinely independent points in $\mathbb{R}^d$, called the *vertices* of σ. Any simplex with vertices at a subset of the vertices of σ is called a *facet* of σ. A *(regular) k-dimensional simplicial complex* in $\mathbb{E}^d$ is a finite set Σ of k-simplices such that, for any pair of distinct simplices $\sigma_1, \sigma_2 \in \Sigma$, either σ_1 and σ_2 are disjoint, or their intersection is the set of facets shared by σ_1 and σ_2. In what follows, a k-simplex will be always called a *cell*, and we will deal only with complexes whose domain is a manifold (also called *subdivided manifolds*).

The intuitive idea behind a *Simplicial Multi-Complex* (SMC) is the following: consider a process that starts with a coarse simplicial complex and progressively refines it by performing a sequence of local updates (see Figure 1). Each *local update* replaces a group of cells with another group of cells at higher resolution. An update U_2 in the sequence *directly depends* on another update U_1 preceding it if U_2 removes some cells introduced with U_1. The *dependency relation* between updates is defined as the transitive closure of the direct dependency relation. Only updates that depend on each other need to be performed in the given order; mutually independent updates can be performed in arbitary order. For instance, in the example of Figure 1, updates 3 and 4 are mutually independent, while update 5 depends on both; thus, we must perform update 4 first, then followed by 3 and 5.

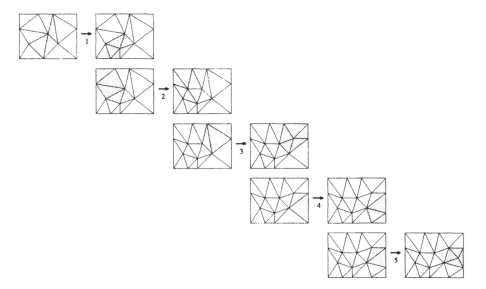

Fig. 1. A sequence of five updates (numbered 1...5) progressively refining an initial coarse triangle mesh. The area affected by each update is shaded.

An SMC abstracts from the totally ordered sequence by encoding a *partial order* describing the *mutual dependencies* between pairs of updates. Updates forming any subset closed with respect to the partial order, when performed in a consistent sequence, generate a valid simplicial complex. Thus, it is possible to perform more updates in some areas, and fewer updates elsewhere, hence obtaining a complex whose resolution is variable in space. Such an operation is known as *selective refinement*, and it is at the basis of LOD management. A few results of selective refinement from an SMC representing a terrain are shown in Figure 2.

An SMC is described by a directed acyclic graph (DAG). Each update is a node of the DAG, while the arcs correspond to direct dependencies between

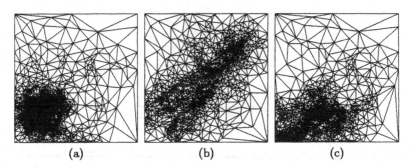

Fig. 2. Three meshes extracted from a two-dimensional SMC representing a terrain (top view). (a) The triangulation has the highest possible resolution inside a rectangular window, and the lowest possible resolution outside it. (b) Resolution inside a view frustum (wedge) is decreasing with the distance from its focus point, while it is arbitrarily low outside it. (c) Resolution is high only in the proximity of a polyline.

updates. Each arc is labeled with the collection of all cells of its source node that are removed by its destination node. For convenience, we introduce two further nodes: a *root* corresponding to the update creating the initial coarse complex, which is connected with an arc to each update that removes some if its cells; and a *drain*, corresponding to the final deletion of the complex obtained by performing all updates, which is connected with an arc from each update that creates some of its cells. Also such arcs are labeled by cells in a consistent way. Figure 3 shows the SMC corresponding to the collection of updates described in Figure 1.

A *front* of an SMC is a set of arcs containing exactly one arc on each directed path from the root (see Figure 3). Since the DAG encodes a partial order, we say that a node is *before* a front if it can be reached from the root without traversing any arc of the front; otherwise the node is said to be *after* the front. Nodes lying before a front define a consistent set of updates, and the corresponding simplicial complex is formed by all cells labeling the arcs of the front [11]. By sweeping a front through the DAG, we obtain a wide range of complexes, each characterized by a different resolution, possibly variable in space.

In the applications, often an SMC is enriched with attribute information associated with its cells. Examples are approximation errors (measuring the distance of a cell from the object portion it approximates), colors, material properties, etc.

3   A Fundamental Query on an SMC

Since an SMC provides several descriptions of a spatial object, a basic query operation consists of selecting a complex which represents the object according to some user-defined resolution requirements. This basic query provides a natural support to variable resolution in many operations, such as:

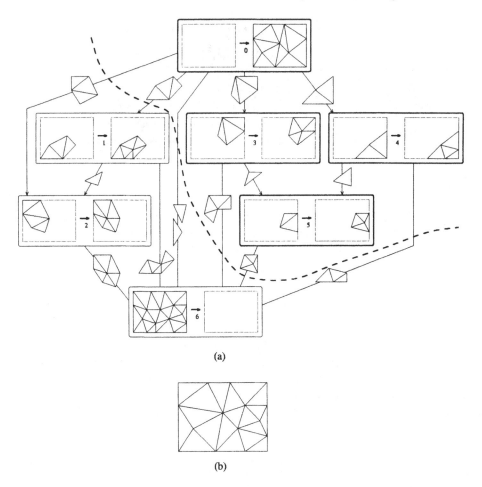

(a)

(b)

Fig. 3. (a) The SMC built over the partially ordered set of mesh updates of Figure 1. Each node represents an update, and it shows the two sets of simplices removed and created in the update. Each arc represents the dependency between two updates, and it is labelled by the triangles created in the first update, which are removed in the second update. A front on the SMC contains the arcs intersected by the thick dashed line; nodes lying before the front are highlighted. (b) The triangle mesh associated with the front.

- *point location*, i.e., finding the cell that contains a given point and such that its resolution meets some user-defined requirements;
- *windowing*, i.e., finding a complex, that represents the portion of the object lying inside a box, at a user-defined resolution;
- *ray casting*, i.e., finding the cells that intersect a given ray at a user-defined resolution;
- *perspective rendering*: in this case, a complex is generated which represents the portion of the object lying inside the view frustum, and whose resolution is higher near the viewpoint and decreases with the distance from it;

– *cut*, i.e., sectioning with a hyperplane: the section is computed by first retrieving the cells that intersect the given hyperplane and have a specific resolution.

In the most general case, resolution requirements are expressed through a *resolution filter*, which is a user-defined function R that assigns to each cell σ of the SMC a real value $R(\sigma)$. Intuitively, a resolution filter measures the "signed difference" between the resolution of a cell and that required by the application: $R(\sigma) > 0$ means that the resolution of σ is not sufficient; $R(\sigma) < 0$ means that the resolution of σ is higher than necessary. A cell such that $R(\sigma) \leq 0$ is said *feasible*.

For example, the meshes depicted in Figure 2 satisfy the following resolution filters: in (a), R is negative for all cells outside the window, zero for all cells inside it that are at the highest resolution, and positive for all others; in (b), R is negative for all cells outside the view frustum, while for a cell σ inside it, R is decreasing with resolution of σ, and with its distance from the focus point; in (c), R is negative for all cells not intersecting the polyline, zero for all cells intersecting it that are at the highest resolution, and positive for all others.

The basic query on an SMC consists of retrieving the simplicial complex of *minimum size* (i.e., composed by the smallest number of cells) which satisfies a given resolution filter R (i.e., such that all its cells are feasible with respect to R). Variants of this query are also described in [11]. The basic query can be easily combined with a culling mechanism, which extracts only the subcomplex intersecting a given *Region Of Interest (ROI)*. This *localized* query permits to implement operations like point location, windowing, etc.

Algorithms for mesh extraction [13, 3, 14, 11] sweep a front through the DAG, until an associated complex formed by feasible cells is found. Minimum size is guaranteed by a front that lies as close as possible to the root of the SMC. In the case of a localized query, spatial culling based on a ROI is incorporated in the DAG traversal, hence using the structure of the SMC also as a sort of spatial index. The key operations used by extraction algorithms consist in either advancing the front after a node, when the resolution of the complex over that area is not sufficient, or moving it before a node when the resolution over that area is higher than required.

The key issues that have impact on the performance of such algorithms are: the evaluation of the resolution function, which is application-dependent; and the evaluation of mutual relations that occur among different entities of the SMC. The cost of computing such relations is highly dependent on the amount of information stored in the data structure.

4   Relations in a Simplicial Multi-complex

In some applications, e.g., in computer graphics, it is often sufficient to represent a simplicial complex by the collection of its cells, where each cell is described by its vertices and its attributes. In other applications, e.g., in GIS, in CAD, or in scientific visualization, topological relations among vertices and cells of the mesh must be maintained as well. A common choice is the *winged* data structure, which stores, for each cell, the $(k+1)$ cells adjacent to it along its $(k-1)$-facets [12]. Building the winged data structure for the mesh produced as the result of a query on the SMC can be more or less expensive, depending on the data structure used to encode the SMC.

In the following, we discuss the relations among the elements of an SMC, which are needed in the traversal algorithms, and in building the winged data structure for the output mesh.

There are essentially three kinds of relations in an SMC:

- *Relations on the DAG*: they define the structure of the DAG describing the SMC by relating its nodes and arcs.
- *Relations between the DAG and the cells of the SMC*: they define the connections between the elements of the DAG (arcs and nodes) and the cells forming the SMC; in the definition given in Section 2, such a connection is defined by labeling each arc of the DAG with the cells created by its source node that are removed by its destination node.
- *Relations between the simplices of the SMC*: they define the relations among vertices and cells in the SMC.

The relations on the DAG are the standard relations in a directed graph: *Node-Arc (NA)*, which associates with a node its incoming and its outgoing arcs; and *Arc-Node (AN)*, which associates with an arc its source and its destination.

The relations between the DAG and the cells of the SMC can be defined as follows:

- *Arc-Cell (AC)* relation, which associates with an arc of the DAG the collection of the cells labeling it.
- *Cell-Arc (CA)* relation, which associates with a cell σ of the SMC the arc of the DAG whose label contains σ.
- *Node-Cell (NC)* relation, which associates with a node U the cells created and deleted by the corresponding update.
- *Cell-Node (CN)* relation, which associates with a cell σ the node U introducing σ in its corresponding update, and the node U' removing σ.

The relations between the simplices in an SMC we are interested in are:

- the relation between a cell and its vertices, that we call *Cell-Vertex (CV)* relation;
- the *adjacency* relation between two cells, which share a $(k-1)$-facet, that we call a *Cell-Cell (CC)* relation.

Since not all cells sharing a $(k-1)$-facet in the SMC can coexist in a cell complex extracted from it, we specialize the CC relation further into four different relations that will be used in the context of data structures and algorithms discussed in the following (see also Figure 4). Given two cells σ_1 and σ_2 that share a $(k-1)$-facet σ'':

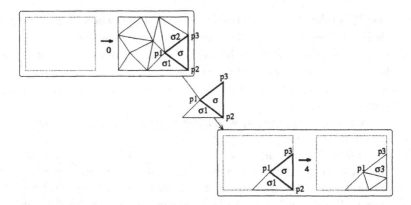

Fig. 4. A fragment of the SMC of Figure 3 and CC relations involving simplex σ. At edge p_1p_2, relation co-CC_1 and co-CC_2 both give simplex σ_1; relations counter-CC_1 and counter-CC_2 are not defined. At edge p_2p_3 no CC relation is defined. At edge p_3p_1, relation co-CC_1 is not defined, relation counter-CC_1 gives σ_3; relation co-CC_2 gives σ_2 and counter-CC_2 is not defined.

1. σ_1 and σ_2 are *co-CC_1* at σ'' if and only if σ_1, σ_2 have been removed by the same update (i.e., they label either the same arc or two arcs entering the same node);
2. σ_1 and σ_2 are *co-CC_2* at σ'' if and only if σ_1, σ_2 have been created by the same update (i.e., they label either the same arc or two arcs leaving the same node);
3. σ_2 is *counter-$CC_{1,2}$* to σ_1 at σ'' if and only if, σ_2 is created by the update that removes σ_1 (i.e., the arc containing σ_1 and that containing σ_2 enter and leave the same node, respectively);
4. σ_2 is *counter-$CC_{2,1}$* to σ_1 at σ'' if and only if σ_2 is removed by the update that creates σ_1 (i.e., the arc containing σ_1 and that containing σ_2 leave and enter the same node, respectively).

Relations co-CC_1 and counter-$CC_{1,2}$ are mutually exclusive: a k-simplex cannot have both a co-CC_1, and a counter-$CC_{1,2}$ cell at the same $(k-1)$-facet. The same property holds for relations co-CC_2 and counter-$CC_{2,1}$. The above four relations do not capture all possible CC relations among cells in an SMC, but they are sufficient to support efficient reconstruction algorithms, as explained in the following.

Relations CV and CC, defined in the context of a mesh extracted from an SMC by the algorithms described in Section 3, also characterize the winged data structure. Now, let us assume that we want to encode our output mesh through such a data structure. We have three options:

1. *Adjacency reconstruction as a post-processing step*: the extraction algorithm returns just a collection of cells and vertices, together with the CV relation; pairs of adjacent (CC) cells in the output mesh are found through a sorting

process. This takes $O(m(k+1)\log(m(k+1)))$ time, where m is the number of cells in the mesh, and k is the dimension of the complex.

2. *Incremental adjacency update*: the pairs of adjacent cells in the output mesh are determined and updated while traversing the SMC, encoded with a data structure that maintains the four relations co-CC_1, co-CC_2, counter-$CC_{1,2}$ and counter-$CC_{2,1}$.

   In the extraction algorithms, when the current front is advanced after a node, the pairs of mutually adjacent new cells introduced in the current mesh are determined by looking at co-CC_2 relations in the SMC; the adjacency relations involving a new cell, and a cell that was already present in the output mesh, are determined by using relation counter-$CC_{2,1}$ and adjacency relations of the old cells replaced by the update. Symmetrically, when the current front is moved before a node, relations co-CC_1 and counter-$CC_{1,2}$ permit updating adjacency relations in the current mesh. The total time is linear in the number of cells swept from one side to the other of the front.

3. *Incremental adjacency reconstruction*: same as approach 2, but without encoding CC relations of the SMC.

   In this case, when sweeping a front through a node (either forward or backward), a process of adjacency reconstruction similar to that used in approach 1 is applied, locally to the part of the current complex formed by the new cells introduced in the current mesh, and the cells adjacent to those deleted by the update operation. The time required is $O(n_{sweep}\log M)$, where n_{sweep} is the number of swept cells, and M is the maximum number of cells removed and created by the update contained in a swept node.

5   Data Structures

Encoding an SMC introduces some overhead with respect to maintaining just the mesh at the highest possible resolution that can be extracted from it. This is indeed the cost of the mechanism for handling multiresolution. However, we can trade-off between the space requirements of a data structure and the performance of the query algorithms that work on it.

From the discussion of previous sections, it follows that basic requirements for a data structure encoding an SMC are to support selective refinement (as outlined in Section 3), and to support the extraction of application-dependent attributes related to vertices and cells. Moreover, a data structure should support the efficient reconstruction of spatial relationships of an output mesh, for those applications that require it.

In the following subsections, we describe and compare some alternative data structures that have different costs and performances. Those described in Sections 5.1 and 5.2 can be used for any SMC, while those described in Section 5.3 can be used only for a class of SMCs built through specific update operations.

5.1   Explicit Data Structures

An *explicit* data structure directly represents the structure of the DAG describing an SMC. It is characterized by the following information:

- For each vertex, its coordinates.
- For each cell: the Cell-Vertex and the Cell-Arc relations, plus possibly a subset of *Cell-Cell* relations (as described below).
- For each node: the Node-Arc relation.
- For each arc: the Arc-Node relation, and the Arc-Cell relation.

Depending on the specific application, additional information may be attached to vertices and/or cells (e.g., approximation errors for simplices). Here, we do not take into account such extra information.

Assuming that any piece of information takes one unit, the space required by this data structure, except for adjacency relations and attributes, is equal to $d\mathbf{v} + (k + 3)\mathbf{s} + 4\mathbf{a}$, where $\mathbf{v}$, $\mathbf{s}$ and $\mathbf{a}$ denote the number of vertices, cells, arcs in the SMC, respectively. Note that $4\mathbf{a}$ is the cost of storing the NA plus the AN relations (i.e., the DAG structure), while $2\mathbf{s}$ is the cost of storing the CA and AC relations, i.e. information connecting the DAG and the cells of the SMC.

We consider three variants of adjacency information that can be stored for each cell σ:

- *Full-adjacency:* all four adjacency relations are maintained: co-CC_1, co-CC_2, counter-$CC_{1,2}$ and counter-$CC_{2,1}$. For each $(k-1)$-facet of σ, co-CC_1 and counter-$CC_{1,2}$ are stored in the same physical link, since they cannot be both defined; similarly, co-CC_2 and counter-$CC_{2,1}$ are stored in the same physical link. Thus, we have $2(k+1)$ links for each simplex.
- *Half-adjacency:* only relations co-CC_2 and counter-$CC_{2,1}$ are stored, by using the same physical link for each edge e of σ, thus requiring $(k+1)$ links.
- *Zero-adjacency:* no adjacency relation is stored.

The version with full-adjacency can support incremental adjacency update (see approach 2 in Section 4). The version with half-adjacency can support incremental adjacency update only when advancing the front after a node. With zero-adjacency, adjacency reconstruction must be performed, either as a postprocessing (approach 1), or incrementally (approach 3).

Figure 5 compares the performance of query algorithms on the zero-adjacency data structure without adjacency generation, and with adjacencies reconstructed through approaches 1 and 3. Adjacency reconstruction increases query times of almost a factor of ten. Therefore, it seems desirable that adjacency information are maintained in the SMC data structure whenever they are necessary in the output mesh, provided that the additional storage cost can be sustained.

5.2   A Data Structure Based on Adjacency Relations

In this section, we describe a data structure that represents the partial order which defines the SMC implicitly, i.e., without encoding the DAG, but only

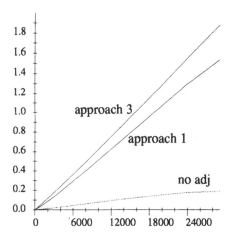

Fig. 5. Query times with adjacency reconstruction on a two-dimensional SMC using approach 1 and 3; the dotted curve represents query time without adjacency generation. The horizontal axis reports the number of triangles in the output mesh, the vertical axis execution times (in seconds).

using adjacency relations. The data structure stores just vertices and cells, and it maintains the following information:

- For each vertex: its coordinates.
- For each cell: the Cell-Vertex relation, and the four Cell-Cell relations, (using $2(k+1)$ links as explained in Section 5.1).

The space required is $d\mathbf{v} + 3(k+1)\mathbf{s}$.

Given a cell σ, removed by an update U, all other cells removed by U are found through co-CC_1 starting at σ. Among such cells, a cell σ'' is found which has at least one counter-$CC_{2,1}$ simplex σ''. Finally, starting from σ'' (which is a cell created by U), all remaining cells created by U are found by using relation co-CC_2. These properties allow us to update the current mesh after any movement of the current front.

The size of the adjacency-based data structure is always larger than that of the explicit structure with zero-adjacency, while it is comparable with that of the explicit data structure encoding some adjacency. Note that the space needed to store adjacency relations tends to explode when the dimension k of the model increases. Implementing query algorithms on the adjacency-based structure is more involved than on the explicit structure, and it requires maintaining larger temporary structures for encoding the internal state (see [11] for details). Therefore, this data structure should be preferred over to the explicit ones only if adjacency relations are fundamental in the output structure, and the dimension of the problem is low. However, the performance of the query algorithms is likely to degrade with respect to the case of explicit data structures. Storage costs for $k = 2, 3$ are compared in Table 1.

k=d=2

structure	space
explicit (zero-adj)	2v + 5s + 4a
explicit (half-adj)	2v + 8s + 4a
explicit (full-adj)	2v + 11s + 4a
adj-based	2v + 9s

k=d=3

structure	space
explicit (zero-adj)	3v + 6s + 4a
explicit (half-adj)	3v + 10s + 4a
explicit (full-adj)	3v + 14s + 4a
adj-based	3v + 12s

Table 1. Space requirements of the explicit and the adjacency-based data structures for $k = 2, 3$.

5.3 Compressed Data Structures

Much of the cost of data structures presented in the previous sections is due to the explicit representation of the cells and of the cell-oriented relations in the SMC. Indeed, the total number of cells is usually quite larger than the number of vertices, arcs, and nodes involved in the model, and relations among cells and vertices are expensive to maintain: for instance, a cell needs $k + 1$ vertex references for representing relation CV.

In some cases, the structure of every update exhibits a specific pattern, which allows us to compress information by representing cells implicitly. Examples of update patterns commonly used in building LOD models for surfaces are: *vertex insertion*, which is performed by inserting a new vertex and retriangulating its surrounding polytope consequently; and *vertex split*, which is performed by expanding a vertex into an edge and warping its surrounding cells consequently. Such update patterns are well defined in any dimension d, and they are depicted in Figure 6 for the two-dimensional case.

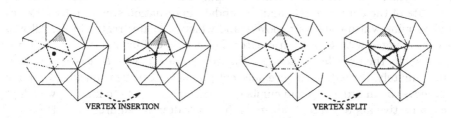

VERTEX INSERTION   VERTEX SPLIT

Fig. 6. Two types of update patterns that allow the design of compressed data structures for an MC. The shaded triangles are those involved in the update.

Since each update U exhibits a predefined pattern, the set of cells it introduces can be encoded by storing just a few parameters within U, that are sufficient to describe how the current complex must be modified when sweeping the front, either forward or backward, through U. The type and number of parameters depend on the specific type of update for which a certain compressed data structure is designed.

The generic scheme for a compressed data structure encodes just the vertices, nodes, and arcs of an SMC. For vertices, it stores the same information as the standard explicit structure; for nodes and arcs it encodes the following information:

- For each node U: the Node-Arc relation, plus an implicit description of the cells defining the update described by U, i.e., an implict encoding of the combination of Node-Cell and Cell-Vertex relations.
- For each arc: the Arc-Node relation.

The space required (except for the implicit encoding of the NC relation combined with the CV one) is equal to $d\mathbf{v} + 4\mathbf{a}$.

Note that, since cells do not exist as individual entities, attribute information for them cannot be encoded explicitly. This means that, while the exact geometry of cells can be obtained through suitable mechanisms associated with update parameters, their attributes can only be approximated through information associated with nodes. In other words, attributes on a node U summarize attributes of the whole group of cells associated with it. In this sense, such a structure is *lossy*, because it cannot discriminate between attributes of different cells in the context of the same node. Since the evaluation of the resolution filter may depend on cell attributes, because of the approximation, a given cell σ may result unfeasible even if it was feasible, or viceversa. This fact may cause the extraction of a mesh that is either over- or under-refined with respect to the input requirements.

Another subtle issue, which affects the performance of the extraction algorithms, is the lack of information on the update that must be applied to refine the mesh at a given cell. This is due to the fact that cells are associated with nodes, rather than with arcs. When sweeping the current front after a node U, the state of the current mesh, and the update information stored in U, allow us to determine which cells are removed, and which cells are created by U. All new cells are tagged with U as their creator. Let σ be a cell introduced by U. If σ is not feasible, then we should advance the front after the node U' that removes σ. Unfortunately, the data structure does not provide information on which child of U removes σ. In order to avoid cumbersome geometric tests to find U', we adopt a conservative approach that advances the front after all children of U. However, such an approach may lead to over-refine the extracted mesh with respect to the output of the same query answered on a general data structure. Similar problems arise when sweeping the current front before a node. See [11] for further details.

In the following, we describe in more detail two specific data structures for the case of vertex insertion. The first data structure applies to SMCs in arbitrary dimension d, while the second structure is specific for two-dimensional SMCs. Similar data structures for the case of vertex split can also be obtained, by combining the ideas presented here with the mechanism described in [8] for a single update.

A Structure for Delaunay SMCs We present a compressed structure designed for d-dimensional SMCs in $\mathbb{R}^d$, such as the ones used to represent the domain of scalar fields (e.g., for $d = 2$, the domain of terrains, or of parametric surfaces), based on Delaunay simplicial complexes. A simplicial complex is called a *Delaunay simplicial complex* if the circumsphere of any of its cells does not contain vertices in its interior. In two dimensions, Delaunay triangulations are widely used in terrain modeling because of the regular shape of their triangles and since efficient algorithms are available to compute them.

In a Delaunay SMC, the initial simplicial complex is a Delaunay one, and every other node U represents the insertion of a new vertex in a Delaunay complex. Thus, every extracted complex is a Delaunay complex.

For a set of points in general positions (no $d + 2$ points are co-spherical), the Delaunay complex is unique; thus, a Delaunay complex is completely determined by the set of its vertices, and the update due to the insertion of a new vertex is completely determined by the vertex being inserted. The data structure encodes cells in the following way:

- at the root, the initial simplicial complex is encoded in the winged data structure;
- for any other node U, the new vertex inserted by U is encoded (this defines an implicit description of the combination of the Node-Cell and the Cell-Vertex relations).

The cost of storing the implicit description is just equal to **v**. It can be reduced to zero by storing vertices directly inside nodes. The total cost of this data structure is equal to $d\mathbf{v} + 4\mathbf{a}$, by considering the space required by the two DAG relations (i.e., NA and AN).

Given a front on the SMC, the vertex stored in a node U is sufficient to determine how the corresponding mesh must be updated when sweeping the front through U, either forward or backward. This operation reduces to vertex insertion or deletion in a Delaunay simplicial complex.

This compression scheme is easily implemented for 2-dimensional SMCs based on vertex insertions in a Delaunay triangulation. For higher values of the dimension d, the algorithms necessary to update the current Delaunay complex become more difficult [6]. Deleting a point from a Delaunay simplicial complex in three or higher dimensions, as required when sweeping backward the front, is not easy; we are not aware of any existing implemented algorithm for such task, even in the three-dimensional case.

A Structure Based on Edge Flips This compression scheme can encode any two-dimensional SMC where nodes represent vertex insertions in a triangle mesh. It is efficient for SMCs where the number of triangles created by each update is bounded by a small constant b. The basic idea is that, for each node U, the corresponding update (which transforms a triangle mesh not containing a vertex p into one containing p) can be performed by first inserting p in a greedy way and then performing a sequence of edge flips. This process, illustrated in

Figure 7, defines an operational and implicit way of encoding the combination of the Node-Cell and Cell-Vertex relations.

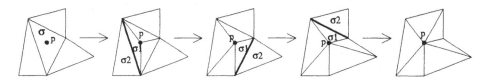

Fig. 7. Performing an update (insertion of a vertex p) through triangle split and edge flips. At each flip, the pair σ_1, σ_2 of triangle sharing the flipped edge is indicated.

First, p is inserted by splitting one of the triangles removed by p, that we call the *reference triangle* for p. This creates three triangles incident at p. Starting from such initial configuration, a sequence of edge flips is performed. Each edge flip deletes a triangle σ_1 incident at p, and the triangle σ_2 adjacent to σ_1 along the edge opposite to p, and replaces them with two new triangles incident at p, by flipping the edge common to σ_1 and σ_2. At the end, p has a fan of incident triangles which are exactly those introduced by update U.

The update represented by a node U is fully described by the new vertex p, a reference triangle σ, and a sequence of edge flips. Edge flips are represented by numerical codes. Let us consider an intermediate situation where a flip replaces the j-th incident triangle of p in a radial order around p (e.g., in counterclockwise order starting from the topmost triangle): then we use the number j to code the flip. The code of the first flip is in the range $0 \ldots 2$ since, at the beginning, there are only three triangles incident at p. Since each edge flip increases the number of these triangles by one unit, the code for the j-th flip is in $0 \ldots j + 1$. The total number of edge flips for a vertex p is $t - 3$, where t is the number of triangles created by the update U. Since $t \leq b$ the flip sequence consists of at most $b - 3$ integers, where the j-th integer is in the range $0 \ldots j + 1$. Therefore, the sequence of flips can be packed in a *flip code* of $\sum_{j=1}^{b-3}(\log_2(j + 2)) = \sum_{i=3}^{b-1}(\log_2(i)) = \log_2((b - 1)!) - 1$ bits.

The reference triangle σ for p is a triangle created by some update U' that is a parent of U in the DAG. Thus, to uniquely define σ, it is sufficient to give a reference to U' and an integer number identifying one of the triangles incident in the central vertex of U' according to a radial order (e.g., counterclockwise). Conventionally, we organize the data structure in such a way that parent U' is the first parent stored for U; thus, there is no need to encode it explicitly. The number identifying σ is in the range $0 \ldots b - 1$. We can pack the flip code and the index of σ together in $\log_2(b!) - 1$ bits. The space required for the implicit encoding of cells in this scheme is $v((\log_2(b!) - 1))$ bits.

Extending this compression scheme to higher dimensions is a non-trivial task. Edelsbrunner and Shah [6] showed that insertion of a point in a Delaunay simplicial complex reduces to a sequence of flips of $(k - 1)$-facets. This result could suggest that a coding based on flips may be possible for k-dimensional Delaunay SMCs built through incremental refinement. However, in k-dimensions it is not clear how the $(k-1)$-facets incident at a vertex could be sorted in such a way to allow the definition of a compact flip code. Moreover, it is difficult to guarantee a bounded degree of vertices in a k-dimensional simplicial mesh and, in any case, the number of flips is not guaranteed to be linear in the degree of the inserted vertex.

Discussion We have compared the sizes of the two compressed structures outlined above with the size of the explicit data structure, for a number of two-dimensional SMCs. On the average, the space occupied by the Delaunay compressed structure, and by the one based on edge flips, is about 1/4 and 1/3, respectively, of the space needed by the explicit structure without adjacencies.

It is interesting to compare the performance of query algorithms on an SMC when it is encoded through an explicit data structure, or through a compressed one, and the quality of the triangle meshes produced by such algorithms. Query algorithms provide the same meshes for the same input parameters with all compressed data structures, but the performances vary depending on the amount of work needed for reconstructing triangles with the specific structure.

Our experiments have shown that, if the given resolution filter does not refer to triangle attributes (e.g., it depends just on the geometry and location of triangles in space), the mesh extracted by a query algorithm using a compressed or an explicit structure are the same. On the contrary, if the resolution filter uses triangle attributes, then the resulting mesh may be quite different due to the approximation of such attributes in the compressed structures.

We have experimented with resolution filters that refer to *approximation errors* associated with triangles of an SMC. The resolution filter imposes an upper bound on the error of triangles that can be accepted in the solution of a query. In the compressed structure, a single error is associated with each node U, defined as the maximum approximation error of the triangles created by U. When such triangles are reconstructed in the current mesh, they receive the approximation error of U, which over-estimates their true error, hence forcing the extraction algorithm to over-refine the solution of the query. In this case, meshes extracted from the compressed SMC may be twice as large as those obtained from the explicit structure.

The performance of query algortihms has been monitored just for the explicit structure, and for the Delaunay-based compressed structure. The compressed structure based on edge flips is still under implementation. The explicit structure supports the extraction of triangle meshes formed by about 10^4 cells from SMCs containing about 10^5 cells, in real-time. Query algorithms on the Delaunay-based compressed structure are much slower. The increase in execution times is due to the on-line computation of a Delaunay triangulation. We expect better results with the structure based on edge flips.

6   Concluding Remarks

We have presented several alternative data structures for encoding a Simplicial Multi-Complex.

General-purpose data structures are characterized by encoding different subsets of the basic relations between the elements of an SMC. Different alternatives can be selected in order to adapt to the needs of a specific task, and to trade-off between space and performance. The SMC has been extended to cell complexes in [11]. However, the main difficulty in extending general-purpose data structures to general cell complexes lies in the intrinsic complexity of data structures for cell complexes, compared with those for simplicial complexes.

Compressed data structures have been defined for SMCs in the two-dimensional case, in which only the DAG structure is stored, and triangles are encoded through an implicit rule. Only the structure for Delaunay SMC extends to three or more dimensions easily, even if the problem of deleting a point from a Delaunay mesh in three or higher dimension is solved only from a theoretical point of view. We plan to investigate more general compressed structures for higher-dimensional SMCs in the future.

Compressed data structures can be much more compact than general-purpose ones, but, on the other hand, the performance of extraction algorithms can be degraded severely, because of additional work necessary to reconstruct the structure of meshes. In the two-dimensional case, it should be remarked that, while the Delaunay-based data structure is more compact than the one based on edge flips, the performance of the extraction algorithms is severely affected by numerical computation necessary to update the Delaunay triangulation.

Based on the data structures presented here, we have developed an object-oriented library for building, manipulating and querying two-dimensional SMCs, which has been designed as an open-ended tool for developing applications that require advanced LOD features [11]. In the current state of development, the library implements both the explicit and the Delaunay-based data structures, the algorithms described in [11], algorithms for building an SMC both for terrains and free form surfaces, application-dependent operations implemented on top of the query operations, mainly for GIS applications (interactive terrain visualization, contour map extraction, visibility computations, etc.). We are currently implementing the structure based on edge flips and a version of the library for dealing with three-dimensional SMCs for representing 3D scalar fields at variable resolution.

An important issue in any application which deals with large data sets is designing effective strategies to use secondary storage. To this aim, we have been studying data structures for handling SMCs on secondary storage. In [11], a disk-based data structure for two-dimensional SMCs is proposed in the context of a terrain modeling application. Such a structure organizes a set of SMCs, each of which describes a subset of a larger area. Individual SMCs reside on separate files, and two or more of them (i.e., the ones contributing to represent a relevant area of space) can be merged into a single SMC when loaded into main memory to answer a query. Future work involves defining and implementing query algorithms having a direct access to large SMCs resident on disk.

Acknowledgments

Part of the work described in this paper has been developed while the first author was on leave from the University of Genova at the University of Maryland Institute for Applied Computer Studies (UMIACS). The support of National Science Foundation (NSF) Grant "The Grand Challenge" under contract BIR9318183 is gratefully acnowledged. This work has been also partially supported by the Coordinated Project "A Library for Applications in Geometric Modeling" of the Italian National Research Council under contract 98.00350.

References

[1] M. de Berg and K. Dobrindt. On levels of detail in terrains. In *Proceedings 11th ACM Symposium on Computational Geometry*, pages C26–C27, Vancouver (Canada), 1995. ACM Press.

[2] L. De Floriani, P. Magillo, and E. Puppo. VARIANT - processing and visualizing terrains at variable resolution. In *Proceedings 5th ACM Workshop on Advances in Geographic Information Systems*, Las Vegas, Nevada, 1997.

[3] L. De Floriani, P. Magillo, and E. Puppo. Efficient implementation of multi-triangulations. In *Proceedings IEEE Visualization 98*, pages 43–50, Research Triangle Park, NC (USA), October 1998.

[4] L. De Floriani and E. Puppo. Hierarchical triangulation for multiresolution surface description. *ACM Transactions on Graphics*, 14(4):363–411, October 1995.

[5] L. De Floriani, E. Puppo, and P. Magillo. A formal approach to multiresolution modeling. In R. Klein, W. Straßer, and R. Rau, editors, *Geometric Modeling: Theory and Practice*. Springer-Verlag, 1997.

[6] H. Edelsbrunner and N. R. Shah. Incremental topological flipping works for regular triangulations. *Algorithmica*, 15:223–241, 1996.

[7] A. Guéziec, G. Taubin, F. Lazarus, and W. Horn. Simplicial maps for progressive transmission of polygonal surfaces. In *Proceeding ACM VRML98*, pages 25–31, 1998.

[8] H. Hoppe. View-dependent refinement of progressive meshes. In *ACM Computer Graphics Proceedings, Annual Conference Series, (SIGGRAPH '97)*, pages 189–198, 1997.

[9] P. Lindstrom, D. Koller, W. Ribarsky, L.F. Hodges, N. Faust, and G.A. Turner. Real-time, continuous level of detail rendering of height fields. In *Comp. Graph. Proc., Annual Conf. Series (SIGGRAPH '96), ACM Press*, pages 109–118, New Orleans, LA, USA, Aug. 6-8 1996.

[10] D. Luebke and C. Erikson. View-dependent simplification of arbitrary polygonal environments. In *ACM Computer Graphics Proceedings, Annual Conference Series, (SIGGRAPH '97)*, pages 199–207, 1997.

[11] P. Magillo. *Spatial Operations on Multiresolution Cell Complexes*. PhD thesis, Dept. of Computer and Information Sciences, University of Genova (Italy), 1999.

[12] A. Paoluzzi, F. Bernardini, C. Cattani, and V. Ferrucci. Dimension-independent modeling with simplicial complexes. *ACM Transactions on Graphics*, 12(1):56–102, January 1993.

[13] E. Puppo. Variable resolution terrain surfaces. In *Proceedings Eight Canadian Conference on Computational Geometry*, pages 202–210, Ottawa, Canada, August 12-15 1996.

[14] E. Puppo. Variable resolution triangulations. *Computational Geometry Theory and Applications*, 11(3-4):219–238, December 1998.

[15] J.C. Xia, J. El-Sana, and A. Varshney. Adaptive real-time level-of-detail-based rendering for polygonal models. *IEEE Transactions on Visualization and Computer Graphics*, 3(2):171–183, 1997.

Spatial Indexing with a Scale Dimension

Mike Hörhammer[1]   Michael Freeston[1,2]

[1]Department of Computer Science, University of California, Santa Barbara
[2]Department of Computing Science, University of Aberdeen, Scotland

horhamm@cs.ucsb.edu   freeston@alexandria.ucsb.edu

Abstract. It is frequently the case that spatial queries require a result set of objects whose scale – however this may be more precisely defined – is the same as that of the query window. In this paper we present an approach which considerably improves query performance in such cases. By adding a scale dimension to the schema we make the index structure explicitly "aware" of the scale of a spatial object. The additional dimension causes the index structure to cluster objects not only by geographic location but also by scale. By matching scales of the query window and the objects, the query then automatically considers only "relevant" objects. Thus, for example, a query window encompassing an entire world map of political boundaries might return only national borders. Note that "scale" is not necessarily synonymous with "size". This approach improves performance by both narrowing the initial selection criteria and by eliminating the need for subsequent filtering of the query result. In our performance measurements on databases with up to 40 million spatial objects, the introduction of a scale dimension decreased I/O by up to 4 orders of magnitude. The performance gain largely depends on the object scale distribution.
We investigate a broad set of parameters that affect performance and show that many typical applications could benefit considerably from this technique. Its scalability is demonstrated by showing that the benefit increases with the size of the query and/or of the database. The technique is simple to apply and can be used with any multidimensional index structure that can index spatial extents and can be efficiently generalized to three or more dimensions. In our tests we have used the BANG index structure.

1 Introduction

Conventionally, an index of spatial objects takes only the physical extent of the objects (or their bounding boxes) into account. Queries on such an index can thus only constrain the location of the result objects. Any further constraints on the type or size of the objects can only be imposed by appending an additional filtering step to the result of the spatial query. This procedure is obviously inefficient in two ways: the initial spatial query may retrieve large numbers of irrelevant objects, which then have to be removed at the second step. As a result, two particular classes of spatial queries are still poorly served in practice:

R.H. Güting, D. Papadias, F. Lochovsky (Eds.): SSD'99, LNCS 1651, pp. 52-71, 1999.
© Springer-Verlag Berlin Heidelberg 1999

1. Window queries which are further constrained to show only features commensurate with the size of the query window; for example, a map display query should retrieve only those features visible at the display scale.
2. Window queries further constrained to show only features of a particular range of subtypes ("range" implying some linear classification of subtypes); for example, a query of political boundaries in a window encompassing the whole world might reasonably only return national and state/province boundaries.

The conventional way of trying to improve the performance of queries of the first class is to break up the database into separate subsets of objects of more-or-less arbitrarily chosen ranges of sizes. For queries of the second class, the familiar "layering" technique of GIS helps to improve performance by correspondingly partitioning the database into subsets of a single type.

But both of these are very inflexible techniques. The first moves away from automated indexing towards manual intervention - which is hardly an advance; and the second suffers from a general objection to layering: it cannot efficiently resolve all the subtleties of subtyping which can be better represented in an object type hierarchy. In addition, the need for a multiplicity of additional files and associated indexes introduces potential new inefficiencies.

In this paper we present a more flexible and integrated technique that can considerably improve query performance under such circumstances. We achieve this by introducing an additional *scale dimension* to the data space of the entire database of spatial objects. This allows the introduction of a new type of query that includes a scale constraint on the objects returned in the query window.

While our approach is applicable to most index structures that can handle multidimensional and spatial data we chose the BANG index structure for our implementation. The ability of this structure to mix spatial and point dimensions reduces the additional space overhead incurred by the scale dimension.

With large data sets of spatial extents (40 million) our performance tests show a reduction in I/O accesses of up to four orders of magnitude. With smaller databases (1 million spatial extents) the improvement is still up to a factor of nearly 600. CPU-time is negligible whether a scale dimension is used or not. As our approach especially favors large queries and large databases it is superscalable.

2 Related Work

There have been several approaches to separating small and large objects for increased performance. One such is the filter tree [SK95], which is something of a misnomer, since it is actually a forest of B-trees. This spatial index structure has a number of layers with one tree for each. Each layer i is assigned a $2^i \times 2^i$ grid layer. Beginning at the top level 0, an inserted spatial object "falls" through the sequence of increasingly dense grid layers until its bounding box hits one of the grid lines of layer n. The according object is then inserted in layer $n-1$, in which it is completely contained by one of the grid squares. Large objects are always in high layers whereas small objects generally *tend* to be in lower levels. The filter tree performs well for spatial joins, but the performance of range queries is reported to be inferior to the *R-tree*. Essentially, the filter tree partitions the set of indexed objects according to a property partially correlated to size, and a separate index structure is maintained for each of the partitions.

A possible way of improving the performance of range queries arises from the fact that some of the layers tend to contain relatively few objects so that those layers could be kept in a cache. The overhead of accessing several separate index structures would then be decreased. Unfortunately, this method does not scale with increasing database size unless the cache can be increased accordingly. The choice of granularity for the smallest grid size is also somewhat arbitrary, and the optimum would depend on the distribution of object sizes. For dynamic data this optimum could change over time, and the introduction or removal of a grid layer would be a multiple update operation. This makes the design unsuitable for dynamic applications with predictable update characteristics.

There are two other approaches that follow a similar pattern: the layering technique in GIS and level-based indexing. In traditional GIS, layering partitions the database in separately indexed groups like "countries", "cities", "streets"... Level-based indexing [Kan98] in contrast separates objects more or less directly based on their extent. The objects are indexed in multiple indexes according to a nesting-based partitioning technique. As with the filter tree, this partitioning is also related to the object size, but uses a different approach. In tests, performance improvements could be achieved if most of the partitions (those with the larger objects) were indexed in main memory or cached. This turns out to be quite advantageous for two reasons: large objects tend to be inefficiently indexed and can benefit tremendously from being located in main memory. Furthermore, large objects tend to be in the minority in geographic databases (with the exception of elevation lines) and thus might reasonably fit into main memory. In the tests only the partition with the smallest objects (mainly indexed as point data) was indexed on disk. If *all* the indexes were kept on disk the method would become inefficient because of access to several separate indexes. This is the same problem as that faced by the filter tree and the fundamental reason why its range query performance is worse than that of the *R-tree*.

A related approach based on a single index is described in [Kan98] where promotion-based indexing is presented for different index structures such as the *R-tree* and *R*-tree*. This approach allows 'large' spatial objects to be promoted to a higher index level. This makes it possible to reduce the extent of a page region at the index level from which the object was promoted. Queries tend to perform better because the query region then intersects less page regions. A performance improvement of up to 45% is reported in tests with different types of data, although in some cases the improvement was considerably less. One possible disadvantage is that the height of the tree may increase as a consequence of the object promotions. However, in contrast to previous promotion methods this approach can be used with intersecting page regions as found in the *R-tree*, the *R*-tree* and other spatial data structures. Promotion can be either extent-based or nesting-based.

– It should be noted that these four approaches have a different target compared to ours: with the exception of the layering technique in GIS they have all been devised to improve performance while returning *all* objects geographically in range. In contrast, our objective is to avoid returning the complete set and to access only the "interesting" scales. The fundamental question is thus not whether our approach is better than the filter tree, level-based or promotion based indexing. Rather, the question is whether it is better for our specific objective of accessing only those objects of sufficiently large-scale to match the scale of a given query box. Although it would be a straightforward variation of all these related techniques to restrict access to those regions of the index structure which contain large objects, there are unfortunately considerable drawbacks to this approach in all cases (including the layering technique):

– None of them is based on scale, but rather more or less directly on size. The scale and the size of an object need not be related (we will make this point using the example of elevation lines). If the four approaches were all changed to use scale instead of their original partitioning criterion then the filter tree would become a special case of level-based indexing; level-based indexing would become fundamentally identical with the layering technique; and promotion-based indexing would loose its ability to reduce page region sizes – its fundamental raison d'être. Level-based indexing could be expected to perform quite well except for the overhead inherent in accessing several separate index structures.

– All four approaches fundamentally partition the objects into a small number of "buckets" (layer, index-layer and containment level respectively). With a scale dimension, in contrast, scale is assigned on a near continuous scale – as continuous as the used data type allows. This allows a more graceful approximation of, for example, differently scaled printed maps.

3 BANG Indexing

We emphasize that the approach described here does not depend on any specific multidimensional indexing method, except that the method must be able to support an additional linear scale dimension in an index of extended spatial objects. For completeness, however, we review below the organization of the BANG index structure used in our tests.

3.1 Multidimensional Point Indexing

A BANG index [F87, F89a] has a balanced tree structure which represents a recursive partitioning of an n-dimensional data space into subspaces. Each subspace corresponds to a disk page, and the partitioning algorithm attempts to ensure that no page is ever less than 1/3 full (but see [F95]). Generally they will, like the B-tree, be approximately 69% full. Tuples are represented by points in the data space, and each data page contains either full tuples (in a clustered index) or index keys with associated pointers to individual tuples stored on a heap (in an unclustered index).

When a page overflows, it is split into two by generating a sequence of regular binary partitions of the corresponding subspace, cycling through the dimensions of the data space in a fixed order, until the content of the two resulting subspaces is distributed as equally as possible. This partition sequence is represented by a Peano code (a variable-length binary string) which uniquely identifies the external boundary of each resulting subspace. The set of subspaces represented within each node of the index tree thus appears as a set of Peano code index keys. Associated with each key k is a pointer to a node at the next lower level of the tree. This node represents the subspace whose external boundary is defined by key k.

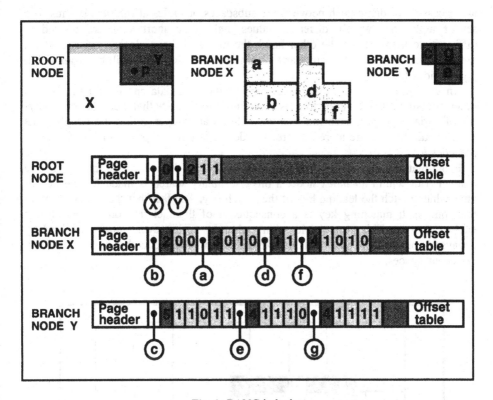

Fig. 1: BANG indexing

Note however that, although an index node containing f key entries represents f subspaces, there is not in general a 1:1 correspondence between entries and subspaces i.e. some subspaces may be defined by a non-singular subset of the f keys. This is a consequence of a unique feature of BANG indexing, whereby the partitioning algorithm allows one subspace to enclose another. In general, therefore, a subspace may have a set of internal boundaries as well as an external boundary. Thus an individual key k in an index node may both represent the external boundary of a subspace s and an element of the set of internal boundaries of a subspace which encloses s.

A further consequence of this enclosure property is that a BANG index is a true prefix tree i.e. any leading bits which are common to all the keys within a node can be deleted and appended to the key representing the node at the index level above. This is because the Peano code of any subspace is a prefix of the Peano code of every subspace or tuple which it encloses. The full key representing an individual subspace or tuple is thus never stored explicitly in the index. It can however always be found by concatenating the keys encountered in the direct descent from the root of the tree to that subspace or tuple.

An additional feature is that it is rarely necessary to employ all the bits of the n index attributes of a tuple to generate its index key. During an exact-match search, the index key of the query tuple is generated dynamically and incrementally as the search proceeds down the tree. At each index level, only as many key bits are generated as are necessary to distinguish between the subspaces or tuples at that level. Thus, if a set of tuples has widely differing values, only very short keys are needed to differentiate them. If, on the other hand, the tuples all have closely similar values, then their index keys will share many common prefix bits which will be promoted to higher index levels.

An example of a simple two-level partitioning of a data space, and the BANG index structure corresponding to it, is given in figure 1. Note that, because index keys are of variable length, each index entry includes an integer giving the length of its Peano code. Entries are stored in prefix order: if key *a* is a prefix of key *b*, then *a* precedes *b*.

Exact-match tuple search proceeds downwards from the root in the familiar B-tree manner. But within an index node, a bit-wise binary search is made for the key or keys which match the leading bits of the search key. The fact that there may be more than one such matching key is a consequence of the fact that one subspace can enclose another. This potential location ambiguity is, however, simply resolved by always choosing the *longest* matching key, since this represents the innermost of any nested subspaces.

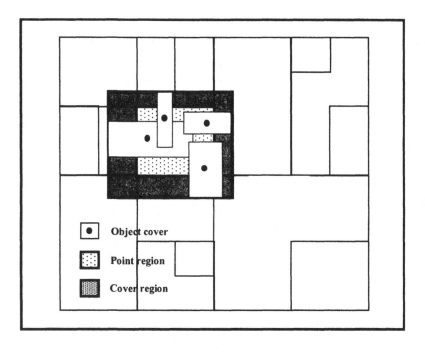

Fig. 2: Object cover indexing

Figure 1 also demonstrates a critically important feature of domain indexing: it is not in general necessary to represent the whole data space in the index, if large tracts of the space are empty. In the example of figure 1, some exact-match queries will fail at the root index level. In general, therefore, the *average path length* traversed to answer an exact-match query *will be less than the height of the index tree*. The more highly correlated the data, the greater the advantage of this feature compared to the conventional range-based indexing of the B-tree. This advantage is reinforced by the extremely compact index due to the (on average) very short index entries. This leads to high fan-out ratios and hence a reduced index tree height and consequent faster access for a given data set.

3.2 Object Cover Indexing

A method of extending BANG multidimensional point indexing to spatial object covers (i.e. rectangular bounding boxes) was originally proposed in [F89b]. The method used here is a variation on that design, which involves a dual representation. Point indexing is used to partition the data space into *point regions* according to the centre points of the object covers. A rectangular *cover region* is then associated with each point region, such that the cover region boundary just encloses the covers of all the objects in the point region. (See figure 2). Additional lower and upper bound coordinate pairs representing this cover region are thus added to each index entry of the point index representation.

The dual is reflected in the queries: containment queries need only check for overlap with point regions, whereas intersection queries must check for overlap with the cover regions.

4   Introducing a Scale Dimension

We propose the introduction of an additional object attribute, which, according to some measure we will define, represents the *scale* of the indexed spatial object.

We define the scale $S(O)$ of an object O as a perceived value of importance or size. It can best be understood by considering maps with different scales: a larger-scale map contains larger-scale objects. In the case of a map there actually is a well-known measure of scale: a length l on the map relates to a multiple m of l in reality. The larger the value of m, the larger the scale of the map. Given a constant physical size of the map this results in a larger geographic area covered[1].

We define:

– $S(M)$ as the scale of a map M. $S(M)$ can be easily measured and is typically given as $1:m$.

– $S(Q)$ as the scale of a query Q. $S(Q)$ is loosely defined based on the scale of a map of the same geographic region.

---

[1] The geographic area covered by a printed map depends on the scale and the physical size of the map. For simplicity we assume the physical size of the map constant. This assumption does not reduce generality but simplifies the discussion.

– $S(O)$ as the scale of an object. This function is based on the perception of which objects should be represented on a map of a given scale. Given a map M_1 with scale $S(M_1)$ we assign the scale $S(O_1) = S(M_1)$ to an object O_1 that is perceived just large-scale enough to be on the map M_1.

For most objects we assume an automatic assignment of scale that sufficiently approximates human perception. For a large class of objects on a map the assignment of scale based on geographic extent is reasonable. Assuming automatic scale assignment, the value for this attribute does not depend on any information that is not already inherent in the original data. The explicit storage of this attribute is thus redundant. The important purpose of the scale value is that it causes the index structure not only to cluster objects geographically, but also by scale. Objects of "similar" scale such as the countries of Germany and Japan would be likely to be clustered close to each other, despite their geographic distance.

A query Q_1 with scale $S(Q_1)$ could (transparently to the user) specify a range of desired object scales $[S(Q_1), \max_{scale}]$. In this query all objects O of scale $S(O) < S(Q_1)$ are ignored.

Of course the introduction of an additional dimension also has a cost, and certain queries might be unfavorably affected by the use of a scale dimension. Specifically, one expects this to happen when a query is spatially so restricted that there are no objects of scale less than $S(Q_1)$. In this case the clustering according to scale does not help and only the incurred cost counts. However, such a small query would need very few disk accesses, and thus any performance decrease would have very little impact. We verify this in the section on "Performance Evaluation".

In the next section we turn our attention to the criteria which should be applied when using the additional scale dimension.

5 Assignment of Scale

5.1 Scale Based on Extent

Several important decisions will influence the success of using a scale dimension:

1. How is the scale of a spatial object assigned?
2. If the scale is a function of the size of an object, should it be logarithmic or linear?
3. If the scale is a function of the size of an object, how much smaller than the query can an object be to still be considered large-scale enough?
4. Which is the domain of sizes or scales we would like to distinguish? Objects outside the domain of interest would be assigned the maximum / minimum scale value.

Point 1: *How is the scale of a spatial object assigned?* Scale should be assigned to objects to represent human perception as closely as possible. For simplicity we based scale on geographic extent in our performance tests. This coincides with scale perception for a large set of objects typically found on maps.

We decided to assign the scale of an object based on the maximum of its extent in the x-dimension and the y-dimension. The longer side of a very long and thin object thus determines its scale. Another option would be to measure the surface covered by the minimal bounding box or the actual object, itself. The latter would have two disadvantages, illustrated by the following examples: lines of latitude and longitude would be assigned a negligibly small scale; and rivers and highways would be assigned a scale value depending on whether they are horizontally/vertically aligned (tiny bounding box) or diagonally (big bounding box).

We thus introduce the scale function:

$$S_1(O) = \max(extent_x(O), extent_y(O))$$

Point 2: *If the scale is a function of the size of an object, should it be logarithmic or linear?* We chose to use a logarithmic scale. To understand our reasoning, consider 3 spatial objects A, B and C.

$$S_1(A) = 2S_1(B) = 4S_1(C)$$

Comparing the relative sizes of A, B and C, A relates to B as B relates to C. Unfortunately, an index data structure works with a linear scale. On such a linear scale, France and Germany would be considered further apart in scale than a city from a village. The user, in contrast, would consider France and Germany very similar in scale because of the relative difference.

We thus do not index $S_1(O)$ for an object O, but rather

$$S_2(O) = k_1 + k_2 \log_b(S_1(O))$$

For simplicity, we assume $b = 2$. k_1 and k_2 will be considered further below.

Point 3: If the scale is a function of the size of an object, how much smaller than the query can an object be to still be considered large-scale enough? In our tests we considered a factor of 256 in each dimension to be appropriate. Thus, an object of size $\frac{1}{256} \times \frac{1}{256}$ of the query window is considered just large enough to be part of the answer. Any other factor could be chosen, but we will base further data on the value 256.

There is no concept of objects too large for a query window. Even a village-scale query will return the according continent.

Point 4: Which is the domain of sizes or scales we would like to distinguish? Objects outside the domain of interest would be assigned the maximum / minimum scale value. Finally, we must decide on a reasonable scale range within which we would like to distinguish objects by scale. Thus, we have to choose k_1 and k_2 in the formula for s_2. The naïve approach would be to index the complete range: suppose the spatial extents of the original object are indexed by 64-bit values. On a logarithmic scale, this results in a domain from 0 to 63. Most objects on a map would typically only use the upper few values. A value of 0 on the logarithmic scale from 0 to 63 would represent an object with side length $\frac{1}{2^{64}}$ of the map.

If such small objects are of interest, the query must obviously be so restricted to a tiny partition of the domain that only very few disk accesses are needed. Thus we are mainly concerned with distinguishing between larger-scale objects. We are willing to consider extremely small-scale objects equal to each other (for indexing purposes) and truncate the lower values in the logarithmic range ([0, 63] in the example). We decided to index a minimum scale value for objects with length $\frac{1}{2^{17}}$ of the corresponding dimension's domain. Smaller objects are indexed with the same scale value.

Should we only truncate on the lower end of the logarithmic scale? Let's consider two objects A and B with sizes $\frac{1}{4} \times \frac{1}{8}$ and $\frac{1}{64} \times \frac{1}{256}$ of the domain, respectively. It is clear that $s_2(A) \neq s_2(B)$. Nevertheless both objects would be returned by any query in whose range they fall. This is because they are not too small-scale even for the largest-scale queries (based on our maximum factor 256 between query and returned object). As a result, we can also truncate the upper few values from the logarithmic scale. The scale we have actually used for our performance tests is:

$$s_2(O) = k_1 + k_2 \log_b(s_1(O))$$

with $b = 2$, $k_1 = -55k_2$, $k_2 = 0x10000000_{(hex)} = 268,435,456_{(dec)}$.

In Table 3 the relation between object size and scale according to this function can be seen in the left-most two columns.

5.2 Scale Based on Type

Up to now, we have considered the scale of an object to depend on its size only. This allows for automatic assignment by the software, transparent to the user. For most types of objects this assignment actually is very similar to human perception of which objects should appear on a map of a given scale.

But how does this scale assignment fare when we index elevation lines? Elevation lines have some special conceptual properties:

- they are contained one in another and thus have a large intersection
- conceptually, they do not represent different objects; rather they represent all geographic points with the same elevation x

Elevation lines (meters)	
1000, 2000, 3000, …	$n+6i$
1500, 2500, 3500, …	$n+5i$
	$n+4i$
1100, 1200, 1300, …	$n+3i$
1050, 1150, 1250, …	$n+2i$
	$n+i$
1010, 1020, 1030, …	n

Table 1: Scale assignment to elevation lines

On a large-scale map elevation lines might be recorded in increments of 500 meters or more. On a small-scale map, in contrast, this would be too coarse and increments of only 1 meter might be used. Obviously, an elevation line at 1000 meters is conceptually considered larger-scale than at 1001 meters or 999 meters. We propose to assign scales to elevation lines based on the according elevations.

Table 1 informally shows one possible algorithm for the scale assignment for elevation lines. Which values exactly are appropriate for n and i can be easily adjusted by comparing with common practice for printed maps. Note that the number of objects of scale x decrease exponentially with increasing x. In Table 1, the number of objects with scale n is roughly 10 times the number of objects with scale $n+3i$. Further, the objects with scale $n+3i$ are roughly 10 times more than objects with scale $n+6i$. In the section on "Performance Evaluation" we show that it is, in fact, advantageous to have more small-scale than large-scale objects. Similar scale assignments can be applied to lines showing levels of precipitation, temperature and other attributes.

5.3 Initial Failures

For the first test results we used a linear scale for the scale dimension and did not truncate the lower or higher range. As a result, the value of the scale attribute did not sufficiently influence clustering. The performance figures were worse than without a scale dimension. Another concern was the type of objects indexed. For an application using data of similarly scaled objects, our approach is clearly not appropriate. But geographic databases often have a wide bandwidth of differently scaled objects. This is shown in the section on "6.2 Relevance of Data and Queries Chosen".

6 Performance Evaluation

We first present the data sets we have used and then the performance results on those sets. We justify the choice of data distributions in terms of typical geographic databases. To see the impact of different parameters we have used synthetic data sets instead of an actual geographic database. Furthermore, many actual geographic databases contain mainly point data that are more efficient to handle than spatial data. Other databases often contain only a rather small number of objects. We want to show that our approach can make many applications scalable, even when millions of truly spatial objects are in the database. Thus, our performance results include measurements on databases with up to 40 million spatial extents – a substantial and demanding database size.

6.1 Type of Data and Queries

We have run performance tests with databases with 1 million and 40 million objects. For each size of database we used 3 object size distributions. The section on the "6.2 Relevance of Data and Queries Chosen" shows that distribution # 1 is most typical for geographic databases.

Database size	Object size distribution		Performance result	
			Without scale dimension	With scale dimension
1 million extents	# 1	Fig 3 left; Table 3	Fig 5 left	Fig 5 right
	# 2	Fig 3 right; Table 3	Fig 6 left	Fig 6 right
	# 3	Fig 4; Table 3	Fig 7 left	Fig 7 right
40 million extents	# 1	Table 3	Fig 8 left	Fig 8 right
	# 2	Table 3	Fig 9 left	Fig 9 right
	# 3	Table 3	Fig 10 left	Fig 10 right

Table 2: The databases created for performance tests

The object centers are randomly distributed. The object size distributions (with 1 million extents) are specified in Fig. 3, Fig. 4 and Table 3.

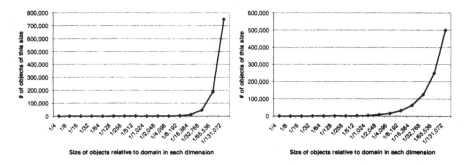

Fig. 3: The object size affects the number of such objects. On the left-hand side the number of objects of a given size is inversely proportional to the area (object size distribution # 1). On the right hand side the number of objects of a given size is inversely proportional to the side length (object size distribution # 2).

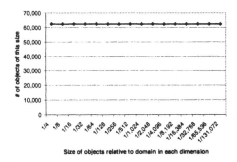

Fig. 4: Object size distribution # 3: The number of objects of each size is the same (bounded by some minimum and maximum size). This means the objects in the database are uniformly distributed according to the scale dimension.

		1,000,000 extents			40,000,000 extents		
		Distribution			Distribution		
		1 (Fig. 5)	2 (Fig. 6)	3 (Fig 7)	1 (Fig. 8)	2 (Fig 9)	3 (Fig 10)
1/4	0x70000000	0	30	62,500	0	1,220	2,500,000
1/8	0x60000000	0	31	62,500	0	1,221	2,500,000
1/16	0x50000000	0	61	62,500	0	2,441	2,500,000
1/32	0x40000000	0	122	62,500	2	4,883	2,500,000
1/64	0x30000000	0	244	62,500	7	9,766	2,500,000
1/128	0x20000000	0	488	62,500	29	19,531	2,500,000
1/256	0x10000000	3	977	62,500	114	39,063	2,500,000
$1/2^9$	0x00000000	12	1,953	62,500	458	78,125	2,500,000
$1/2^{10}$	0xf0000000	46	3,906	62,500	1,831	156,250	2,500,000
$1/2^{11}$	0xe0000000	183	7,813	62,500	7,324	312,500	2,500,000
$1/2^{12}$	0xd0000000	732	15,625	62,500	29,297	625,000	2,500,000
$1/2^{13}$	0xc0000000	2,930	31,250	62,500	117,188	1,250,000	2,500,000
$1/2^{14}$	0xb0000000	11,719	62,500	62,500	468,750	2,500,000	2,500,000
$1/2^{15}$	0xa0000000	46,875	125,000	62,500	1,875,000	5,000,000	2,500,000
$1/2^{16}$	0x90000000	187,500	250,000	62,500	7,500,000	10,000,000	2,500,000
$1/2^{17}$	0x80000000	750,000	500,000	62,500	30,000,000	20,000,000	2,500,000
Object Size	Scale Value (signed int)						

Table 3: Object size distributions with 1 million an 40 million spatial extents.

6.2 Relevance of Data and Queries Chosen

We have chosen the data and the queries to represent applications with geographic databases as faithfully as possible. Which parameter of the data distributions – if inappropriately chosen – could make the performance results (partially) irrelevant for typical applications? There are several candidates, which we consider subsequently to show that our performance results are, in fact, relevant to typical applications:

– Choice of queries	(A)
– Choice of object locations (centers of represented objects)	(B)
– Choice of object sizes	(C)
– Assignment of object scales	(D)

(A) We have chosen a wide range of queries for each test. The query locations are randomly distributed, which we consider an appropriate choice. For each query size between 100% and $\frac{1}{256} = 0.4\%$ of each dimension (between $(100\%)^2 = 100\%$ and $(\frac{1}{256})^2 = 0.0015\%$ of the 2-dimensional domain) we have run 100 random queries (10 random queries when 100 queries took too long[2]). The performance results in Fig 5 through Fig. 10 show the minimum, maximum and average values for all queries. With respect to an application using only a narrow range of query sizes the relevant range can be easily extracted from the figures.

---

[2] The results based on 10 queries of each size are marked by gray backgrounds in the graphs (in Fig 8, Fig. 9 and Fig. 10).

(B) The object centers are randomly distributed in our data sets. Considering that there are only very few "empty" areas on maps, this seems an appropriate distribution. Note that also the oceans contain objects such as lines of depth, temperature and others. As we apply the same object distribution to the tests with and without scale dimension the object center distribution should not have any great effect on our performance results.

(C) The object size distribution is comparable to the object center distribution considering its impact on our performance results. We consider the broad range of object sizes we have used typical for geographic applications. Nevertheless, whatever object size distribution is chosen impacts both the results with and without scale dimension equally. The performance difference between tests with and without scale dimension depends on the scales of objects, not their sizes (note the difference between size and scale).

(D) The object scale distribution finally is the concept with the actual influence on how much our approach can help increase performance. If we have chosen non-typical distributions for it our performance results are of limited relevance to actual applications. We will thus compare our distributions with the distribution found in a very large digital map. Note that the object scale distribution is solely chosen to influence what should be returned upon a query of a given scale. Our claim is that exactly object scale distribution 1 will typically result in query answers similar to most existing printed maps.

Informal proof: Let us assume a geographic database containing all objects that can be found on *any* existing printed map. When we ask a query on such a database we would perfectly expect to be returned exactly the same objects that would actually be on a printed map of the same area. We thus consider exactly those objects both within the specified geographic region *and* the appropriate scale interval.

The important fact is that most printed maps we typically work with, show a very similar number of objects. Let's only consider the detail information and not for example the country borders that are only partially visible. This detail information in a query Q_i has scale $S(Q_i)$. Had a map much more objects than typical, it would be considered too detailed. Consequently, some objects should be assigned a smaller scale so that they only appear in smaller-scale maps. Had a map much less objects, it would be considered not detailed enough. Thus some objects should be assigned a larger scale so that they also appear in otherwise "empty" larger-scale maps. Obviously a scale assignment to objects is considered reasonable by cartographers if maps of different scales contain a roughly similar number of objects.

What does this mean for the object scale distribution? We consider a query window Q_i that covers a proportion a_i of the complete domain. Independent of the scale we expect a typical number n of objects of scale $S(Q_i)$ as result to this query. If a query Q_i covering a proportion a_i of the domain contains n objects of scale $S_i = S(Q_i)$ then the complete domain can typically be expected to contain roughly $\frac{n}{a_i}$ such objects. Thus the overall number of objects of scale $S(O) = S_i$ tends to be inversely proportional to the area covered by a map (query) with scale $S(Q_i) = S_i$.

This distribution is exactly object scale distribution #1 presented in Fig. 3 (left side) and Table 3. In fact, this distribution shows the best performance results among the three distributions we have used. We have added the other two distributions to show how our method performs with adverse object scale distributions.

If a geographic database has an object scale distribution like distribution #3 or even "worse" then it contains only little small-scale detail. Serious applications with large geographic databases arguably can be expected to contain considerable small-scale detail.

6.3 Performance Measurements

The following figures present our performance measurements (I/O) for the different data sets. We have run 100 random queries of each query size (figures with white background). With those tests involving large numbers of disk accesses per query we only performed 10 random queries of each size (figures with gray background). The query sizes can be seen on the x-coordinate of the figures.

Thus we have varied
- The object size distribution (each row with a pair of figures relates to one object size distribution)
- The database size (1 million spatial extents in the first 3 rows, 40 million in the last 3 rows)
- The query size (varied on the x-axis in each figure)
- The use of a scale dimension (without a scale dimension on the left-hand side and with a scale dimension on the right-hand side)

The vertical bars in each figure show the minimum and maximum values for single queries (among 10 or 100 random queries each). The curve crosses each vertical bar at the average number of disk accesses per random query.

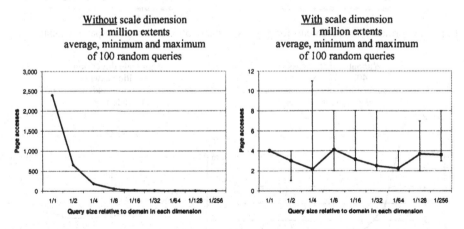

Fig. 5: Performance measurements for object size distribution 1 (1 million spatial extents) – according to the proof in section 0 this is the typical distribution for large geographic databases.

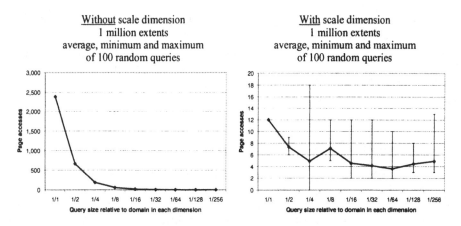

Fig. 6: Performance measurements for object size distribution 2 (1 million spatial extents)

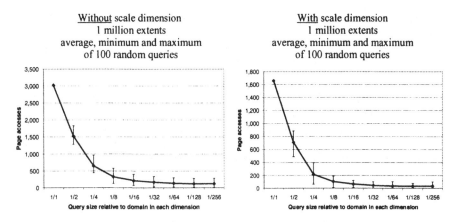

Fig. 7: Performance measurements for object size distribution 3 (1 million spatial extents)

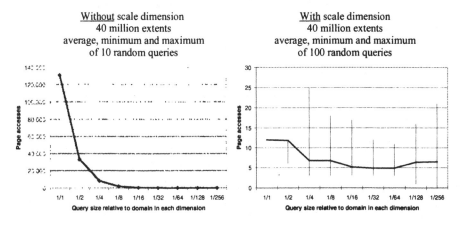

Fig. 8: Performance measurements for object size distribution 1 (40 million spatial extents) – according to the proof in section 0 this is the typical distribution for large geographic databases.

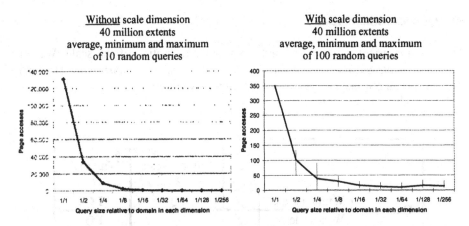

Fig. 9: Performance measurements for object size distribution 2 (40 million spatial extents)

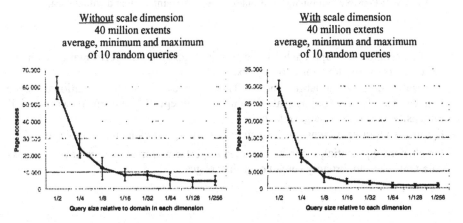

Fig. 10: Performance measurements for object size distribution 3 (40 million spatial extents). Note that the query on the *complete* domain (returning 40 million objects) could not be measured in this case, because main memory was exhausted.

6.4 Analysis

In Fig. 5 through Fig. 10 it becomes clear that the scale dimension can considerably improve performance. Especially in the typical[3] case that there are far more small-scale than large-scale objects (object size distribution 1, Fig. 5, Fig. 8) the improvement is substantial. CPU-time is negligible with and without scale dimension.

With increasing database size (from 1 million to 40 million extents) the improvement factor due to the scale dimension even increases. Our technique thus scales very well with the size of the database.

---

[3] As proven in the section "Relevance Of Data And Queries Chosen" this is typical for a large geographic database.

6.5 Queries without Scale

The question arises how queries perform *without* restriction of the scale. We thus *index with scale* but make the conscious decision to *query without scale*. The query then returns all objects geographically in the query window.

We thus have three approaches:

| | | | Indexing | |
| | | | Without scale | With scale |
			Γ	Γ^+
Querying	Without scale	Q^-	ΓQ^-	$\Gamma^+ Q^-$
	With scale	Q^+	–	$\Gamma^+ Q^+$

Table 4: Possible combinations of indexing and querying with and without scale

It can be assumed that approach $\Gamma^+ Q^-$ results in slightly lower performance than the initial ΓQ^-. This is obvious for two reasons:

- Γ^+ results in a 25% larger index. This is because of the additional space requirements for the (point) scale dimension (1 integer). The original two (spatial) dimensions need two integers each.
- Γ^+ clusters by three dimensions instead of two. As Q^- fundamentally only uses $\frac{2}{3}$ of the clustering potential, it can be expected to perform worse: queries have to access data that are more broadly spread over disk pages.

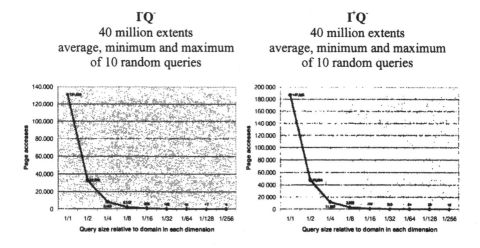

ΓQ^-
40 million extents
average, minimum and maximum
of 10 random queries

$\Gamma^+ Q^-$
40 million extents
average, minimum and maximum
of 10 random queries

Fig. 11: Performance comparison between ΓQ^- and $\Gamma^+ Q^-$

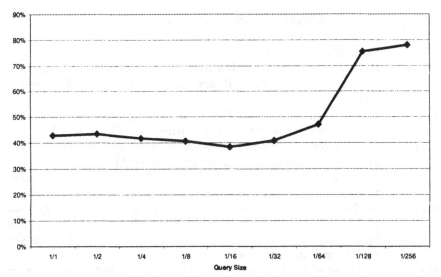

Performance Penalty of I+Q- vs. I-Q-

Fig. 12: Relative performance penalty of I^+Q^- compared to I^-Q^-

Fig. 12 shows a performance comparison between I^-Q^- and I^+Q^-. The performance penalty for using I^+Q^- is between 40% and 45% for the larger queries. In case of the smaller queries, the penalty increases to nearly 80%. Note that the difference is only a few disk accesses in absolute terms: the small queries are geographically very restricted.

7 Conclusion

We have introduced the application of an additional scale dimension to spatial data. Although the additional dimension increases the size of the index slightly, it can considerably speed up spatial queries. Specifically, it results in better scalability with respect to the query size: especially very large-scale queries that generally need many disk accesses benefit from our approach. Extremely small-scale queries that generally only need very few disk accesses might be slightly inhibited.

We have presented tuning considerations that can have a considerable effect on performance, and have pointed out applications that typically may not benefit from our approach. Our approach is not suited for spatial data sets with only minimal variations in object scales or more large-scale than small-scale objects.

With a scale dimension, the index structure itself is given the ability to consider only spatial objects that are sufficiently large-scale with respect to the query scale. This not only increases indexing performance considerably, but also makes a filtering step superfluous. It is an integrated approach to both

– keep the query answer restricted to relevant objects
– considerably improve performance

References

[BBK98]   Stefan Berchtold, Christian Böhm and Hans-Peter Kriegel: *"The Pyramid Technique: Towards Breaking the Curse of Dimensionality"*, SIGMOD 98, pages 142 – 153.

[DKL+94]   David J. DeWitt, Navin Kabra, Jun Luo, Jignesh M. Patel, Jie-Bing Yu: *"Client Server Paradise"*, Proceedings of the 20th VLDB Conference, 1994, pages 558 – 569.

[F87]   M. Freeston: *"The BANG file: a new kind of grid file"*, Proceedings ACM SIGMOD Conference, San Francisco, California, May 1987.

[F89a]   M. Freeston: *"Advances in the Design of the BANG File"*, 3rd International Conference on Foundations of Data Organization and Algorithms (FODO), Paris, June 1989. [Lecture Notes in Computer Science No. 367, Springer-Verlag, 1989].

[F89b]   M. Freeston: *"A well-behaved file structure for the storage of spatial objects"*, Symposium on the Design and Implementation of Large Spatial Databases, Santa Barbara, California, July 1989. [Lecture Notes in Computer Science No. 409, Springer-Verlag, 1989].

[F95]   M. Freeston: *"A General Solution of the n-dimensional B-tree Problem"*, ACM SIGMOD Conference, San Jose, California, May 1995.

[HKP97]   J. M. Hellerstein, E. Koutsoupias and C. H. Papadimitriou: *"On the Analysis of Indexing Schemes"*, Proceedings of the 16th ACM SIGACT-SIGMOD-SIGART Symposium on Principles of Database Systems, 1997.

[Kan98]   Ravi Kanth V. Kothuri: *"Indexing Multidimensional Data"*, Ph.D. Thesis, UCSB, 1998.

[KT98]   Elias Koutsoupias and David Scot Taylor: *"Tight Bounds for 2-Dimensional Indexing Schemes"*, PODS 98, pages 52 – 58.

[SK95]   Kenneth C. Sevcik and Nikos Koudas: *"Filter Trees for Managing Spatial Data Over a Range of Size Granularities"*, Technical Report, University of Toronto, November 1995.

[SM98]   Vasilis Samoladas and Daniel P. Miranker: *"A Lower Bound Theorem for Indexing Schemes and its Application to Multidimensional Range Queries"*, PODS 98, pages 44 – 51.

[Vit98]   Jeffrey Scott Vitter: *"External Memory Algorithms"*, PODS 98.

Indexing

XZ-Ordering: A Space-Filling Curve
for Objects with Spatial Extension

Christian Böhm[1], Gerald Klump[1] and Hans-Peter Kriegel[1]

[1] University of Munich, Computer Science Institute, Oettingenstr. 67,
D-80538 Munich, Germany
{boehm,klump,kriegel}@informatik.uni-muenchen.de

Abstract. There is an increasing need to integrate spatial index structures into commercial database management systems. In geographic information systems (GIS), huge amounts of information involving both, spatial and thematic attributes, have to be managed. Whereas relational databases are adequate for handling thematic attributes, they fail to manage spatial information efficiently. In this paper, we point out that neither a hybrid solution using relational databases and a separate spatial index nor the approach of existing object-relational database systems provide a satisfying solution to this problem. Therefore, it is necessary to map the spatial information into the relational model. Promising approaches to this mapping are based on space-filling curves such as Z-ordering or the Hilbert curve. These approaches perform an embedding of the multidimensional space into the one-dimensional space. Unfortunately, the techniques are very sensitive to the suitable choice of an underlying resolution parameter if objects with a spatial extension such as rectangles or polygons are stored. The performance usually deteriorates drastically if the resolution is chosen too high or too low. Therefore, we present a new kind of ordering which allows an arbitrary high resolution without performance degeneration. This robustness is achieved by avoiding object duplication, allowing overlapping Z-elements, by a novel coding scheme for the Z-elements and an optimized algorithm for query processing. The superiority of our technique is shown both, theoretically as well as practically with a comprehensive experimental evaluation.

1. Motivation

Index structures for spatial database systems have been extensively investigated during the last decade. A great variety of index structures and query processing techniques has been proposed [Güt 94, GG 98]. Most techniques are based on hierarchical tree-structures such as the R-tree [Gut 84] and its variants [BKSS 90, SRF 87, BKK 97]. In these approaches, each node corresponds to a page of the background storage and to a region of the data space.

There is an increasing interest in integrating spatial data into commercial database management systems. Geographic information systems (GIS) are data-intensive applications involving both, spatial and thematic attributes. Thematic attributes are usually best represented in the relational model, where powerful and adequate tools for evaluation and management are available. Relational databases, however, fail to manage spatial attributes efficiently. Therefore, it is common to store thematic attributes in a relational database system and spatial attributes outside the database in file-based multidimensional index structures (*hybrid solution*).

R.H. Güting, D. Papadias, F. Lochovsky (Eds.): SSD'99, LNCS 1651, pp. 75-90, 1999.

The hybrid solution bears various disadvantages. Especially the integrity of data stored in two ways, inside and outside the database system, is difficult to maintain. If an update operation involving both, spatial and thematic attributes fails in the relational database (e.g. due to concurrency conflicts), the corresponding update in the spatial index must be undone to guarantee consistency. Vice versa, if the spatial update fails, the corresponding update to the relational database must be aborted. For this purpose, a distributed commit protocol for heterogeneous database systems must be implemented, a time-consuming task which requires a deep knowledge of the participating systems. The hybrid solution involves further problems. File systems and database systems have usually different approaches for data security, backup and concurrent access. File-based storage does not guarantee physical and logical data independence. Thus, changes in running applications are complicated.

A promising approach to overcome these disadvantages is based on object-relational database systems. Object-relational database systems are relational database systems which can be extended by application-specific data types (called *data cartridges* or *data blades*). The general idea is to define data cartridges for spatial attributes and to manage spatial attributes in the database. For data-intensive GIS applications it is necessary to implement the multidimensional index structures in the database. This requires the access to the block-manager of the database system, which is not granted by most commercial database systems. For instance the current universal servers by ORACLE and INFORMIX do not provide any documentation of a block-oriented interface to the database. Data blades/cartridges are only allowed to access relations via the SQL interface. Thus, current object-relational database systems are not very helpful for our integration problem.

We can summarize that anyway, using current object-relational database systems or pure relational database systems, the only possible way to store spatial attributes inside the database is to map them into the relational model. An early solution for the management of multidimensional data in relations is based on space-filling curves. Space-filling curves map points of a multidimensional space to one-dimensional values. The mapping is distance preserving in the sense that points which are close to each other in the multidimensional space, are *likely* to be close to each other in the one-dimensional space. Although distance-preservation is not strict in this concept, the search for matching objects is usually restricted to a limited area in the embedding space.

The concept of space-filling curves has been extended to handle polygons. This idea is based on the decomposition of the polygons according to the space-filling curve. We will discuss this approach in section 2 and reveal its major disadvantage that it is very sensitive to a suitable choice of the resolution parameter. We will present a new method for applying space-filling curves to spatially extended objects which is not based on decomposition and avoids the associated problems.

For concreteness, we concentrate us in this paper on the implementation of the first filter step for queries with a given query region such as window queries or range queries. Further filter steps and the refinement step are beyond the scope of this paper. The rest of this paper is organized as follows: In section 2, we introduce space-filling curves and review the related work. Section 3 explains the general idea and gives an overview of our solution. The following sections show how operations such as insert, delete and search are handled. In section 5, a comprehensive experimental evaluation of our technique using the relational database management system ORACLE 8 is performed, showing the superiority of our approach over standard query processing techniques and competitive approaches.

2. Z-Ordering

Z-Ordering is based on a recursive decomposition of the data space as it is provided by a space-filling curve [Sag 94, Sam 89] called *Z-ordering* [OM 84], *Peano/Morton Curve* [Mor 66], *Quad Codes* [FB 74] or *Locational Codes* [AS 83].

2.1 Z-Ordering in Point Databases

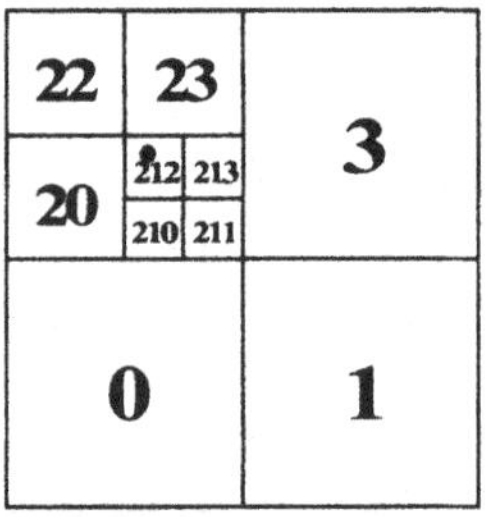

Figure 1: Z-Ordering.

Assume a point taken from the two-dimensional unit square $[0..1]^2$. The algorithm partitions the unit square into 4 quadrants of equal size (we change the description of Z-ordering here slightly to make it more comparable to our approach), which are canonically numbered from 0 to 3 (cf. figure 1). We note the number of the quadrant and partition this quadrant into its four sub-quadrants. This is recursively repeated until a certain basic resolution is reached. The fixed number of recursive iterations is called the resolution level g. Then we stop and use the obtained sequence of g digits (called *quadrant sequence*) as ordering key for the points (we order lexicographically). Each quadrant sequence represents a region of the data space called *element*. For instance, the sequence <00> stands for an element with side length 0.25 touching the lower left corner of the data space. Elements at the basic resolution which are represented by quadrant sequences of length g are called *cells*. If an element e_1 is contained in another element e_2, then the corresponding quadrant sequence $Q(e_2)$ is a prefix of $Q(e_1)$. The longer a quadrant sequence is, the smaller is the corresponding element. In the unit square, the area of an element represented by a sequence of length l is $(1/4)^l$. In a point database, only cells at the basic resolution are used. Therefore, all quadrant sequences have the same lengths and we can interpret the quadrant sequences as numbers represented in the quaternary system (i.e. base 4). Interpreting sequences as numbers facilitates their management in the index and does not change ordering of the points, because the lexicographical order corresponds to the less-equal relation of numbers. The points are managed in an order-preserving one-dimensional index structure such as a B^+-tree.

2.2 Query Processing in Z-Ordering

Assume a window query with a specified window. The data space is decomposed into its four quadrants. Each quadrant is tested for intersection with the query window. If the quadrant does not intersect the query window, nothing has to be done. If the quadrant is completely enclosed in the query window, we have to retrieve all points from the database having the quadrant sequence of this element as a prefix of their keys. If the keys are represented as integer numbers (cf. section 3.2), we have to retrieve an interval of subsequent numbers. All remaining quadrants which are intersected by the window but not completely enclosed in the window (i.e. "real" intersections) are decomposed recursively until the basic resolution is reached.

Figure 2: The one-value-representation.

2.3 A Naive Approach for Polygon Databases

To extend the concept of Z-ordering for the management of objects with a spatial exten-sion (e.g. rectangles or polygons), we face the problem that a given polygon intersects with many cells. A naive approach could be to store every cell covered by the object in the database. Obviously, this method causes a huge storage overhead unless the basic grid is very coarse. Therefore, several methods have been proposed to reduce the over-head when using a finer grid.

2.4 One-Value-Approximation

The objects are approximated by the smallest element which encloses the complete object (cf. figure 2). In this case our recursive algorithm for the determination of the quadrant sequence is modified as follows: Partition the current data space into four quadrants. If exactly one quadrant is intersected by the object, proceed recursively with this quadrant. If more than one quadrant is intersected, then stop. Use the quadrant sequence obtained up to that point as the ordering key. This method has the obvious advantage that each object is represented by a single key, not by a set of keys as in our naive approach. But this method yields also several disadvantages. The first disadvan-tage is that the quadrant sequences in this approach have different lengths, depending on the resolution of the smallest enclosing quadrant. Thus, our simple interpretation as a numerical value is not possible. Keys must be stored as strings with variable length and compared lexicographically, which is less efficient than numerical comparisons. The second problem is that objects may be represented very poorly. For instance any polygon intersecting one of the axis-parallel lines in the middle of the data space (the line $x = 0.5$ and the line $y = 0.5$) can only be approximated by the empty quadrant sequence. If the polygon to be approximated is very large, an approximation by the empty sequence or by very short sequences seems to be justified. For small polygons, the relative approxi-mation error is too large. The relative space overhead of the object approximation is thus unlimited. In fact, objects approximated by the empty quadrant sequence are candidates to every query a user asks. The more objects with short quadrant sequences are stored in the database, the worse is the selectivity of the index.

2.5 Optimized Redundancy

To avoid the unlimited approximation overhead, Orenstein proposes a combination of the naive approach and the one-sequence representation [Ore 89a, Ore 89b]. He adopts the idea of the object decomposition in the naive approach, but does not necessarily decompose the object until the basic resolution is reached. Instead, he proposes two different criteria, called *size-bound* and *error-bound* to control the number of quadrants into which an object is decomposed. Each subobject is stored in the index by using its quadrant sequence, e.g. represented as a string. Although this concept involves object duplication (which is called redundancy by Orenstein), the number of records stored in the index is not directly determined by the grid resolution as in the naive approach. Unlike in the one-sequence approach, it is not necessary to represent small objects by the empty sequence or by very short sequences. According to Orenstein, typically a decomposition into 2-4 parts is sufficient for a satisfactory search performance.

Orenstein's approach alleviates the problems of the two previous approaches, but a duplicate elimination is still required and the keys are sequences with varying length. Orenstein determines an optimal degree of redundancy only experimentally. An analytical solution was proposed by Gaede [Gae 95] who identified the complexity of the stored polygons, described by their perimeter and their fractal dimension, as the main parameters for optimization. A further problem when redundancy is allowed arises in connection with secondary filters in a multi-step environment. Information which can be exploited for fast filtering of false hits, such as additional conservative approximations (e.g. *minimum bounding rectangles MBR*), should not be subject to duplication due to its high storage requirement. To avoid duplication of such information, it must be stored in a separate table which implies additional joins in query processing.

A further consequence of Gaede's analysis [Gae 95] is that the number of intervals which are generated from the query window is proportional to the number of grid cells intersected by the boundary of the query window (i.e. its perimeter). This means that a too fine resolution of the grid leads to a large number of intervals and thus to deteriorated performance behavior when a relational database system is used. The reason is that the intervals must be transferred to and processed by the database server, which is not negligible, if the number of intervals is very high (e.g. in the thousands).

2.6 Alternative Techniques

Several improvements of the Z-ordering concept are well-known (cf. figure 3). Some authors propose the use of different curves such as Gray Codes [Fal 86, Fal 88], the Hilbert Curve [FR 89, Jag 90] or other variations [Kum 94]. Many studies [Oos 90, Jag 90, FR 91] prefer the Hilbert curve among the proposals, due to its best distance preservation properties (also called *spatial clustering properties*). [Klu 98] proposes a great variety of space-filling curves and makes a comprehensive performance study using a relational implementation. As this performance evaluation [Klu 98] does not yield a substantial performance improvement of the Hilbert curve or other space-filling curves over Z-ordering, we use the Peano/Morton curve because it is easier to compute.

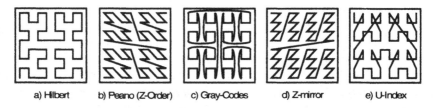

a) Hilbert   b) Peano (Z-Order)   c) Gray-Codes   d) Z-mirror   e) U-Index

Figure 3: Various Space-Filling Curves.

3. A Space-Filling Curve for Spatially Extended Objects

In contrast to the previous approaches, we propose a solution which avoids the disadvantages of object duplication and variable-length quadrant sequences. Our method is thus completely insensitive against a too fine grid resolution. There is no need to optimize the resolution parameter. It can always be taken as fine as possible, i.e. the full bit precision of the CPU can be exploited. Three ideas are applied to achieve this robustness: The first idea presented in section 3.1 is to incorporate overlap into the concept of elements. We will define the elements such that adjacent elements at the same resolution level l overlap each other up to 50%. This method enables us to store objects without redundancy (i.e. object duplication) and without uncontrolled approximation error. In particular, it is impossible that a very small object must be represented by a very short sequence or even by the empty sequence.

The second idea is to use a sophisticated coding scheme for the quadrant sequences which maps quadrant sequences with varying length into the integer domain in a distance-preserving way. The coding algorithm is presented in section 3.2. The third idea (cf. section 4.3) is an efficient algorithm for interval generation in query processing. The goal of the algorithm is to close small gaps between adjacent intervals if the overhead of processing an additional interval is larger than the cost for overreading the interval gap.

We call our technique which maps polygons into integer-values *extended Z-ordering* or *XZ-ordering*. The integer values forming the keys for search are called *XZ-values*. For each polygon in the database, we store one record which contains its XZ-value and a pointer to the exact geometry representation of the polygon. As we avoid object duplication, further information such as thematic attributes or information for secondary filters (the *MBR* or other conservative and progressive approximations, cf. [BKS 93]) can be stored in the same table.

3.1 Overlapping Cells and Elements

The most important problem of the *one-value representation* is that several objects are approximated very poorly. Every object intersecting with one of the axis-parallel lines $x = 0.5$ and $y = 0.5$ is represented by the empty quadrant sequence which characterizes the element comprising of the complete data space. If the object extension is very small (close to 0), the relative approximation error diverges to infinity.

In fact, every technique which decomposes the space into disjoint cells gets into trouble if an object is located on the boundary between large elements. Therefore, we modify our definition of elements such that overlap among elements on the same resolution level l is allowed.

The easiest way to envisage a definition of overlapping elements is to take the original elements as obtained by Z-ordering and to enlarge the height and width by a factor 2 upwards and to the right, as depicted in figure 4. Then, two adjacent cells overlap each other by 50%. The special advantage is, that this definition contains also small elements for objects intersecting with the middle axis.

Definition 1: Enlarged elements

The lower left corner of an enlarged element corresponds to the lower left corner of Z-ordering. Let s be the quadrant sequence of this lower left corner and let $|s|$ denote its length. The upper right corner is translated such that the height and width of the element is $0.5^{|s|-1}$.

It is even possible to guarantee bounds for a minimal length of the quadrant sequence (and thus of the approximation quality) based on the extension of the object in x- and y-direction.

Lemma 1. Minimum and Maximum Length of the Quadrant Sequence

The length $|s|$ of the quadrant sequence s for an object with height h and width w is bounded by the following limits:

$$l_1 \leq |s| < l_2 \text{ with } l_1 = \left\lfloor \log_{0.5}(\max\{w, h\}) \right\rfloor \text{ and } l_2 = l_1 + 2$$

Proof (Lemma 1)

Without loss of generality, we assume $w \geq h$. We consider the two disjoint space decompositions into elements at the resolution levels l_1 and l_2 and call the arising decomposition grids the l_1-grid and the l_2-grid, respectively. The distances w_1 and w_2 between the grid-lines are equal to the widths of the elements at the corresponding decomposition levels l_1 and l_2.

(1) As the distance w_1 between two lines of the l_1-grid is greater than or equal to w, because

$$w_1 = 0.5^{l_1} = 0.5^{\left\lfloor \log_{0.5}(w) \right\rfloor} \geq 0.5^{\log_{0.5}(w)} = w,$$

the object can at most be intersected by one grid line parallel to the y-axis and by one grid line parallel to the x-axis. If the lower left element among the intersecting elements is enlarged as in definition 1, the object must be completely contained in the enlargement.

(2) As the distance w_2 between two lines of the l_2-grid is smaller than $w/2$, because

$$w_2 = 0.5^{l_2} = 0.5^{\left\lfloor \log_{0.5}(w) \right\rfloor + 2} < 0.5^{\log_{0.5}(w) + 1} = w/2,$$

the object is intersected at least by two y-axis parallel lines of the l_2-grid. Therefore, there is no element at the l_2-level which can be enlarged according to definition 1 such that the object is contained.

❑

Lemma 1 can be exploited to provide boundaries for the relative approximation error of objects. As polygons can be arbitrary complex, it is not possible for any approximation technique with restricted complexity (such as MBRs or our technique) to provide error boundaries. However, we can guarantee a maximum relative error for square objects which are not smaller than the basic resolution:

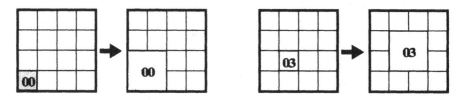

Figure 4: Enlarged Regions in XZ-Ordering.

Lemma 2. Maximum approximation error

The relative approximation error for square objects is limited.

Proof (Lemma 2)

According to lemma 1, the quadrant sequence for a square object of width w has at least the length $l_1 = \lfloor \log_{0.5}(w) \rfloor$. The width of the corresponding cell is

$$w_1 = 0.5^{l_1} = 0.5^{\lfloor \log_{0.5}(w) \rfloor} < 0.5^{\log_{0.5}(w) - 1} = 2 \cdot w.$$

The enlarged element has the area $A = 4 \cdot w_1^2 < 16 \cdot w^2$.

The maximum approximation error is limited by the following value:

$$E_{rel} = \frac{A - w^2}{w^2} < 15 \quad .$$

❏

Although a relative approximation error of 15 seems rather large, it is an important advantage of our technique over the one-value representation that the approximation error is limited at all. Our experimental evaluation will show that also the average approximation error of our technique is smaller than that of the one-value representation.

3.2 Numbering Quadrant Sequences

We are given a quadrant sequence with a length varying from 0 to the maximum length g determined by the basic resolution. Our problem is to assign numbers to the sequences in an order-preserving way, i.e. the less-equal-order of the numbers must correspond to the lexicographical order of the quadrant sequences. Let the length of the quadrant sequence s be l. The following lemma is used to determine the number of cells and elements contained in the corresponding region $R(s)$:

Lemma 3. Number of cells and elements inside of region $R(s)$

The number of cells of resolution g contained in a region described by the quadrant sequence s with $|s| = l$ is

$$N_{cell}(l) = 4^{g-l}.$$

The corresponding number of elements (including element $R(s)$ and the cells) is

$$N_{elem}(l) = \frac{4^{g-l+1} - 1}{3}.$$

Proof (Lemma 3)

(1) There is a total of 4^l elements with length l and a total of 4^g cells in the data space. As both cells and elements of length l cover the data space in a complete and overlap-free way, the area of an element is $(1/4)^l/(1/4)^g = 4^{g-l}$ times larger than the area of a cell.

(2) The number of elements of length i contained in an element of length l corresponds to 4^{i-l}. For obtaining the number of all elements, we have to summarize over all i ranging from l to g:

$$\sum_{l \le i \le g} 4^{g-i} = \frac{4^{g-l+1} - 1}{3}$$

❑

For the definition of our numbering scheme, we have to make sure that between the codes of two subsequent strings s_1 and s_2 of length l there are enough numbers for all strings which are ordered between s_1 and s_2. These are exactly the strings having s_1 as prefix, and their number is thus $(4^{g-l} - 1)/3$. We therefore multiply each quadrant number q_i in the sequence $s = <q_0\, q_1\, ...\, q_i\, ...\, q_{l-1}>$ with $(4^{g-i} - 1)/3$.

Definition 2: Sequence Code

The sequence code $C(s)$ of a quadrant sequence $s = <q_0\, q_1\, ...\, q_i\, ...\, q_{l-1}>$ corresponds to

$$C(s) = \sum_{0 \le i < l} q_i \cdot \frac{4^{g-i} - 1}{3} + 1$$

Lemma 4. Ordering Preservation of the Sequence Code

The less-equal order of sequence codes corresponds to the lexicographical order of the quadrant sequences:

$$s_1 <_{\text{lex}} s_2 \Leftrightarrow C(s_1) < C(s_2)$$

Proof (Lemma 4)

'$\Rightarrow$': Suppose $s_1 <_{\text{lex}} s_2$ with $s_1 = <q_0...q_{l-1}>$ and $s_2 = <p_0...p_{m-1}>$. Then, one of the following predicates must be true according to the definition of the lexicographical order:

(1) there exists an i ($0 \le i < \min\{m, l\} - 1$) such that $q_i < p_i$ and $q_j = p_j$ for all $j < i$
(2) $m > l$ and $q_j = p_j$ for all $j < l$.

In case (1), we know that the i-th term of the sum of $C(s_1)$ is at least by $(4^{g-i} - 1)/3$ less than the i-th term in the sum of $C(s_2)$. The summands for all $j < i$ are equal. We have to show that the difference D of the sums of all remaining terms is smaller than $N_{\text{elem}}(g - i)$ to guarantee $C(s_1) < C(s_2)$. The difference D is maximal if $q_j = 3$ and $p_j = 0$ for all $j > i$. In this case, we can determine D as follows:

$$D = \sum_{i < j < g} 3 \cdot \frac{4^{g-j} - 1}{3} = \sum_{0 < j < g-i} 4^j - 1 = \frac{4^{g-i} - 1}{3} - (g - i + 1) < \frac{4^{g-i} - 1}{3}$$

In case (2), $C(s_2) > C(s_1)$, because the first i summands are equal, and $C(s_2)$ has additional positive summands which are not available in the sum of $C(s_1)$.

'$\Leftarrow$': We can rewrite the condition $s_1 <_{\text{lex}} s_2 \Leftarrow C(s_1) < C(s_2)$ to $s_1 \ge_{\text{lex}} s_2 \Rightarrow C(s_1) \ge C(s_2)$. The proof is then analogue to the '$\Rightarrow$'-direction.

❑

Lemma 5. Minimality of the Sequence Code

There exists no mapping from the set of quadrant sequences to the set of natural numbers which requires a smaller interval of the natural numbers than $C(s)$.

Proof (Lemma 5)

From lemma 3 it follows that there are $N_{elem}(0) = (4^{g+1} - 1)/3$ different elements. $C(s)$ is maximal if s has the form $<3^g>$. In this case, $C(s)$ evaluates to the term

$$C(s) = \sum_{0 \le i < g} 3 \cdot N_{elem}(g - i) + 1 = \sum_{0 \le i < g} 3 \cdot \frac{4^{i+1} - 1}{3} + 1 = \left(\sum_{0 \le i \le g} 4^i \right) - 1$$

$$= \frac{4^{g+1} - 1}{3} - 1 = N_{elem}(0) - 1$$

The coding of the empty string $C(<>) = 0$. Therefore, the $N_{elem}(0)$ different strings are mapped exactly to the interval $[0, N_{elem}(0) - 1]$.

❑

From lemma 5, it also follows that C is a surjective mapping onto the set $\{0, ..., N_{elem}(0) - 1\}$. We note without a formal proof that C is also injective. The quadrant sequence can be reconstructed from the coding in an efficient way, but this is not necessary for our purposes.

4. Query Processing

4.1 Insert and Delete

We know from lemma 1 that the quadrant sequence has the length $l_1 = \lfloor \log_{0.5}(\max\{w, h\}) \rfloor$ or $l_1 + 1$. We can decide this by the following predicate which tests whether the object is intersected by one or two grid lines (x_l denotes the lower boundary of the object in x direction; the same criterion must be applied for y_l):

$$\left\lfloor \frac{x_l}{l_1} + 2 \right\rfloor \cdot l_1 \le x_l + w$$

Once the length l of the quadrant sequence is determined, the corresponding quadrant sequence s is determined for the lower left corner of the bounding box of the object, as described in section 2.1. This sequence s is clipped to the length l and coded according to definition 2. The obtained value is used as a key for storage and management of the object in a relational index. Our actual algorithm performs the operations of recursive descent into the quadrants and coding according to definition 2 simultaneously without explicitly generating the quadrant sequence. The algorithm runs in $O(l)$ time.

4.2 From Window Queries to Interval Sets

For query processing, we can proceed in a recursive way similar to the algorithm presented in section 2.2. We determine which quadrants are intersected by the query. Those which are not intersected are ignored. If a quadrant is completely contained in the query window, all elements having the corresponding quadrant sequence as prefix are completely contained in the query. Therefore, the interval of the corresponding XZ-values is generated and marked for a later retrieval from the database. If a quadrant is intersected,

```
FOREACH interval i in list {
        WHILE succ (i).lower - i.upper < maxgap {
                i.upper := succ (i).upper ;
                delete succ (i) ;
}       }
```

Figure 5: The simple algorithm for gap closing.

the corresponding XZ-value is determined and marked (it is handled as a one-value interval). Then the algorithm is called recursively for all its sub-quadrants. Finally, we have marked a set of intervals of XZ-values which are to be retrieved from the database. This set is translated into an SQL statement which is transferred to the DBMS.

Problems arise if the resolution of the basic grid is chosen too fine, because in this case, typically many elements are partially intersected. In the average case, the number of generated intervals is in the order $O(2^g)$. It is very costly to transfer and compile such a complex query. To alleviate this problem, one could apply the simple densification algorithm depicted in figure 5 to the set of intervals. As only small gaps between subsequent intervals are closed, it is unlikely that this densification of intervals causes additional disk accesses in query processing. The densification after the interval generation optimizes query compilation cost and related cost factors, but does not change the general complexity of the interval generation. For this purpose, an algorithm must be devised which generates the intervals directly in a way that closes the gaps. This algorithm is described in the subsequent section.

4.3 An Efficient Algorithm for the Interval Generation

This algorithm allows the user to specify the number of intervals n_{int} which are generated. We exploit a general property of XZ-ordering and Z-ordering, which we note here without a formal proof: Let us consider interval sets which come up if we restrict our interval generation to a certain length l ($0 \leq l \leq q$) of the corresponding quadrant sequences. If l is increased, it is possible that more intervals are generated. The factor by which the number of intervals in the interval set grows when increasing the length l by one, is restricted by 4. All intervals of the larger interval set are contained in an interval of the smaller set. The additional gaps between the intervals in an interval set generated when l is increased by 1, can only be smaller than all gaps which are visible when l is decreased by 1. If we would descend the recursion tree of the algorithm in section 2.2 in a breadth-first fashion and stop if we have found 4 times as many intervals as demanded ($4n_{int}$), we know that we can densify this set to a set of n_{int} intervals with the largest possible gaps between them.

Instead of the breadth-first traversal, we have implemented the following algorithm: In a first phase, depicted in figure 6, the algorithm performs a depth-first traversal for determining the number of intervals in each recursion level. The clue is, that the recursive descent is avoided, if the corresponding level has reached a number of $4n_{int}$ intervals, because we are only interested in the first level having more than $4n_{int}$ intervals. In each level, the algorithm measures the number of transitions in the XZ-order, where the extended elements of the space-filling curve cross the query region. As the algorithm is

```
VAR num_ch: ARRAY [0..g] OF INTEGER ;

PROCEDURE det_num_changes (element, query: REGION; cur_num_ch: INTEGER;
                                      c_depth: INTEGER; VAR inside: BOOLEAN) {
    IF (NOT intersect (element, query) ) {
        IF (inside) {
            inside := FALSE; INCREMENT (num_ch [c_depth]);
    }   }
    ELSE IF (contains (query, element) ) {
        IF (NOT inside) {
            inside := TRUE; INCREMENT (num_ch [c_depth] ) ;
    }   }
    ELSE {
        IF (NOT inside) {
            inside := TRUE; INCREMENT (num_ch [c_depth] ) ;
        }
        IF (current_depth < g AND c_num_ch + num_ch[c_depth] < n_int) {
            FOREACH subquadrant {
                det_num_changes (   subquadrant, query, c_num_ch + num_ch [c_depth],
                                    c_depth + 1, inside) ;
}   }   }   }

FUNCTION suitable_level (query:REGION): INTEGER {
    INITIALIZE num_ch := {0, 0, ..., 0} ;
    det_num_changes (dataspace, query, 0, 0, FALSE) ;
    suitable_level := first i where num_ch[i] >= n_int ;
}
```

Figure 6: The Algorithm for Determining the Suitable Recursion Level ("Phase 1").

in each level called $4n_{int}$ times maximum, the complexity of the algorithm is bounded by $O(n_{int} \cdot g)$.

In the second phase, we generate the corresponding interval set by a depth-first traversal which is limited to the recursion depth obtained in phase one. Whenever the number of intervals becomes greater than n_{int}, the two neighboring intervals with the smallest gap between them are merged. In a third phase, the upper bounds of the intervals are investigated. It is possible that the upper bounds can be slightly improved (i.e. decreased) by a deeper descent in the recursion tree. All three phases yield a linear complexity in g.

5. Experimental Evaluation

In order to verify our claims that an implementation of spatial index structures does not only provide advantages from a software engineering point of view but also in terms of performance, we actually implemented the XZ-Ordering technique on top of ORACLE-8 and performed a comprehensive experimental evaluation using data from a GIS application. Our database contains 324,000 polygons from a map of the European Union. We also generated smaller data sets from the EU map to investigate the behavior with vary-

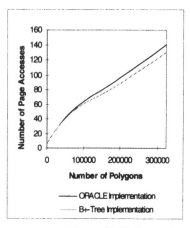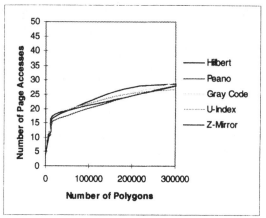

Figure 7: (l.) Comparison between ORACLE and file-based B+-tree
(r.) Comparison between various Space-Filling Curves.

ing database size. Our application was implemented in Embedded SQL/C (dynamic SQL level 3) on HP C-160 workstations. The database server and the client application run on separate machines, connected by a 100MBit Fast Ethernet.

We also had an implementation on top of a file-system based B+-tree which was used for comparison purposes. The reason for this test was to show that our technique is implementable on top of a commercial database system and that the additional overhead induced by the database system is small. In both cases, we varied the database size from 1,000 to 324,000 and executed window queries with a selectivity of 1% (with respect to the number of polygons in the query result). All experiments presented in this section were repeated ten times. The presented results are averages over all trials. The number of page accesses (cf. figure 7, left diagram) is in Oracle only by up to 8% higher than in the file-based B+-tree implementation. We used a comparable node capacity in both implementations. A second experiment depicted on the right side of figure 7 shows that the application of a specific space filling curve has no strong influence on the performance of our technique. Using the same settings, we tested various curves including Z-Ordering (Peano), the Hilbert-curve and Gray-Codes. There was no consequent trend observable which could suggest the superiority of one of the curves. Therefore, we decided to perform all subsequent experiments using the Peano-/Morton-curve, because the implementation is facilitated.

The purpose of the next series of experiments is to show the superiority of our approach over competitive techniques such as Orenstein's Z-Ordering. First, we demonstrate in figure 8 that our technique, in contrast to Z-Ordering, is not subject to a performance deterioration when the grid resolution is chosen too high. We constructed several indexes with varying resolution parameter g for both techniques, XZ-Ordering and Z-Ordering, where we applied the size-bound decomposition strategy resulting in a redundancy of 4. In this experiment, we stored 81,000 polygons in the database and retrieved 1% of them in window queries. Z-Ordering has a clear minimum at a resolution of $g = 8$ with a satisfying query performance. When the resolution is slightly increased or decreased, the query performance deteriorates. For instance, if g is chosen to 10, the num-

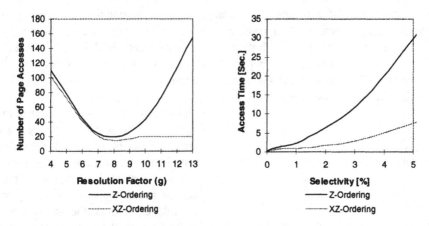

Figure 8: (l.) Z-Ordering is sensitive to the resolution
(r.) Performance comparison with varying selectivity.

ber of page accesses is by 80% higher than at the optimum. At the maximum resolution factor $g = 13$, the number of page accesses was by a factor more than 6.4 higher than the optimum. In contrast, our technique shows a different behavior. If the grid resolution parameter g is too small, the performance is similar to the performance of Z-ordering. A too coarse grid leads obviously to a bad index selectivity, because many objects are mapped to the same Z-values or XZ-values, respectively. Both techniques have the same point of optimum. Beyond this point, XZ-Ordering yields a constant number of page accesses. Therefore, it is possible to avoid the optimization of this parameter which is difficult and depends on dynamically changing information such as the fractal dimension of the stored objects and their number. This trend is maintained and even intensified if the selectivity of the query is increased, as depicted on the right side of figure 8. Here, the number of polygons was 324,000 and the resolution was fixed to $g = 10$.

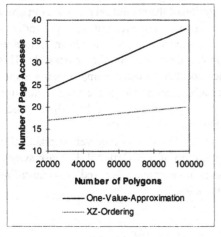

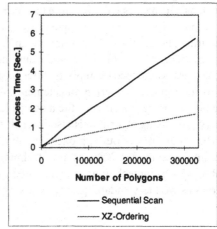

Figure 9: (l.) Comparison with the One-Value-Approximation
(r.) Comparison with the Sequential Scan.

In a further series of experiments, we compared our technique with the One-Value-Approximation (cf. section 2.4) and with the sequential scan of the data set. The remaining test parameters correspond with the previous experiments. Both competitive techniques are clearly outperformed as depicted in figure 9. The One-Value-Approximation yields up to 90% more page accesses. The sequential scan needs up to 326% as much processing time as XZ-Ordering.

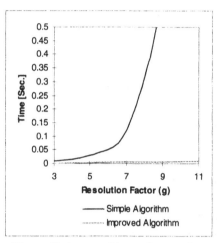

Figure 10: Improved interval generation.

In a last experiment, we determined the influence of the query processing algorithm presented in section 4.3. We varied again the resolution parameter g from 3 to 11 and generated interval sets of 20 intervals according to a window query with side length 0.03 (the unit square is the data space). We measured the CPU time which is required for the generation of the intervals and of the corresponding string for the dynamic SQL statement. We compared our algorithm with the simple algorithm (cf. section 2.2) extended by the gap closing algorithm (cf. figure 5). The results are presented in figure 10. The simple algorithm has an exponential complexity with respect to the resolution, whereas the improved algorithm is linear. For the finest resolution ($g = 11$), the improved algorithm is by the factor 286 faster than the simple algorithm.

6. Conclusion

In this paper, we have proposed XZ-Ordering, a new technique for a one-dimensional embedding of extended spatial objects such as rectangles or polygons. In contrast to previous approaches which require the optimization of the resolution parameter in order to become efficient, our technique is insensitive against a fine resolution of the underlying grid. Therefore, the resolution parameter is only restricted by hardware constants (the number of bits for an integer value) and can be chosen as fine as possible. This robustness is achieved by applying three basic concepts. Object decomposition is avoided by the concept of overlapping elements. A sophisticated order-preserving coding into the integer domain facilitates the management of the search keys for the DBMS. A new query processing algorithm is also insensitive to the grid resolution.

The superiority of our approach was shown both, theoretically as well as practically. We implemented the XZ-Ordering technique on top of the relational database system ORACLE-8. Our technique outperformed competitive techniques based on space-filling curves as well as standard query processing techniques.

References

[AS 83]   Abel D.J., Smith J.L.: 'A Data Structure and Algorithm Based on a Linear Key for a Rectangle Retrieval Problem', Computer Vision 24, pp. 1-13.

[BKK 97]   Berchtold S., Keim D., Kriegel H.-P.: 'Using Extended Feature Objects for Partial Similarity Retrieval', VLDB Journal Vol. 6, No. 4, pp. 333-348, 1997.

[BKS 93]   Brinkhoff T., Kriegel H.-P., Schneider R.: 'Comparison of Approximations of Complex Objects Used for Approximation-based Query Processing in Spatial Database Systems', ICDE 1993, pp. 40-49.

[BKSS 90] Beckmann N., Kriegel H.-P., Schneider R., Seeger B.: 'The R*-tree: An Efficient and Robust Access Method for Points and Rectangles', Proc. ACM SIGMOD Int. Conf. on Management of Data, Atlantic City, NJ, 1990, pp. 322-331.

[Gae 95]   Gaede V.: 'Optimal Redundancy in Spatial Database Systems', Proc. 4th Int. Symposium on Advances in Spatial Databases, SSD'95, Portland, Maine, USA, 1995, Lecture Notes in Computer Science Vol. 951, pp. 96-116.

[GG 98]   Gaede V., Günther O.: 'Multidimensional Access Methods', ACM Computing Surveys, Vol. 30, No. 2, 1998, pp. 170-231.

[Güt 94]   Güting R. H.: 'An Introduction to Spatial Database Systems'. VLDB Journal , Vol. 3, No. 4, 1994.

[Gut 84]   Guttman A.: 'R-trees: A Dynamic Index Structure for Spatial Searching', Proc. ACM SIGMOD Int. Conf. on Management of Data, Boston, MA, 1984, pp. 47-57.

[Hil 91]   Hilbert D.: 'Über die stetige Abbildung einer Linie auf ein Flächenstück', Math. Annln., Vol. 38, 1891, pp. 459-460.

[Klu 98]   Klump G.: 'Development, Implementation and Evaluation of Strategies for Geometric Query Processing Under Oracle 8' (in German), master thesis, University of Munich.

[Kum 94]   Kumar A.: 'A Study of Spatial Clustering techniques', DEXA 1994, pp. 57-71.

[Fal 86]   Faloutsos C.: 'Multiattribute Hashing Using Gray Codes', Proc. ACM SIGMOD Int. Conf. on Management of Data, Washington D.C., 1986, pp. 227-238.

[Fal 88]   Faloutsos C.: 'Gray Codes for Partial Match and Range Queries', IEEE Transactions on Software Engineering (TSE) , Vol. 14, No. 10, 1988, pp. 1381-1393.

[FB 74]   Finkel R, Bentley J.L. 'Quad Trees: A Data Structure for Retrieval of Composite Keys', Acta Informatica , Vol. 4, No. 1, 1974, pp. 1-9.

[FR 91]   Faloutsos C., Rong Y.: 'DOT: A Spatial Access Method Using Fractals', Proc. 7th Int. Conf. on Data Engineering, Kobe, Japan, 1991, pp. 152-159.

[FR 89]   Faloutsos C., Roseman S.: 'Fractals for Secondary Key Retrieval', Proc. 8th ACM PODS, Philadelphia, PA, 1989, pp. 247-252.

[Jag 90]   Jagadish H. V.: 'Linear Clustering of Objects with Multiple Atributes', Proc. ACM SIGMOD Int. Conf. on Managment of Data, Atlantic City, NJ, 1990, pp. 332-342.

[Mor 66]   Morton G.: 'A Computer Oriented Geodetic Data BAse and a New Technique in File Sequencing', IBM Ltd., 1966.

[Oos 90]   Oosterom P.: 'Reactive Data Structures for Geographic Iinformation Systems'. Ph.D. thesis, University of Leiden, The Netherlands.

[Ore 89a] Orenstein J. A.: 'Redundancy in Spatial Databases', Proc. ACM SIGMOD Int. Conf. on Management of Data, Portland, OR, 1989, pp. 294-305.

[Ore 89b] Orenstein J. A.: 'Strategies for Optimizing the Use of Redundancy in Spatial Databases', Proc. 1st Symposium on Large Spatial Databases, Santa Barbara, CA, pp. 115-134.

[OM 84]   Orenstein J. A., Merrett T. H.: 'A Class of Data Structures for Associative Searching', Proc. 3rd ACM PODS, Waterloo, Ontario, Canada, 1984, pp. 181-190.

[Sag 94]   Sagan H.: 'Space-Filling Curves', Berlin/Heidelberg/New York: Springer-Verlag, 1994.

[Sam 89]   Samet H.: 'The design and analysis of spatial data structures'. Reading, MA: Addison-Wesley.

[SRF 87]   Sellis T. K., Roussopoulos N., Faloutsos C.: 'The R+-Tree: A Dynamic Index for Multi-Dimensional Objects', Proc. 13th Int. Conf. on Very Large Data Bases, Brighton, England, 1987, pp. 507-518.

GBI: A Generalized R-Tree Bulk-Insertion Strategy*

Rupesh Choubey, Li Chen, and Elke A. Rundensteiner

Department of Computer Science
Worcester Polytechnic Institute
Worcester, MA 01609-2280
{rupesh|lichen|rundenst}@cs.wpi.edu

Abstract. A lot of recent work has studied strategies related to bulk loading of large data sets into multidimensional index structures. In this paper, we address the problem of *bulk insertions* into *existing* index structures with particular focus on R-trees – which are an important class of index structures used widely in commercial database systems. We propose a new technique, which as opposed to the current technique of inserting data one by one, bulk inserts entire new incoming datasets into an active R-tree. This technique, called GBI (for Generalized Bulk Insertion), partitions the new datasets into sets of clusters and outliers, constructs an R-tree (small tree) from each cluster, identifies and prepares suitable locations in the original R-tree (large tree) for insertion, and lastly performs the insertions of the small trees and the outliers into the large tree in bulk. Our experimental studies demonstrate that GBI does especially well (over 200% better than the existing technique) for randomly located data as well as for real datasets that contain few natural clusters, while also consistently outperforming the alternate technique in all other circumstances.

Index Terms — Bulk-insertion, Bulk-loading, Clustering, R-Tree, Index Structures, Query Performance.

1 Introduction

1.1 Background and Motivation

Spatial data can commonly be found in diverse applications including Cartography, Computer-Aided Design, computer vision, robotics and many others. The amount of available spatial data is of an ever increasing magnitudes. For example, the amount of data generated by satellites is said to be several gigabytes

---

* This work was supported in part by the University of Michigan ITS Research Center of Excellence grant (DTFH61-93-X-00017-Sub) sponsored by the U.S. Dept. of Transportation and by the Michigan Dept. of Transportation. Dr. Rundensteiner thanks IBM for the Corporate IBM partnership award, and Li Chen thanks IBM for the Corporate IBM fellowship as well as mentoring from the IBM Toronto Labora.

R.H. Güting, D. Papadias, F. Lochovsky (Eds.): SSD'99, LNCS 1651, pp. 91–108, 1999.

per minute. Hence, efficient storage and indexing techniques for such data are needed.

Among many index structures proposed for spatial data, R-trees remain a popular index structure employed by numerous commercial DBMS systems [Inf] [Map]. The R-tree structure was initially proposed by Guttman [Gut84] and various variations and improvements over the original structure have been suggested ever since [BKSS90] [AT97] [TS94].

Generally, index structures need to be set up by loading data into them before they can be utilized for query processing. Initially, the basic *insert* operation proposed in [Gut84] was used to load sets of data into an R-tree, i.e., each object was inserted one by one into the index structure. In this paper, we will refer to this traditional technique of data sequential loading as the OBO (one-by-one) technique. The insertion of data sequentially into the R-tree has been found to be inefficient in the case when the entire tree needs to be set up [TS94]. Thus as an improvement, various *bulk loading* strategies have been proposed in recent years [LLG97, LLE97, BWS97, KF93]. These bulk loading strategies were aimed at creating a complete R-tree *from scratch*. These techniques thus assumed that totally new data was being collected (or existing data files were set up to be utilized by some application), and thus an index structure had to be constructed from scratch for this new dataset.

An upsurge of interest in spatial databases is how to efficiently manipulate existing massive amounts of spatial data, especially the problem of *bulk insertions* of new data assuming an already existing R-tree. The importance of this problem for numerous real-world examples is apparent. For example, new data obtained by satellites needs to be loaded into existing index structures. Active applications using the spatial data should continue functioning while being minimally impacted by the insertion of new data and in addition should be given the opportunity to make use of the new data as quickly as possible. The construction of a new index structure each time from scratch to contain both the old as well as the new data is not likely to scale well with an increasing size of the existing original index structure. Instead, new techniques specially tuned to this problem at hand are needed.

1.2   The Proposed Bulk-Insertion Approach

In our earlier work [CCR98a], we focussed on the problem of bulk insertion when the incoming dataset was skewed to a certain subregion of the original data. By skewed, we mean that the dataset to be inserted is localized to some portion of the region covered by the R-tree instead of being spread out over the whole region. In this paper, we extend the work and now provide a solution that deals with the general problem of bulk insertion of datasets of any nature instead of just skewed datasets. Both works look into the problem of bulk insertion of new datasets into an *existing active* R-tree. By active, we mean an R-tree which already contains (large) datasets, may have been used for some amount of time, and which has currently active applications that forbid the possibility of scheduling a long down-time for the index structure construction process.

However, in contrast to the prior STLT (Small-Tree-Large-Tree) technique [CCR98a], we now propose a new technique that is designed to handle bulk-insertion cases for different characteristics of the incoming datasets both in terms of the size of the new versus existing datasets as well as completely localized, partially skewed, somewhat clustered, or even completely uniform datasets. This technique, called GBI (for Generalized-Bulk-Insertions), partitions the new dataset into a set of clusters, constructs R-trees from the clusters (small trees), identifies and prepares suitable locations in the original R-tree (large tree) for insertions, and lastly performs the insertions of the small trees into the large tree.

Extensive experiments are conducted both with synthesized datasets as well as with real world datasets from the Sequoia 2000 storage benchmark to test applicability of our new technique. The results for both are comparable, thus indicating the appropriateness of the synthesized data for our tests. In our experiments, we find that GBI does especially well (in some cases even more than 200% better than the existing technique) for non-skewed large datasets as well as for large ratios of large tree to small tree data insertion sizes, while consistently outperforming the alternate technique in practically all other circumstances. Our experimental results also indicate that the GBI not only reduces the time taken for loading the data, but it also provides reasonable query performance on the resultant tree. The quality of the resulting tree constructed by GBI in terms of query performance is comparable to that created by the traditional tree insertion approach.

In summary, the contributions of this work include:

1. Design a general solution approach, GBI, to address the problem of *bulk insertions* of spatial data into *existing* index structures - in contrast to recent work on bulk loading data from scratch [LLE97] [LLG97] [BWS97] [KF93] and our previous work that deals with skewed datasets only [CCR98a].

2. Select and modify McQueen's k-means clustering algorithm to achieve the clustering desired by GBI.

3. Implement GBI and the conventional insertion technique in an UNIX environment using C++ in order to establish a uniform testbed for evaluating and comparing them.

4. Conduct experimental studies both using real-world as well as synthetic datasets. These experiments demonstrate that GBI is a winner, both in terms of substantial cost savings in insertion times as well as in keeping retrieval costs down.

The paper is organized as follows. Section 2 reviews related work. Section 3 defines the bulk insertion problem. Section 4 discusses our solution approach. Performance results for query tree loading and query-handling are presented in Section 5. This is followed by conclusions in Section 6.

2   Related Work

The general topic of bulk loading data into an initially empty structure has been the focus of much recent work. Two distinct categories of bulk loading algorithms have been proposed.

The first class of algorithms involves the bottom-up construction of the R-tree. Kamel and Faloutsos [KF93] use the Hilbert sorting technique to first order data and then build the R-tree. Leutenegger et al. [LLE97] proposed the STR (Sort-Tile-Recursive) technique in which a k-dimensional set of rectangles is sorted on an attribute and then divided into slabs. Both techniques are concerned with bulk loading and not with bulk insertion.

The other class of algorithms focuses on a top-down approach to build the R-tree. Bercken et al. [BWS97] adapt a strategy using a memory-based temporary structure called the buffer-tree. This technique is not likely to be very applicable when inserting new data over time – unless the temporary structure was continuously to be maintained along with the actual R-tree. This would be expensive, as buffer pages would be wasted. Arge et al. [AHVV99] presented a new buffer strategy for performing bulk operations on dynamic R-trees. Their method uses the buffer tree idea, which takes advantage of the available main memory and the page size of the operating system, and their buffering algorithms can be combined with any of the existing index implementations. However, their bulk insertion strategy is conceptually identical to the repeated insertion algorithm while we will present in this paper a conceptually unique bulk insertion strategy that can potentially be combined with their buffering algorithms.

[RRK97] discuss bulk incremental updates in the data cube. A portion of their work deals with bulk insertions of data which is collected over a period of time into R-trees. Their approach uses the sort-merge-pack strategy in which the incoming data is first sorted, then merged with the existing data from the R-tree and then a new R-tree is built from scratch. The strategy resolves back to eventually loading the tree up from scratch, whereas our approach avoids this for large existing trees prohibitly expensive step. [Moi93] suggests batching of data and sorting it prior to insertion. However, sorting phase is expensive and requires the data to be collected beforehand, while our algorithm tries to avoid the sorting phase. Bulk updates and bulk loading have been studied for various structures and in various scenarios [Che97, LRS93, LN97, CP98]. These techniques are typically specific to the structure in question and are not directly applicable to our problem.

Ciaccia et al. [CP98] proposed methods of bulk loading the M-tree which is possibly the work closest to ours. The proposed bulk loading algorithm performs a clustering of data objects into a number of sets, obtains sub-trees from the sets, and then performs reorganization to obtain the final M-tree. Our initial STLT algorithm is similar to the proposed algorithm in that STLT also constructs trees from subsets of data objects. However, in STLT, the choice of an appropriate location removes the need for reorganization of the tree in order to re-establish the balance of the tree. Again, the problem handled here was of bulk loading data and not of bulk insertion which is the focus of this paper. Heuristics of sizes of subtrees to insert were not developed, as done in our work [CCR98c].

Clustering algorithms have been the focus of extensive studies for long. Innumerable clustering algorithms with varying characteristics have been developed, such as [DO74, Spa82, War63, Rom84]. [Spa82] classifies clustering algorithms into the two classes of hierarchical and non-hierarchical algorithms. We do not attempt to improve any of the clustering algorithms nor suggest a new one. On the contrary, we simply select one of the algorithms based on our needs and apply it in our GBI solution.

3   Four Issues of Bulk Insertion

There are two conflicting goals for our proposed technique of bulk inserting data into an R-tree, the first being that the quality of the structure should be as good as possible and the second being that the time to load the new data into the R-tree should be minimized. This observation raises several issues to be addressed by our work:

- First, what is an effective strategy for inserting sets of data in bigger chunks rather than one-by-one so as to minimize down time for applications that use the R-tree structure?
- Second, which characteristics should these sets of to-be-inserted data objects (insertion sets) ideally possess so that the bulk-insertion strategy works most efficiently (in terms of both insertion time as well as resulting tree quality)?
- Third, how to group incoming possibly continuous streams of spatial objects into insertion sets so that each of them possesses desirable characteristics?
- Lastly, is it possible to design a framework which allows multiple solutions which have different balances between data loading times and the resultant tree quality?

For the first issue, we propose a new bulk-insertion strategy called the 'Generalized Bulk Insertion' (GBI) algorithm (see Section 4). GBI not only demonstrates efficient insertion times, but it also does have minimal impact on existing applications (1) by requiring no down time of the existing index by avoiding to build a new R-tree from scratch, and (2) by locking as few portions as possible of the R-tree for as short as possible a time (often only one single index node).

For the second issue, we conduct an extensive study that identifies numerous possible characteristics of the to-be-inserted data set in order to determine their impact on the insertion time as well as final tree quality (see Section 5). Characteristics of particular interest are (1) the number of objects in one insertion set (size of insertion set versus size of existing tree) and the spatial distribution of the new objects (skewness).

For the third issue, we could simply chop input streams into equally sized sets or as we do in our work employ a suitable clustering technique that takes data distributions into account (see Section 4). A new dataset is fed into a clustering tool that allows tuning of parameters to control the desired compactness and density of the clusters to be generated. Then clusters as well as outliers are ready for the next bulk insert phase, using STLT and OBO, respectively.

For the fourth issue, we provide a solution framework representing a mechanism for realizing multiple strategies for bulk inserting the new data. These parameters allow solutions which range from insertion of all the data in one shot to the insertion of data items one at a time. The middle portion of this range where data is inserted in multiple sets provides a compromise between loading time and tree quality. We provide a set of heuristics for selecting among the strategies for insertion [CCR98c].

4   GBI: A Generalized Bulk Insertion Strategy

4.1   The GBI Framework

As depicted in Figure 1, the 'Input Feed' module of GBI takes all the data to be inserted into the existing tree and gives it to the 'Input Feed Analyzer'. Analysis of the input data is done by the latter module, which passes the result of its analysis (number of incoming data, area of coverage, etc.) to the 'Cluster Detector'. The latter identifies clusters of suitable dimensions and accordingly separates the data into different clusters and a set of outliers. The generated clusters are used to construct a series of separate R-trees (small trees) by the 'Small Tree Generator' module. Note that this step can be done in parallel in order to improve performance. The 'Strategy Selector' module determines what strategy of insertion should be adopted, i.e., whether to gather all the data into one large single set and insert it in one shot or to group the data into multiple sets. This choice reflects the tradeoff between fast insertion of data and higher quality of the resultant tree. The small trees constructed are inserted into the existing large tree by the 'STLT Insertion' module. The outliers are inserted into the existing R-tree using the traditional (OBO) insert function.

The main steps of our Generalized Bulk Insertion (GBI) framework are the following:

1. Using our proposed heuristics, based on incoming data size and its area of coverage, determine values of the clustering modules' parameters.
2. Execute the clustering algorithm on the new dataset and obtain a set of clusters and outliers.
3. Build an R-tree (small tree) from one of the generated clusters.
4. Find a suitable position in the original tree (big tree) for insertion of the newly built small tree.
5. Handle unavailability of entry slot space in large tree using different heuristic techniques.
6. Insert the small tree into the identified location (or created as a result of Step 5).
7. Repeat steps 3 through 6 for each of the generated clusters.
8. Insert the outliers (also generated from step 2) using a traditional insertion algorithm.

A detailed explanation of the key steps of the GBI framework follows.

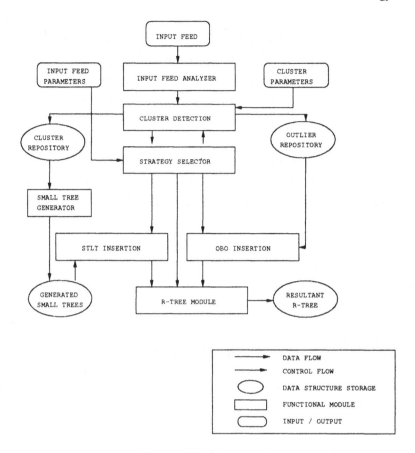

Fig. 1. GBI Framework

4.2   The GBI Terminology and Assumptions

For the following, we assume the terminologies and notations below :

$NumClusters$	= number of clusters generated by cluster generator
ST	= newly built small tree
LT	= original R-tree before insertion
ST_Root	= root node of the small tree
LT_Root	= root node of the large tree
$IndexNode$	= an index node of the large tree
M	= maximum number of slot entries in a node
$DiffHeight$	= difference in heights between large and small trees

An explanation of each of the parameters employed in the clustering algorithm is as follows :

k : Initial number of clusters. This may increase or decrease depending on the number of clusters actually present and the ordering in which input data arrives or is considered. The final number of generated clusters is mainly dependent on the input dataset.

C : Coarsening value which determines how close one cluster can be to another. Any two clusters whose centroids are at a distance less then this value will be merged.

R : Refining value which determines how close a data point should be to a cluster to make the cluster a candidate for insertion of the data point. Any data to be inserted that is at distances greater than R from all the clusters will be inserted in its own separate cluster. If there are more than one clusters at distance less than R, then the data will be inserted in the closest cluster. The cluster where the data is inserted has its centroid recomputed and the possibility of cluster merges is examined.

fmin : Minimum flush value which determines how large a cluster should be (in terms of the number of data elements present) in order to consider it a candidate for a small tree. If the number of elements in a cluster is less than fmin, then the data elements are inserted into the outlier list.

fmax : Maximum flush value which determines the maximum size allowed for a cluster.

```
ALGORITHM GBI(LT_Root)
{
    SetClusteringParameters(f, C, R, K); // Initial tuning values
    while a pause ( > setted time period ) of inputing data {
        NumClusters = InvokeClusteringAlgo(DatasetFromInputFeed);
        repeat (for NumCluster clusters) {
            ST_Root = BuildSmallTree(next ClusterFile);
            DiffHeight = LTHeight - STHeight; // Insertion level
            if (DiffHeight < 1)                 // Large tree <= small tree
                adopt OBO insertion;    // not suitable for GBI bulk insertion
            else                                // Insert small tree
                InsertSTintoLT(LT_Root, ST_Root, DiffHeight);
        }
        InsertUsingOBO(OutliersSet);
    }
}
```

Fig. 2. The GBI Framework: Bulk Insertion of New Data into Large Tree

4.3 GBI Framework Description

First, the entire new dataset is fed to the clustering tool that then breaks up the dataset into appropriate clusters and a set of outliers. The generated clusters are controlled by a few parameters which determine the number of data items in a cluster and also the compactness or density of the generated clusters. The GBI algorithm given in Figure 2 and as visually depicted in Figure 3 first identifies clusters and outliers from the given input dataset and then builds a small tree (denoted by ST) from each of the clusters, as indicated by the function *BuildSmallTree()* invoked in the algorithm. Next, considering one small tree at a time, we compute the difference in heights of the large tree and small tree as this would determine how many levels we need to go down in the large tree in order to locate the appropriate place for insertion of the small tree. If the new data is larger than the existing data, then the proposed technique is not meaningful. Then we instead simply apply one of the bulk-loading strategies to build a new tree containing both old and new data from scratch. Otherwise, we invoke the *InsertSTintoLT()* function to insert ST into an appropriate IndexNode in the big tree. The previous step is repeated for each of the small trees. Finally, once all clusters are exhausted, the outliers are inserted into the large tree one-by-one.

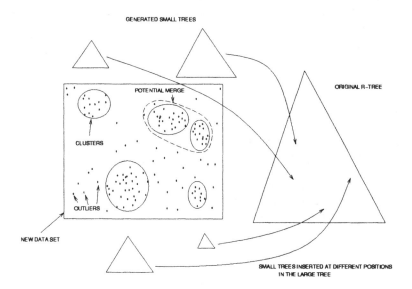

Fig. 3. Graphical Depiction of GBI Process

4.4   GBI Clustering Module

For the clustering, we use a variant of the MACQueen's k-Means Method [And73] with a suitable extension. This algorithm is chosen because it can be easily modified to allow for clusters of fixed maximum or minimum sizes. It can also be made to adapt for continuous incoming data. Figure 3 shows how clusters and outliers are formed from a given set of input data elements and how some of the formed clusters can be potentially merged (based on the values of the parameters of the clustering algorithm).

1. Select the proper values of the parameters : number of clusters k, coarsening value C, refining value R, and flush values *fmin* and *fmax*.
2. For the input, let each of the first k data unit be a cluster of size one.

3. Determine the minimal distance C between clusters. Merge the clusters with the distance of their centroids less than C until all the clusters are at a distance greater than C from one another.
4. Take the next data element and determine the cluster closest to it.
5. Decide to insert the data into its closest cluster if it is at a distance of less than R to that cluster, then recompute its centroid and merge clusters that their distance become less than C. If the closest cluster is at a distance greater than R, take the data as a new cluster of one member.
6. Repeat Step 4 and 5 for the remaining data given by the Input Feed module.
7. Take each of the cluster centroids as fixed seed points and reallocate each data to its nearest seed point.
8. For each cluster, if the number of data items it contains is greater than f, then consider it as a cluster, otherwise insert it in the outlier list.

4.5 GBI - STLT Module

A detailed description of the STLT (small-tree-large-tree) strategy, including precise algorithms for the module to insert one small tree into an existing large tree, is given in [CCR98a], while below we give a brief summary of its basic ideas below. Let the height of the newly built small R-tree be h_r and the height of the original R-tree be H_R. We consider the root rectangle of the small R-tree (enclosing rectangle of all new data rectangles) as a data rectangle. In other words, we use the standard *insert* operation to find a suitable place to insert the newly built R-tree into the existing R-tree (referred to as the *large tree*) as if it were an individual data item. We try to insert it into the level $l = H_R - h_r$ of the original R-tree. Our goal here is to assure that the bottom level of the small R-tree is on the same level as that of the original R-tree as seen in Figure 4. This is in order to ensure that the resultant tree remains balanced which is a fundamental requirement for the structure to be an R-tree.

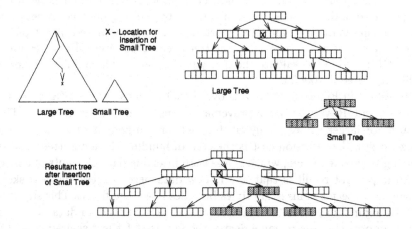

Fig. 4. Insertion of One Small Tree into the Large Tree

5   Experimental Evaluation of the GBI Framework

5.1   Experimental Setup

Testbed Environment: Our performance studies are conducted on a testbed on a SUN Sparc-20 workstation running the UNIX operating system. The testbed includes the original R-tree with Quadratic splitting [Gut84], an I/O buffer manager, modified MACQueen's k-means clustering module, input feed analyzer, strategy selector module and other supporting data structures and procedures.
Test Data: The test data comprised both real datasets and synthetic datasets. Most of our tests are based on synthetically generated testing datasets to be able to verify the usefulness of GBI under different extreme settings. The synthetic data was generated with varying parameters to control the distribution and the nature of the objects. The real data from the TIGER/Line files distributed by the US Census Bureau consists of a dataset of streets (131,461 objects) and a dataset of rivers and railway tracts (128,971 objects) from an area in California.
Test Types: We carried out two major classes of tests. The first type of experiments were conducted to compare the *I/O insertion costs* of GBI with OBO for different parameters. The tests were designed to evaluate the performance of GBI and compare that of OBO in terms of I/O costs for different parameters and determine the usefulness and limitations of GBI. The second type of experiments focussed on evaluating the *quality* of the resultant trees formed by GBI or OBO style of insertions. The tests comprised asking queries on the resultant trees to measure the number of nodes visited to answer the queries and the I/O cost incurred in answering the queries.

5.2   Experiments Measuring Insertion Cost and Query Cost for Different Data Area Ratios

This experiment is used to evaluate the performance of GBI as compared to OBO, when the area of the new dataset is varied from a small percentage of the large tree until it equals the area covered by the original tree. A set of 5000 data elements is inserted and the insertion times are measured. A set of 50000 random queries is asked on each of the trees generated after OBO insertion and after GBI insertion to evaluate the query performance. The number of elements in the large tree is 100000.

As shown in Figure 5, GBI wins over OBO for most 'C' and 'R' values. For larger 'C' and 'R' values, the improvement in insertion times is maximal. This is because, for larger values, many clusters are formed and there are few outliers. In such cases, the insertion cost is the sum of building the small trees and then inserting them one by one, which is less than inserting data elements one by one.

An important result is that as the new data becomes less and less skewed, the savings in terms of insertion times becomes more significant. This shows that the GBI algorithm proves to be useful for non-skewed random data. The reason for this improvement is to some degree the fact that for less skewed data, OBO is not able to exploit the locality of pages in the buffer as most elements belong

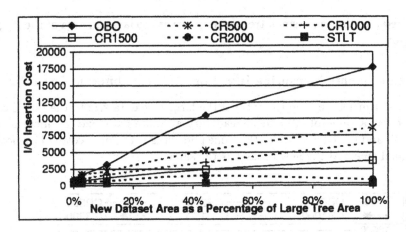

Fig. 5. I/O Insertion Cost for Different Area Ratio of Original Tree Dataset to New Dataset.

to different pages. Hence, the I/O cost of insertion for OBO increases as the new dataset becomes less localized.

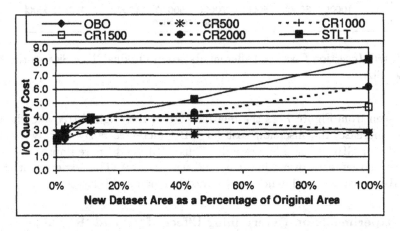

Fig. 6. Query Cost for Different Area Ratio of Original Tree Dataset to New Dataset.

On the other hand, Figure 6 shows that the query performance of the GBI algorithm is comparable to the OBO query performance for smaller 'C' and 'R' values. This is because the clusters formed are much tighter if the values of 'C' and 'R' are lower. The figure also shows that GBI improves on the STLT [CCR98a] query performance while yielding significant savings in the insertion time. This is done by ensuring that the data to be inserted into the original tree is clustered and thus each small tree does not cover too large an area of the large tree region.

5.3   Experiments Measuring Insertion Cost for Different Data Sizes

In this experiment, we keep the ratio of the large tree data size to the new data size fixed (at a value of 50) and instead vary absolute data sizes by increasing both small and large tree sizes at the same percentage. The area covered by both the original tree and the new dataset is the same.

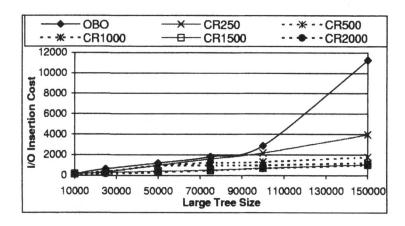

Fig. 7. I/O Insertion Cost for Different Ratios of Original Tree Size to the New Dataset Size

As seen from Figure 7, as the relative sizes of the original tree and the new dataset increase, the insertion cost for OBO increases fairly rapidly whereas the costs for GBI increase less rapidly. Thus, GBI yields more improvement for relatively larger sizes of original tree and new dataset. This experiment shows that GBI is scalable and is not constrained to small sized datasets.

5.4   Experiments on Determining Effect of 'C' and 'R' Values

In this experiment, we analyze the effect of varying 'C' and 'R' values on the insertion times and the resulting tree quality in terms of query performance. We insert a set of 5000 data elements randomly located over the region of the large tree and measure the insertion costs and query costs for 50000 random queries. By randomly located, we mean that the data elements are randomly spread out all over the area enclosed by the original tree elements which is 30000 by 30000.

As can be seen from Figure 8, as the 'C' and 'R' values increase, the insertion time decreases. This is because for larger 'C' and 'R' values, larger size clusters are formed and there are fewer data elements which are inserted individually thus improving on the insertion times tremendously (100-fold improvement for

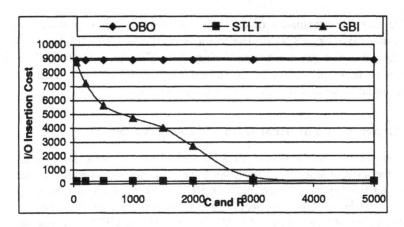

Fig. 8. I/O Insertion Cost for Different 'C' and 'R' Values

'C' and 'R' values greater than 3000). For low 'C' and 'R' values, there are fewer clusters which are used to construct small trees and most of the data not being part of a cluster but being an outlier is inserted using the OBO technique. For very small values of 'C' and 'R', the technique yields no small trees and hence there is no difference in OBO and the GBI insertion costs.

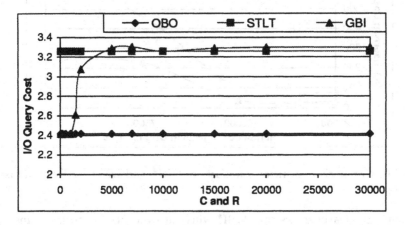

Fig. 9. I/O Query Cost for Different 'C' and 'R' Values

The query performance corresponding to the resultant trees from the above experiment is shown in Figure 9. This shows clearly that GBI yields trees of good quality when the 'C' and 'R' values are kept low. This is because low 'C' and 'R' values yield more dense clusters which keep the query costs low. For higher 'C' and 'R' values, the clusters formed are less dense and thus the trees generated using them are of slightly lower quality (in terms of intersecting MBRs) and thus the query performance of GBI becomes worse. This experiment yields a maximum of one extra I/O on the average for the tree constructed using GBI as compared to the one constructed using OBO. Clearly, settings for 'C and 'R' values must be empirically selected to find a trade-off between maximizing tree retrieval quality while minimizing tree insertion times.

5.5   Experiments on Real Datasets Using Tuned GBI

Based on the experiments listed thus far and additional ones that can be found in our technical report [CCR98c], we have empirically determined values for initializing the key parameters of the GBI framework, such as values for C, R, *flushmin*, *flushmax*, etc. (see Section 4.1 for explanation of these parameters). A justification of this tuning of GBI, then called the GBI* heuristic strategy, can be found in [Cho99], while below we give some insight into the performance achieved by this tuned GBI*. The experiments are done with real data extracted from the Sequioa 2000 storage benchmark. The experiment uses different sizes of the large tree and new dataset but the ratio is kept fixed at 50.

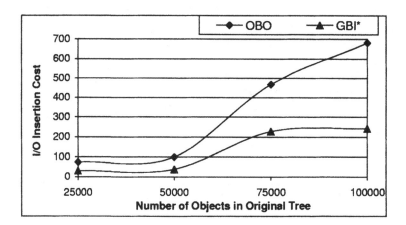

Fig. 10. Comparison of Insertion Costs of GBI* and OBO for Real Datasets

Figure 10 clearly shows that GBI* wins out over OBO in terms of insertion costs. As the sizes of the datasets increase, the savings in insertion cost increase as well. Figure 11 displays the query costs for GBI* and OBO which are very similar thus indicating that the GBI* generated tree is of acceptable quality.

6   Conclusions

In contrast to previous work on bulk loading data which primarily focussed on building index structures from scratch, in this paper we tackle the problem of bulk insertions into existing active index structures. We extend our earlier work

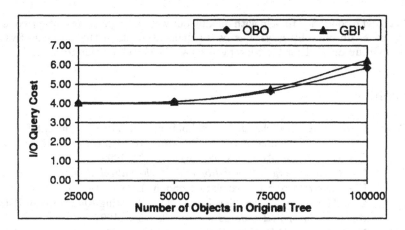

Fig. 11. Comparison of Query Costs of GBI* and OBO for Real Datasets

[CCR98b] where we addressed the problem for skewed datasets only. While our current work focuses on R-tree, we contend that the overall strategy of group-first and then insert-as-bulk is a general approach and thus could also be explored for other multi-dimensional index structures.

The proposed framework, called GBI (for Generalized Bulk Insertion), considers the new dataset as a set of clusters and outliers, constructs from the clusters a set of R-trees (small trees), identifies and prepares suitable locations in the original R-tree (large tree) for insertion of each of the small trees, and lastly bulk-inserts each small tree into the large tree.

We have reported experimental studies designed to not only compare GBI against the conventional technique, but also to evaluate the suitability and limitations of the GBI framework under different conditions. We have found that GBI does especially well (over 200% better than the existing technique) for non-skewed datasets as well for large ratios of large tree sizes as compared to sizes of the new data insertions (see Figure 7), while consistently outperforming the alternate technique in all other circumstances. Our experimental results also indicate that the quality of the resulting tree constructed by GBI in terms of query performance is acceptable when compared to the resulting tree created by the traditional tree insertion approach. All in all, the GBI *bulk insertion* strategy has thus been found to be a viable and significant optimization over the conventional one-by-one insertion approach, gaining both in terms of update costs as well as preserving the resulting index access quality.

Possible future tasks include application of the general approach of GBI to other multi-dimensional index structures as well as experimental evaluation of alternate approaches against GBI such as from-scratch bulk-loading techniques.

Acknowledgments. The authors would like to thank first and fore-most Yun-Wu Huang, Ning Jing, and Matthew Jones who have developed most of the experimental system that served as platform for our work, including the R-tree index structure, storage manager, and buffer management modules. We are especially grateful to Yun-

Wu Huang who continued to help with making our way through the prototype code and in many other uncountable ways. Lastly, we thank students at the Database Systems Research Group at WPI for their interactions and feedback on this research.

References

[AHVV99]   L. Arge, K. Hinrichs, J. Vahrenhold, and J. Vitter. Efficient Bulk Operations on Dynamic R-trees. *Workshop on Algorithm Engineering and Experimentation ALENEX 99*, pages 92–103, 1999.

[And73]   M. R. Anderberg. *Probability and Mathematical Statistics*. Academic Press, New York, San Francisco, London, 1973.

[AT97]   C.H. Ang and T.C. Tan. New Linear Node Splitting Algorithm for R-trees. *Advances in Spatial Databases*, pages 339–349, 1997.

[BKSS90]   N. Beckmann, H.P. Kriegel, R. Schneider, and B. Seeger. The R*-tree : An Efficient and Robust Access Method for Points and Rectangles. *Proceedings of SIGMOD*, pages 322–331, 1990.

[BWS97]   J. Bercken, P. Widmayer, and B. Seeger. A Generic Approach to Bulk Loading Multidimensional Index Structures. *International Conference on Very Large Data Bases*, pages 406–415, 1997.

[CCR98a]   L. Chen, R. Choubey, and E. A. Rundensteiner. Bulk Insertions into R-trees using the Small-Tree-Large-Tree Approach. *Proceedings of ACM GIS Workshop*, pages 161–162, 1998.

[CCR98b]   L. Chen, R. Choubey, and E.A. Rundensteiner. Bulk Insertions into R-Trees. *WPI, Tech. Rep. CS-WPI-98-05*, February 1998.

[CCR98c]   R. Choubey, L. Chen, and E. A. Rundensteiner. GBI: A Generalized R-Tree Bulk Insertion Strategy. *WPI Technical Report TR-98-15-STLT*, 1998.

[Che97]   W. Chen. Programming with Logical Queries, Bulk Updates, and Hypothetical Reasoning. *IEEE Transactions on Knowledge and Data Engineering*, pages 587–599, July 1997.

[Cho99]   R. Choubey. R-Tree Bulk Insertion Strategies. *Master Thesis in progress, Worcester Polytechnic Institute*, 1999.

[CP98]   P. Ciaccia and M. Patella. Bulk Loading the M-tree. *Proceedings of the Australasian Database Conference*, February 1998.

[DO74]   B. Duran and P. Odell. *Cluster Analysis - A Survey*. Springer-Verlag, Berlin, Heidelberg, New York, 1974.

[Gut84]   A. Guttman. R-trees: A Dynamic Index Structure for Spatial Searching. *Proceedings of SIGMOD*, pages 47–57, 1984.

[HJR97a]   Y.W. Huang, N. Jing, and E.A. Rundensteinder. Spatial Joins Using R-trees: Breadth-First Traversal with Global Optimizations. *International Conference on Very Large Data Bases*, pages 396–405, 1997.

[HJR97b]   Y.W. Huang, M. Jones, and E.A. Rundensteiner. Symbolic Intersect Detection: A Method for Improving Spatial Intersect Joins. *Proc. of the International Symposium on Spatial Databases*, pages 165–177, 1997.

[HNR97]   Y.W. Huang, N.Jing, and E.A. Rundensteiner. A cost model for estimating the performance of spatial joins using R-tree. *International Working Conference on Scientific and Statistical Database Management*, pages 30–38, August 1997.

[Inf]   Informix Corporation ("http://www.informix.com"). *Informix*.

[KF93]   I. Kamel and C. Faloutsos. On Packing R-trees. *Proceedings of International Conference on Information and Knowledge Management*, pages 490–499, November 1993.

[LLE97]   S. Leutenegger, M. Lopez, and J. Edgigton. STR: A Simple and Efficient Algorithm for R-tree Packing. *Proceedings of IEEE International Conference on Data Engineering*, pages 497–506, 1997.

[LLG97]   S. Leutenegger, M. Lopez, and Y. Garcia. A Greedy Algorithm for Bulk Loading R-trees. Technical report, University of Denver Computer Science Technical Report # 97-02, 1997.

[LN97]   S. Leutenegger and D. Nicol. Efficient Bulk-Loading of Gridfiles. *IEEE Transactions on Knowledge and Data Engineering*, pages 410–420, May 1997.

[LRS93]   J. Li, D. Rotem, and J. Srivastave. Algorithms for Loading Parallel Gridfiles. *Proceedings of SIGMOD*, pages 347–356, 1993.

[Map]   MapInfo Corporation.
            SpatialWare - http://www.mapinfo.com/spatialware/spatial20.html.

[Moi93]   A. Moitra. Spatio-Temporal Data Management Using R-trees. *International Journal of Geographic Information Systems*, 1993.

[Rom84]   H. Charles Romesburg. *Cluster Analysis for Researchers.* Lifetime Learning Publications, Belmont, California, 1984.

[RRK97]   N. Roussopoulos, M. Roussopoulos, and Y. Kotidis. Cubetree : Organization of and Bulk Incremental Updates on the Data Cube. *Proceedings of SIGMOD*, pages 89–99, 1997.

[Spa82]   H. Spath. *Cluster Analysis Algorithms for Data Reduction and Classification of Objects.* Ellis Horwook Publishers, Chichester, 1982.

[TS94]   Y. Theodoridis and T. Sellis. Optimization Issues in R-tree Construction (extended abstract). *Lecture Notes in Computer Science*, pages 270–273, 1994.

[War63]   J. H. Ward. Hierarchical Grouping Analysis For Applications. *Journal of American Statistics Association*, 58:236–244, 1963.

Moving Objects and Spatio-temporal Data

Capturing the Uncertainty of Moving-Object Representations

Dieter Pfoser and Christian S. Jensen

Department of Computer Science, Aalborg University, DENMARK
{pfoser|csj}@cs.auc.dk

Abstract. Spatiotemporal applications, such as fleet management and air traffic control, involving continuously moving objects are increasingly at the focus of research efforts. The representation of the continuously changing positions of the objects is fundamentally important in these applications. This paper reports on on-going research in the representation of the positions of moving-point objects. More specifically, object positions are sampled using the Global Positioning System, and interpolation is applied to determine positions in-between the samples. Special attention is given in the representation to the quantification of the position uncertainty introduced by the sampling technique and the interpolation. In addition, the paper considers the use for query processing of the proposed representation in conjunction with indexing. It is demonstrated how queries involving uncertainty may be answered using the standard filter-and-refine approach known from spatial query processing.

1 Introduction

A relatively new research area, spatiotemporal databases concerns the management of objects with spatiotemporal extents, and real-world objects with continuously changing spatial extents are attracting substantial attention. The variety of applications suggests that there is not just one prototypical type of spatiotemporal application.

Spatiotemporal applications may be distinguished based on the data they manage, which may pertain to the past, the present, and the future, or a combination of these. For example, applications managing past data often conduct analyses of movements over time, answering queries such as, "What were the movements of the Vikings in the North Sea between year 1000 and year 1200?" Applications dealing with present and future data capture the current spatial extents of objects in the database and typically make predictions about the future extents of the objects. Sample queries include, "What is the position of flight SAS 286?" and "Where will flight SAS 286 be in 20 minutes?" Next, a specific type of application concerns real-world objects that move continuously and disregards the spatial extents of the objects, representing instead their positions as points. Candidate applications include fleet management, air traffic control, military command-and-control systems, and people tracking. This paper focuses

R.H. Güting, D. Papadias, F. Lochovsky (Eds.): SSD'99, LNCS 1651, pp. 111–131, 1999.

on the representation of the past and present positions of such moving-point objects.

Fundamental issues in these applications include the acquisition and representation of the movements of objects, including the inherent imprecision in the representation. For example, when representing the positions of vehicles based on sampling, the sampled positions are inherently imprecise, as are the interpolated positions in-between the samples. As a result, the record of the movements of objects as stored in the database differs from the actual movement. The imprecisions due to the measurements and caused by the use of sampling are inherently quite different. It is highly relevant to understand the nature of these imprecisions because this makes it possible to decide on their relative importance.

This paper's contributions are three-fold. First, it offers a proposal for representing the positions of moving-point objects in databases. Second, it quantifies the imprecisions in the proposed representation. The representation is modular, allowing the imprecision to be captured or not, depending on the application requirements. Third, the paper illustrates how the representation may be used in conjunction with indices to answer queries involving uncertainty. The two-step filter-and-refinement process known from spatial query processing is used together with error information.

Past database research has focussed on spatiotemporal applications where only the present and future positions of moving-point objects are relevant. In the context of applications that predict the movements of objects based on their current positions, speeds, and directions, Wolfson et al. (16) address position update policies and the imprecision involved in the database-representation of the positions. Next, Moreira et al. (9) present a data model for moving-point objects that is based on the decomposition of the trajectories of the objects into sections. In addition, so-called superset and subset semantics are proposed that aim to address uncertainty issues. A maximum error occurs when linearly approximating the movement of an object in-between samples, and this error is used in the process of query processing. However, this work is not connected to any specific application or technological context and thus does not cover the ranges of errors and the relationships between different error measures. The query processing aspects also do not consider the availability of indices. Güting et al. (5) present a comprehensive framework of abstract data types for moving objects. This work, however, does not address representation issues, nor does it accommodate uncertainty.

The outline of the paper is as follows. Section 2 presents an application scenario and describes a particular technological context for the application, the Global Positioning System. Section 3 proceeds to describe, quantify, and relate the measurement and the sampling errors in the context of the application scenario and accommodates also error information in the representation. This sets the stage for a proposal for a database representation for moving-point objects, presented in Section 4. Section 5 considers the utilization of this representation in query processing using indices. Finally, Section 6 concludes and offers directions for future research.

2   An Application Scenario—GPS-Based Fleet Management

This section presents a sample spatiotemporal application scenario, fleet management, and briefly introduces the Global Positioning System (GPS), the technology that is assumed used for sampling the positions of moving objects.

2.1   Fleet Management

The optimization of transportation, especially in highly populated areas, is a very challenging task that may be supported by an information system. An example fleet management project, conducted by the Department of Transportation of the State of California, Caltrans (3), aims to design what is termed the Advanced Transportation System. In this application, vehicles equipped with GPS devices transmit their positions to a central computer using either radio communication links or cellular phones. At the central site, the data is processed and utilized. Example queries occurring in such an application are as follows.

- Which taxi is closest to customer A?
- What is optimal taxi distribution over the area (somewhat related to pickups per area)?
- Compute the optimal route for a ride, considering road characteristics such as the actual and theoretical speed limits, congestions, accidents, etc.

Taking uncertainty into account, more sophisticated queries may be formulated.

- Which taxis were, with a 50% probability, within 100 meters of the Ritz hotel at 14.20 on April 22, 1999?
- How likely is it that taxi 1234 had visual contact with (was within 100 meters of) taxi 4321 between 9.00 and 13.00 on April 22, 1999?
- Which taxis were with 50% probability in Central Park at 10.00 on April 22, 1999.

2.2   Global Positioning System

The Global Positioning System is able to determine exact positions on Earth anytime, in any weather, and anywhere. The system consists of 24 satellites that orbit Earth at 20000 km. The satellites transmit signals that can be detected by GPS receivers, which then are able to determine their locations with great precision.

The principle behind the GPS is the measurement of the distances between a receiver and several satellites. A total of four distances, and thus signals from four satellites, are needed to solve a set of four equations that expresses the latitude, longitude, height, and time (Magellan Corporation 8). The distance from the satellite to the receiver can be calculated by multiplying the time it takes for the signal to arrive by the speed at which it travels–the speed of light.

Although four visible satellites are enough to compute a position, the more satellites that are visible, the more precise the computed position becomes.

More information about the GPS can be found in, e.g., Magellan Corporation (8) and Leick (7).

3   Sampling and Uncertainty

This section covers how to acquire and represent the movement of point objects. We first give the technical means of how to determine the time-varying positions of moving point objects, and subsequently give a suitable way to represent the entire movement. An important part of the representation is the uncertainty caused by the acquisition process. The section describes the uncertainty caused by the measurement error and the sampling error, and it concludes with a discussion of the relative importance of these errors.

3.1   Acquiring Movement—Measuring Position in Time

In order to record the movement of an object, we would have to know the position at all times, i.e., on a continuous basis. However GPS and telecommunications technologies only allows us to sample an object's position, i.e., to obtain the position at discrete instances of time such as every few seconds.

The solid line in Fig. 1(a) represents the movement of a point object. Space (x and y axes) and time (t axis) are combined to form one coordinate system. The dashed line shows the projection of the movement in two-dimensional space (x and y coordinates).

A first approach to represent the movements of objects would be to store the position samples. For our database, this would mean we could not answer queries about the objects' movements at times in-between sampled positions. Rather, to obtain the entire movement we have to interpolate. The simplest approach is to use linear interpolation, as opposed to other methods such as polynomial splines (Bartels et al. 1). The sampled positions then become the end points of line segments of polylines, and the movement of an object is represented by an entire polyline in three-dimensional space. In geometrical terms, the movement of an object is termed a *trajectory* (we will use "movement" and "trajectory" interchangeably).

Fig. 1(b) shows the spatiotemporal space (the cube in solid lines) and several trajectories (the solid lines). Time moves in the upward direction, and the top of the cube is the time of the most recent position sample. The wavy-dotted lines at the top symbolize the growth of the cube with time.

3.2   Quantifying Uncertainty

The research on uncertainty in geospatial information is concerned with all sources of incorrectness and incompleteness in the measurement, analysis, and interpretation of digitally-represented, Earth-referenced phenomena (Unwin 13).

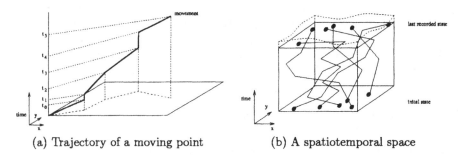

(a) Trajectory of a moving point   (b) A spatiotemporal space

Fig. 1. Movements and spaces

A representation of moving-point trajectories is inherently imprecise: imprecision is introduced by the measurement process used in the sampling of positions and by the sampling approach itself. A useful representation of moving points must take these uncertainties into account.

In this paper, we make the following assumptions.

- We will not consider any error connected to the times of measurements. We assume that we know precisely the time a position sample was taken. This assumption seems to be justified when using the GPS and its precise clocks as a measuring device.
- Within one application, we will only consider objects with similar movement characteristics, such as speed and range. Typical examples of objects with different characteristics include people, cars, and planes.

A first step in incorporating uncertainty into a representation of trajectories is to quantify it. We thus proceed to describe the errors introduced by the trajectory acquisition process.

3.3   Measurement Error

Generally, an error can be introduced by inaccurate measurements (Leick 7). The accuracy and thus the quality of the measurement depends largely on the technique used. This paper assumes that the GPS is used for the sampling of positions.

Two assumptions are generally made when talking about the accuracy of the GPS. First, the error distribution, i.e., the error in each of the three dimensions and the error in time, is assumed to be *Gaussian*. Second, we can assume that the horizontal error distribution, i.e., the distribution in the x-y plane, is circular (van Diggelen 14).

The error in a positional GPS measurement can be described by the probability function in Equation (1). The probability function is composed of two normal distributions in the two respective spatial dimensions. The mean of the distribution is the origin of the coordinate system. Fig. 2 visualizes the error distribution. In addition to the mean, the standard deviation, σ, is a characteristic parameter of a normal distribution . Within the range of $\pm\sigma$ of the mean, in a bivariate normal distribution (2-dimensional), 39.35% of the probability is concentrated.

$$P_1(x,y) = \frac{1}{2\pi\sigma^2} e^{-\frac{x^2+y^2}{2\sigma^2}} \tag{1}$$

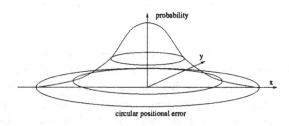

Fig. 2. Positional error in the GPS

Example 1. A typical GPS module used in vehicle navigation systems is the CrossCheck AMPS Cellular from Trimble Navigation Ltd. This GPS/cellular phone system has an error of $2m$ (equal to 1 σ) (Trimble Navigation Ltd. 12). This measure refers to the standard deviation of a bivariate normal distribution centered at the receiver's true antenna position.

3.4   Uncertainty in Sampling

We capture the movement of an object by sampling its position using a GPS receiver at regular time intervals. This introduces uncertainty about the position of the object in-between the measurements. In this section, we give a model for the uncertainty introduced by the sampling, based on the sampling rate and the maximum speed of the object.

Sampling Error The uncertainty of the representation of an object's movement is affected by the frequency with which position samples are taken, the *sampling rate*. This, in turn, may be set by considering the speed of the object and the desired maximum distance between consecutive samples. Let us consider the running example, in which we want to record the movements of taxis.

Example 2. As a requirement to the application, the distance between two consecutive samples should be maximally 10 meters. If the maximum speed of a taxi is $150km/h$, this means that we would need to sample the position at least 4.2 times per second. If a taxi moves slower than its maximum speed, the distance between samples is less than 10 meters.

We proceed to consider how the position samples resemble the true movement of the object. Consider the three trajectories shown in Fig. 3. Each is possible given the three measured positions P_1 through P_3. However, by just "looking" at the

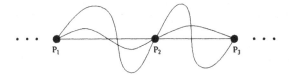

Fig. 3. Possible trajectories of a moving object

three positions, one would assume that the straight line best resembles the actual trajectory of the object. Since we did not measure the positions in-between two consecutive position samples, the best we can do is to *limit the possibilities of where the moving object could have been.* We have to constrain the trajectory of the object by what we know about the object's actual movement. Considering the trajectory in a time interval $[t_1, t_2]$, delimited by consecutive samples, we know two positions, P_1 and P_2, as well as the object's maximum speed, v_m; see Fig. 4. If the object moves at maximum speed v_m from P_1 and its trajectory is a straight line, its position at time t_x will be on a circle of radius $r_1 = v_m(t_1 + t_x)$ around P_1 (the smaller dotted circle in Fig. 4). Thus, the points on the circle represent the furthest away from P_1 the object can gotten at time t_x. If the object's speed is lower than v_m, or its trajectory is not a straight line, the object's position at time t_x will be somewhere within the area bounded by the circle of radius r_1. Next, we know that the object will be at position P_2 at time t_2. Thus, applying

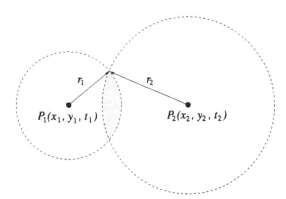

Fig. 4. Uncertainty between samples

the same assumptions again, the object's position at time t_x is on the circle with radius $r_2 = v_m(t_2 - t_x)$ around P_2. If the object moves slower or its trajectory is not a straight line, it is somewhere within the area bounded by this circle.

The above constraints on the position of the object mean that the object can be anywhere in the intersection of the two circular areas at time t_x. This intersection is shown by the shaded area in Fig. 4, and we use the term *lens* for this intersection. Since we do not have any further information, we assume a uniform distribution for the position within the lens.

Thus, the sampling error at time t_x for a particular position can be described by the probability function shown in (2), where r_1 and r_2 are the two radii described above, s is the distance between the measured positions P_1 and P_2, and A denotes the area of the intersection of the two circles.

$$\mathcal{P}_2(x, y) = \begin{cases} \frac{1}{A} & \text{for } x^2 + y^2 \le r_1^2 \wedge (x - s)^2 + y^2 \le r_2^2 \\ 0 & \text{otherwise} \end{cases} \quad (2)$$

Substituting $v_m(t_1 + t_x)$ and $v_m(t_2 - t_x)$ for the radii r_1 and r_2, respectively, the probability function shown in Equation (3) results. Its parameters are described in Table 1.

$$\mathcal{P}_2(x, y) = \begin{cases} \frac{1}{A} & \text{for } x^2 + y^2 \le (v_m(t_1 + t_x))^2 \wedge \\ & (x - s)^2 + y^2 \le (v_m(t_2 - t_x))^2 \\ 0 & \text{otherwise} \end{cases} \quad (3)$$

For a visualization of a sampling error, refer to Fig. 5(a), in which the two horizontal axes depict x and y coordinates, and the vertical axis the positional probability.

Table 1. Parameters of the probability function, $\mathcal{P}_2$, describing the sampling error

v_m	maximum speed of the moving object
t_x	time for which the error distribution is computed
t_1	time of the first measured position
t_2	time of the second measured position
s	distance between the two positions, i.e., the length of the line segment
A	lens area, i.e., the area of the intersection of the two circles

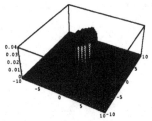
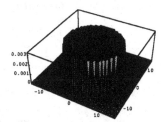

(a) Normal-case sampling error   (b) Worst-case sampling error

Fig. 5. Probability functions for sampling errors

Sampling Error Across Time So far, we have quantified the sampling error for the position at a single point in time. To determine the error across time, as a first step, we compute the lens for various $t_x \in [t_1; t_2]$ as shown in Figs. 6(a)–(c).

The circle around the first point, P_1, measured at time t_1, is initially a point and grows as time advances, and the circle around the second point, P_2, shrinks with the advancement of time and eventually becomes a point. In the first situation in Fig. 6(a), the circle around P_2 contains the one around P_1, meaning that the constraint on how far away the object can be from P_1 at t_x is more restrictive than the constraint on how close it has to be to P_2. The area of intersection is the total circle or radius r_1. In the second situation, Fig. 6(b), the two circles start intersecting, and in Fig. 6(c) they show a clear intersection.

We observe that the intersection points of the two circles over time, i.e., for the cases the circles do actually intersect, lie on an *error ellipse* with positions P_1 and P_2 as its foci (cf. Fig. 7). The length of the semi-major axis is $2a = r_1 + r_2$. This is not surprising if we consider the definition of an ellipse. An ellipse is

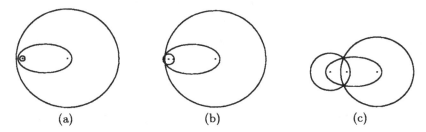

(a)   (b)   (c)

Fig. 6. Evolving sampling error

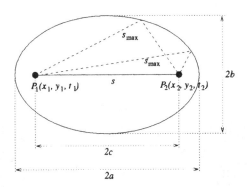

Fig. 7. Error ellipse

a curve consisting of all points in the plane whose sum of distances, r_1 and r_2, from two fixed points, P_1 and P_2 (the foci) separated by a distance of $2c$, is a given constant, $2a$. The measure $2c$ can be interpreted as the observed distance between P_1 and P_2, whereas $2a$ is the maximum distance the object can travel. The "thickness" of the ellipse, $2b$, is determined by the equation $b^2 = a^2 - c^2$.

This means that the smaller the difference between the observed distance, $2c$, and the maximum distance, $2a$, the "thinner" the ellipse. In the extreme case, the ellipse degrades to a line segment. In the worst case, where the object does not move between consecutive position samples, the ellipse becomes a circle.

Sampling Rate Having derived the general principle behind the sampling error, we give an example of how an increased sampling rate affects the error size. To illustrate the underlying principle, we use the error ellipse given in Fig. 7 as a measure for the size of the sampling error per line segment.

Example 3. In Fig. 8, we show the actual trajectory of a moving object as a bold line. As a first step, we sample the movement of the object at position P_1 and P_2. The time in-between the samples is 10 seconds. The shortest distance from P_1 to P_2 is 300 meters. Thus, to the best of our knowledge the object travels at a speed v of $30m/s$. If we further know the maximum speed of the object to be $42m/s$, we can draw an error ellipse around the line approximating the movement. The error ellipse has an eccentricity $2c = 300m$, a major axis $2a = v_{max} \cdot \Delta t = 42m/s \times 10s = 420m$, and a minor axis $2b = \sqrt{(2a)^2 - (2c)^2} = \sqrt{420^2 - 300^2} = 294m$. This rather large error ellipse means that the position of the object in-between samples is quite uncertain. Quadrupling the sampling rate, i.e., sampling the position every 2.5 seconds, leads to an error ellipse that has an eccentricity $2c = 80m$, a major axis $2a = v_{max} \cdot \Delta t = 42m/s \times 2.5s = 105m$, and a minor axis $2b = \sqrt{(2a)^2 - (2c)^2} = \sqrt{105^2 - 80^2} = 68m$.

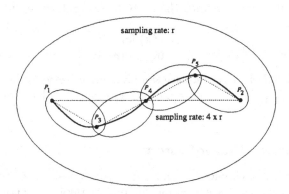

Fig. 8. Varying sampling rate

If we increase the sampling rate, the sample positions better approximate the movement, and the error introduced by sampling is decrease.

Maximum Speed Challenged An underlying assumptions so far has been that the maximum speed of a moving object is fixed at v_{max}. However, the more we know about the object in question, the further we can narrow down v_{max} and thus reduce the uncertainty. For example, if we know that a taxi can reach $200km/h$, but regulations of the company set $120km/h$ as the upper limit, we may decide to assume that v_{max} is $150km/h$. Further examples of such additional information are local speed limits and road conditions; thus the maximum speed can vary depending on the area the taxi is in. Traffic volumes, which are time dependent, may also be taken into account. Further, there might be individual speed limits for drivers and cars. Generally, the more information we have about an object, the better we can adjust the sampling rate, reduce the error, and, consequently, minimize the uncertainty attached to its polyline trajectory.

Worst-Case Sampling Error Previously we identified the size and extent of the sampling error for a particular line segment and time. However, for use in Section 5, we also need an error measure in the situation where an object does not move between consecutive samples. In this case, the sampling error is determined by a circle of radius, r, equal to half the sampling interval multiplied by the maximum speed.

Example 4. Consider again the taxi from Example 3, whose position is sampled every 2.5 seconds. If the taxi is stopped, the eccentricity is $2c = 0$ (the foci coincide) and the error ellipse degrades to a circle. The major axis, $2a = v_{max} \cdot \Delta t = 42m/s \cdot 2.5s = 105m$, is equal to the minor axis. The radius of the circle then is 52.5 meters.

If we have no further information about the position of the object, all positions within the circle have the same possibility, yielding a circular uniform, worst-case error distribution, for which the probability function is given below, where r is the radius.

$$P_3(x, y) = \begin{cases} 1/(r^2\pi) & \text{for } \sqrt{x^2 + y^2} \leq r \\ 0 & \text{otherwise} \end{cases} \qquad (4)$$

For a visualization of the worst-case sampling error, refer to Fig. 5(b), in which the two horizontal axes depict x and y coordinates and the vertical axis the positional probability.

3.5 Comparison of Error Sources

With current GPS technology, a moving object's position can be determined instantaneously with an accuracy of $2m$ (cf. Example 1), and this error will be reduced further with the advancement of GPS technology. How frequently position samples are taken depends on the particular application. In fleet management, determining the position every 2.5 seconds leads to a worst-case error of roughly $50m$. This is the radius of a circular distribution assuming that the maximum speed of the objects is $150km/h$, cf. Example 4). In practice the sampling rates will be much lower, thus allowing for worst-case errors of $200m$ or more.

It follows that the measurement error is small compared to the sampling error in fleet management. Therefore, we will consider only the uncertainty that stems from the sampling, and disregard the uncertainty caused by the measurement technique, in the remainder of the paper.

4   A Representation for Moving Point Objects

Section 3 proposed a technique for capturing the movement of point objects that utilized polylines, and the section characterized the error introduced by this technique, this way also revealing the uncertainty inherent to the polyline representation. This section's objective is to provide a format for representing the history of the positions of continuously moving point objects, along with the uncertainty associated with our records of their positions. For this, we propose a relational database schema that incorporates all the spatiotemporal and error information previously presented in this paper.

Specifically, the schema in Table 2 defines relations for objects, for the line segments constituting the trajectories of the objects, and for the error information associated with the recorded trajectories. Relation **Object** has attributes

Table 2. Relational schema for capturing moving-point objects, their trajectories, and associated error information

Object	$<$ object_id, max_speed, *etc.* $>$
Line_segment	$<$ line_id, object_id, t_1, t_2, x_1, x_2, y_1, y_2, error_id $>$
Error	$<$ error_id, error_type, param1, param2 $>$

object_id, which is the key attribute, and max_speed, which determines the maximum speed at which the object can move. In addition this relation may include any number of attributes unrelated to the objects' spatial extents. Relation **Line_segment** captures the line segments that compose the trajectories of the objects. Attribute line_id is the key attribute; object_id is a foreign key referencing relation **Object**; and t_1 and t_2 are the times when the two positions, (x_1, y_1) and (x_2, y_2), constituting the line segment, were measured. Finally, relation **Error** contains the error information associated with the line segments. Attribute error_id is the key; error_type specifies the type of error that a tuple refers to, and thus specifies how parameters param1 and param2 are to be interpreted. In the current schema, there is only one type of error. However, if we consider more error sources in our application, additional types of errors may occur.

The domains of the attributes are as follows. Define $dom(x)$ to be a function that returns the domain of its argument attribute x. Then $dom(\text{object\_id}) = dom((\text{line\_id}) = dom(\text{error\_id}) = dom(\text{max\_speed}) = dom(t_1) = dom(t_2) = \mathcal{N}$, where $\mathcal{N}$ is the natural numbers, $dom(\text{param1}) = dom(\text{param2}) = \mathcal{N} \cup \text{NIL}$, $dom(x_1) = dom(x_2) = dom(y_1) = dom(y_2) = \mathcal{Z}$, where $\mathcal{Z}$ is the integers, and $dom(\text{error\_type}) = \{\text{worst-case\_sampling}\}$.

The following example illustrates how the above schema can be put to use.

Example 5. Our taxi company operates a number of taxis in a city. The database in Fig. 9 captures the movement of the taxis together with the associated errors.

This database permits the company to reconstruct the trajectories of its taxis and to compute the associated error information. All taxis are recorded in relation **Object**, and their trajectories are kept in **Line_segment** and are referenced through the foreign key object_id.

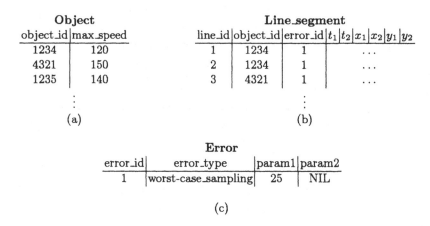

(a)

(b)

(c)

Fig. 9. An example database containing positional and error information connected to a fleet management application of a taxi company

To utilize the error information, the parameters for the various probability functions are recorded. The parameter of the worst-case sampling error, $\mathcal{P}_3$, is the radius r, which in our database is stored as the param1 attribute value (25) of the only tuple in relation **Error**.

The parameters of the sampling error, $\mathcal{P}_2$, as shown in Table 1, are the distance s, which is computed from the attribute values for x_1, x_2, y_1, and y_2 in relation **Line_segment**, together with the times t_1, and t_2. The maximum speed v_m is stored in relation **Object**. Finally, t_x is not a static parameter that can be stored in a relation, but is an input parameter from a query. Thus, the intersection area A, which is different for each position in time contained in a particular line segment, can also only be computed once t_x is known.

5   Query Processing and Indexing

The objective of this section is to explore the use of error information when using indices for processing queries involving the positions of moving objects. The section first sets the context within spatiotemporal indexing for its contribution. Subsequently, it shows how a moving-point index may be put to use in the processing of spatiotemporal range queries involving positional uncertainty. A discussion of what types of queries that can be answered in the given framework is given. The section ends with a summary of the section's proposed approach.

5.1   Context

The purpose of spatiotemporal indexing is to efficiently support the retrieval of those objects, from a large set of objects, with spatiotemporal extents that satisfy a specified query predicate. The most commonly considered predicate is intersection with a specified region.

Substantial research is currently ongoing in spatiotemporal indexing, and a number of spatiotemporal indices have already been proposed; see Theodoridis et al. (11) for an overview. Although an index well suited for indexing the trajectories of the kinds of moving-point objects considered here still does not exist, it is expected that such an index will be invented.

In terms of the representation proposed in this paper, this means that we can expect to be able to index the polyline segments that represent trajectories. However, taking the uncertainty of the trajectories into consideration corresponds to the indexing of (non-point) objects with spatial extents, and the envisioned moving-point indices are no longer readily applicable.

Based on the assumption that it will be substantially more attractive to index the trajectories of moving-point objects than to index the trajectories of objects with spatial extents, which are more complex, this section offers an approach to using moving-point trajectory indices while taking into account the uncertainty of the trajectories and also taking into account query predicates relating to the uncertainty.

The approach employs the fundamental technique from spatial indexing of using approximations for the spatial extents to be indexed (Güting 4). For instance, R-trees generally use minimum bounding boxes. This use leads to a filter-and-refine strategy for query processing. First, based on the approximations, a filtering step is executed that returns a superset of the objects fulfilling the query predicate. Second, in the refinement step, the exact extents of the objects resulting from the first step are checked against the query predicate (Brinkhoff et al. 2).

5.2   Processing Uncertainty Queries

The goal here is to be able to use a moving-point index to answer queries such as "Retrieve the positions of taxis that were inside area A (specified as a rectangle) between times B and C with a probability of at least 30%?"

The first step is to specify the meaning of an object's position being within an area A with a probability of 30%. An object's position is described by means of a probability function centered around the positional mean (e.g., recall the probability function of the worst-case sampling error in Fig. 5(b)).

If all of an object's positional probability is within an area A, we say that the object is within area A for certain. We can determine if this is the case by integrating over the probability function with area A as the limit. If the result is 1, this is the case.

If the object is within area A with a probability of at least 30%, at least 30% positional probability has be concentrated within area A. This case is shown in

Fig. 10, where the circle represents the probability function of the worst-case sampling error, the rectangular shape is the query rectangle, and the shaded region represents the probability in the query window. The result of integrating over the probability function with rectangle $A = ([x_{min}, x_{max}], [y_{min}, y_{max}])$ as the limit thus has to be 0.3 or higher. Further, if the positional error is rota-

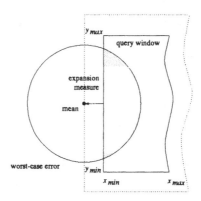

Fig. 10. Summing up the probability

tionally symmetric around the positional mean, as is the case for the worst-case sampling error, we can determine the maximum distance of the query window to the positional mean such that the probability of the position to be within the query window is 30% or higher. We term this computed distance the *expansion measure*. In Fig. 10 this distance is indicated by the arrow from the edge of the query window to the center of the probability function (the mean). The expansion measure can be interpreted as the measure by which the query window has to be expanded to contain the positional mean (the dotted query window in the figure).

Note that we are assuming that the query window is longer and wider than $2r$, the diameter of the error distribution; later in this section, we will revisit this assumption. However, for now, we proceed to show how the expansion measure can be used in the filter step.

The Filter Step We record the positions in time of the moving objects by means of line segments. The points on these segments are the mean values of the positional probability functions Our objective for the filter step is to retrieve those line segments that contain positions in time qualifying for the query result, i.e., those positions that, with the probability specified in the query, are in the query window. To retrieve these line segments, we intersect the expanded query window with the indexed line segments. Expanding the query window means that all positions with a probability higher or equal the one specified in the query (the one used to compute expansion measure) are contained in the query window.

For the filter step, the error measure used to determine the expansion measure can be coarse, but has to be universal so that it applies for all positions in the database. This is true for the worst-case sampling error described in Section 3.4.

As we shall see next, this method can only be applied if the probability specified in the query is less than 50%.

Consider again the above query, but with a probability of 60%. Using the worst-case sampling error leads to a negative query expansion measure (cf. Fig. 11(b)). If we use this smaller query window, we would retrieve a subset rather than a superset of the qualifying objects, since, e.g., positions that have no error (or a small error) and lie on (or close within) the borders of the query window would be disregarded. An example here is position P that has no error associated in Fig. 11(b). Shrinking the query window by the size of the negative expansion measure would eliminate this position from our set of candidate solutions.

This problem is solved by simply using the original query window with no expansion (shrinking) for probabilities higher than 50%. This means that we retrieve a superset of the qualifying objects.

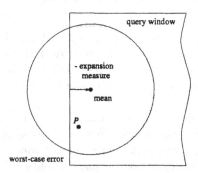

Fig. 11. Query window expansion: high probability

The Refinement Step To determine the final result, we have to evaluate the query predicate on all objects identified during the filtering step. In our case we have identified line segments that intersect with the transformed query region. As the final answer, we would like to have a set of positions, and so the refinement step extracts those parts of the line segments that qualify for this set.

In the filter step the intersection of line segments with the query window was determined with the help of the worst-case sampling error. To evaluate positions in time in the refinement step we will use the sampling error, unique for every position. A very straightforward way to achieve this is to apply the brute-force method of computing the probability functions in turn for all positions in time contained in a line segment (cf. Section 3.4) and check whether at least 30% probability is concentrated within area A. Fig. 12 shows two positions in time, P_1 and P_2, and their respective sampling errors (depicted by dotted lines). For each of the positions, the probability concentrated within the query window is depicted by a shaded area. The set of solutions after the refinement step

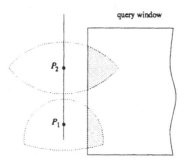

Fig. 12. Refinement step

comprises all positions in time whose positional probability within a given query window is at least as high as specified in the query.

On the Size of Query Windows Some query types deserve special attention within the presented framework. Point queries such as "Which taxis were in location A (point) at time B with 50% probability?" cannot be answered within this framework, since we cannot compute how much probability is concentrated within a point.

However, some "point" queries actually might be translated into window queries, e.g., location A might refer to a road crossing or a waiting area for taxis. In this case we are confronted with a small-window query.

Consider the above query where the query window of location A has an extent of, e.g., a taxi stand. If the sampling rate of the taxis' positions was very coarse, the positions have a high degree of uncertainty, and the sampling error is very large. To find an answer to our query, we have to determine the positions for which at least 50% of the probability is concentrated within the query window. If the query window is too small with respect to the error measure, no positions will qualify, e.g., consider query window QW_1 in Fig. 13.

On Non-Empty Query Results To derive a first minimum size of a query window for which the result is not guaranteed to be empty, we assume the worst case for both error and query. The largest possible error is the worst-case sampling error. The worst case of a query is to specify 100% probability, e.g., "Which taxis were in location A (point) at time B with 100% probability?" The smallest square query window we can consider has side length $2r$, the diameter of the worst-case sampling error, e.g., QW_2 shown in Fig. 13. If the query window is smaller, the probability of the worst-case sampling error cannot by contained entirely within the query window any more, i.e., less than 100% probability is concentrated within the query window, and the result is guaranteed to be empty.

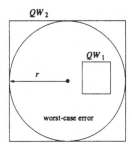

Fig. 13. Small-window query

However, this only depicts the worst-case scenario. For query windows with side length larger than $2r$, queries specifying varying degrees of uncertainty may have non-empty answers. If the query window is smaller and the query specifies 100% probability, certain positions are eliminated because their associated error is too large for them to be for certain (100%) in the query window. Although this seems only to eliminate solutions, it has significant consequences for the use of the worst-case error with the query window in the filter step, as will be explained next.

The Filtering Step Revisited In the case the query window is smaller than $2r$, the filtering as outlined earlier might eliminate positions that satisfy the query predicate. Consider the example shown in Fig. 14(a). First, we determine the expansion measure for a probability of 30% and the query window (the rectangle). The shaded area symbolizes the intersection of the error measure and the query window. The size of the area corresponds to the positional probability concentrated within the query window.

Using this expansion measure, however, would exclude qualifying positions, e.g., position P would be discarded in the filter step, although 30% or more of its actual positional probability (dotted lens shape of the sampling error) is concentrated within the query window.

To avoid the elimination of qualifying positions, we will initially expand the sides of small query windows that are smaller than $2r$ to be of size $2r$ and then use the resulting window to determine the expansion measure as describe earlier in this section. This is illustrated in Fig. 14(b), where the probability concentrated in the window of height $2r$ is symbolized by shading, and where the expansion measure is symbolized by the longer arrow. With this measure, position P will be in the set of candidate solutions. To recap, small window queries are addressed as follows. A query window can be arbitrarily small. In connection with databases considering spatial uncertainty, the size of the query window is also determined by the uncertainty specified in the query. Further, the "extent" of a spatial position stored in the database is determined by the associated error measure. Consequently, when specifying a query, one has to keep all these measures in mind not to retrieve an unwantingly small result.

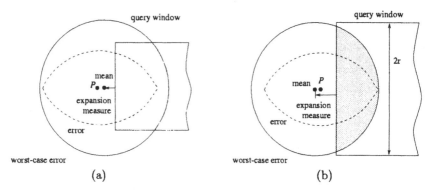

Fig. 14. Query window expansion: small query window

5.3   Summary of Approach

Or goal for this section was to give a method of how to use a moving point-index
to process queries of the form "Retrieve the moving-object positions that were
inside query rectangle A at some time between times B and C with a probability
of at least X%." The trajectories are indexed using a moving point index that
supports range queries. A superset of the qualifying positions are retrieved in a
filtering step, in which we expand the query window to retrieve all line segments
containing positions that are in the query window with probability at least X%.
The expansion is determined using probability X and the worst-case sampling
error, stored in the database.

In the refinement step, the positions contained in the retrieved line segments
that actually are within query rectangle A with probability at least X% are
identified. Here we use the sampling error, which is distinct for all positions. The
following pseudo-algorithm summarizes the full retrieval procedure. We assume
that the diameter of the worst-case sampling error is $2r$.

IF query window A has either height or width less than $2r$ THEN
    Increase the smaller side(s) to be of size $2r$;
IF X < 50% THEN
    Apply query window expansion to A;
Let S contain the result of searching the index with A;
Apply refinement to the line segments in S and return the resulting points;

6   Conclusions and Future Research

The paper investigates the representation of moving-point objects in databases.
First, a set of queries derived from requirements to an application managing
moving-point objects is presented. The Global Positioning System is the technology used for obtaining samples of the positions of these objects.

The paper proposes a method for acquiring and representing the movements
of point objects. The positions of objects are sampled at selected points in time,

and the positions in-between these points in time are obtained using interpolation, thus capturing the complete movement.

The representation of movements is inherently imprecise, and the paper considers two types of errors, the measurement error and the sampling error. Two measures were derived for the sampling error, one pertaining to each position in time, and a global worst-case error. It is further shown that the measurement error can be ignored in the application context considered.

A database schema is proposed that incorporates both the polyline representation of movements as well as the parameters of the various error distributions associated with the polyline representation. The schema is illustrated by an example database suited for the taxi management application.

Finally, the paper shows how to use this database to answer spatiotemporal queries derived from the example application. The error information is used in connection with an arbitrary moving-point index to answer spatiotemporal queries using the standard filter-and-refinement process.

This work points to several directions for future research. First, for the representation of the movement, we chose to linearly interpolate in-between measured positions. More advanced techniques may be used for this purpose as well, e.g., polynomial splines. Second, two types of error measures were considered, namely the measurement error and the sampling error. Additionally manipulating the measured positions before storing them in the database introduces another error that needs to be considered. Thus, generalizing the approach to an arbitrary number of error measures poses an interesting challenge. Third, in our work we only consider uncertainty in the spatial dimensions (cf. Section 3.2). This is partly because of the high precision with respect to time of the positional measurement device, GPS, we use. Using motion sensors or other techniques instead poses the question of quantifying uncertainty with respect to time as well. Fourth, one of the underlying assumptions in our work is that objects are not restricted in their movements through space. In reality, the space considered will typically contain roads, railroad tracks, walls, floors, mountains, lakes, or other "infrastructure" that facilitate or inhibit movement. This infrastructure may be taken into account to yield a reduced overall uncertainty and error in the database, as well as other benefits.

Acknowledgements

This research was supported in part by the CHOROCHRONOS project, funded by the European Commission, contract no. FMRX-CT96-0056, by the Danish Technical Research Council through grant 9700780, and by the Nykredit Corporation.

The Mathematica software package (Wolfram 15) was used to compute the probability functions shown in Figs. 5(a) and (b).

References

Bartels, R. H., Beatty, J. C., and Barsky, B. A.: An Introduction to Splines for Use in Computer Graphics & Geometric Modeling. Morgan Kaufmann Publishers, Inc., 1987.

Brinkhoff, T., Kriegel, H. P., and Seeger, B.: Efficient Processing of Spatial Joins Using R-trees. *SIGMOD Record*, 22(2):237–246, 1993.

Caltrans New Technology and Research Program. Advanced Transportation Systems Program Plan. URL: <http://www.dot.ca.gov/hq/newtech/nt_page.html>.

Güting, R. H.. An Introduction to Spatial Database Systems. *VLDB Journal*, 3:357–399, 1994.

Güting, R.H., Böhlen, M., Erwig, M., Jensen, C. S., Lorentzos, N. A., Schneider, M., and Vazirgiannis, M.: A Foundation for Representing and Querying Moving Objects. Technical Report, Informatik-Bericht 238, FernUniversität Hagen, Germany, 1998.

Greenwalt, C. R. and Shultz, M. E.: Principals of Error Theory and Cartographic Applications. Technical Report, ACIC Technical Report No. 96, Aeronautical Chart and Information Center, St. Louis, MO, 1962.

Leick, A.: GPS Satellite Surveying. John Wiley & Sons, Inc., 1995.

Magellan Corporation, The : About Global Positioning - the Basics of GPS and GLONASS. URL: <http://www.ashtech.com/Pages/gpsndx.html>.

Moreira, J., Ribeiro, C., and Saglio, J.: Representation and Manipulation of Moving Points: An Extended Data Model for Location Estimation. *Cartography and Geographical Information Systems*, to appear.

Nascimento, M. A., Silva, J. R. O., and Theodoridis, Y.: Access Structures for Moving Points. Technical Report TR-33, TIMECENTER, Aalborg University, Denmark, 1998.

Theodoridis, Y., Sellis, T., Papadopoulos, A., and Manolopoulos, Y.: Specifications for Efficient Indexing in Spatiotemporal Databases. In *Proceedings of the 10th International Conference on Scientific and Statistical Database Management*, 1998.

Trimble Navigation Ltd.: CrossCheck AMPS Cellular. Product Datasheet, 1998.

Unwin, D. J.: Geographical Information Systems and the Problem of Error and Uncertainty. In *Progress in Human Geography 19*, 549–558, 1995.

van Diggelen, F.: GPS Accuracy: Lies, Damn Lies, and Statistics. In *GPS World*, 5(8), 1998.

Wolfram, S.: The Mathematica Book. Cambridge University Press, 1996.

Wolfson, O., Chamberlain, S., Dao, S., Jiang, L., and Mendez, G.: Cost and Imprecision in Modeling the Position of Moving Objects. In *Proceedings of the Fourteenth International Conference on Data Engineering*, 1998.

Dynamic Spatial Clustering for Intelligent Mobile Information Sharing and Dissemination*

Eddie C. Shek, Giovanni Giuffrida, Suhas Joshi, and Son K. Dao

Information Sciences Laboratory
HRL Laboratories
3011 Malibu Canyon Road
Malibu, CA 90265, USA
shek@hrl.com, giovanni@wins.hrl.com, suhas@wins.hrl.com, skdao@hrl.com

Abstract. Intelligent Mobile Information Systems support information-centered applications that require support for a large number of distributed mobile users collaborating on a common mission and with interests in a common situation domain. A mobile user operating in the field changes location, consumes resources, investigates situations "on the horizon," and performs other incrementally evolving activities. A mobile user's information needs are therefore continually evolving in a neighborhood of interrelated data centered on the user's current location. Broadcast data dissemination is most effective when each broadcast information packet has multiple interested parties. To maximize the value of multicast dissemination, we dynamically cluster similar user profiles into aggregate user classifications that are served by independent multicast channels of custom information packets. Mobile user locations are also continuously tracked and mapped onto a cartographic representation of the real scenario. Spatial proximity between users is then computed by taking into account real boundaries as described in the cartographic map. Spatial information and spatial relationships among mobile users are then provided to the clustering algorithm with an eventual improved quality of the disseminated data.

1 Introduction

Intelligent Mobile Information Systems [19,18] support information-centered applications that require support for a large number of distributed mobile users collaborating on a common mission and with interests in common situation domains. Most existing work focus on access methods to spatial databases that contain pre-recorded situation information [17]. On the other hand, we aim to provide support for proactive and adaptive multicast-based dissemination of real-time situation information. The underlying basis behind creating a multicast

---

* This research was funded in part by the Defense Advanced Research Project Agency (DARPA) under the Battlefield Awareness Data Dissemination (BADD) Program. The view and conclusions contained in this document are those of the authors, and should not be interpreted as representing the official policies, either expressed or implied, of DARPA or the US Government.

R.H. Güting, D. Papadias, F. Lochovsky (Eds.): SSD'99, LNCS 1651, pp. 132–146, 1999.

service for sharing mobile information is to efficiently disseminate relevant information to users that are in close proximity, and hence likely share a common interest in transportation information.

A mobile user operating in the field changes location, consumes resources, investigates situations "on the horizon," and performs other incrementally evolving activities. A mobile user's information needs are therefore continually evolving in a neighborhood of interrelated data centered on the user's current location. This need can be effectively satisfied by listening to the broadcast of situation update by fixed stations as well as other mobile users in the user's neighborhood. Broadcast data dissemination is most effective when each broadcast information packet has multiple interested, or receiving, parties. To maximize the value of multicast dissemination, we dynamically cluster similar user profiles into aggregate group profiles that are served by independent multicast channels of custom information packets. In other words, given a collection of information interests of mobile users against a collection of information streams, a clustering process is performed to organize users into disjoint groups. Multicast communication channels are then created for user groups whose membership changes as mobile users migrate. Through the multicast channel assigned to the group he/she belongs to, a user can make announcements to his/her fellow group members and obtain information from sources servicing the area covered by the group.

One of the most important issues in realizing the information dissemination framework for mobile information sharing and dissemination is to balance the often antagonistic goals of maintaining accurate clustering, so that users receive highly relevant information, and economical utilization network bandwidth. At the same time, care has to be taken to accommodate constraints imposed by mobile wireless networks that form the communication infrastructure. In particular, existing terrestrial wireless networks have limited bandwidth as well as support only a limited number of allowable multicast channels (if multicasting supported at all). Such a framework is further complicated by the mobile nature of users which move around in the real world while participating at a multicast session. This implies that user's interests, correlated to their current geographical location, vary across space and time. As a result, clustering have to be continuously modified to maintain its quality. Furthermore, mobile users' move are generally constrained in the real world by natural and/or artificial boundaries such as roads, rivers, tracks, etc. Such boundaries should be exploited to help guide decisions on whether users are "similar" and hence improve the quality of clustering.

In this paper, we present a framework that combines knowledge discovery and spatial database techniques to realizing effective mobile information sharing given the real-life limitations of existing mobile wireless networks. In Section 2, we describe the fundamental cartographic map representation and its usage to dynamically compute real-world distance among mobile users in the system. In Section 3, we describe an algorithm to dynamically cluster mobile users based on their spatial domain of interest and compare it with previous clustering algorithms appeared in the literature. Users' locations are continuously mapped onto the cartographic representation of the actual scenario. Sections 4 and 5 present simulation results and the current prototype implementation of our intelligent mobile information system respectively. Finally, Section 6 concludes the paper by summarizing the contributions of this research.

2   Geographic Information Modeling

As already mentioned, under very many circumstances users move around in the real world following well defined natural and/or artificial routes such as roads, rivers, tracks, air channels, etc. Often *spatial clustering* has been solely based on Euclidean distance among moving entities. This measure may be ineffective in real world situations where, rarely, actual distance between pairs of entities moving along defined paths coincides with their Euclidean distance. For instance, in Figure 1 user a is moving along the same route of b, while c is moving on another route. According to the actual road distance, a is closer to b than to c (even though the Euclidean distance computation would suggest the opposite), therefore, a would be more likely clustered together with b since his/her interest in, say, road information may be closer to b's.

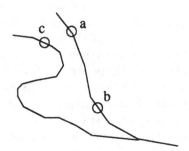

Fig. 1. Road Distance between Users

Identifying user's location on a cartographic map, and relating this location to the one of other users, improve *user profile prediction* with a subsequent more relevant information being multicasted to specific users. In order to include spatial information in our profile clustering algorithm, we superimposed a cartographic representation of the real scenario to our system. Mobile users' locations are then *mapped* onto such a cartography with a subsequent richer set of information being provided to the clustering algorithm.

Henceforth, in our system user profiles are augmented with information such as: user's location, movement direction, speed. Furthermore, distance among entities is the real-world physical distance based on the map representation by taking into account the actual shape and length of real routes. These variables are then included in the objective function of the profile clustering algorithm discussed above.

2.1   Cartographic Map Representation

In our system, users' locations are continuously mapped onto a cartographic map. Distance among moving objects corresponds to their physical distance computed in the map by following the real routes reproduced in the map. Hence, we assume that entities move along such routes and these latter ones are properly represented in the cartography at hand. Routes in the map are represented as *polylines* in a vector based representation of a GIS. Each polyline is described by the set of its vertices coordinate pairs: $(< x_0, y_0 >, < x_1, y_1 >, \ldots, < x_n, y_n >)$.

The cartography needs to be modeled by a data structure that can guarantee an efficient computation of the distance between each pair of objects (the spatial clustering objective function is based upon such a distance. To satisfy such a requirement we adopt a modified version of a *spatial network* [17] (alternatively called *Spatial Graph*). Spatial networks have been exploited in many different domains as the kernel data representation; examples of these domains are: transportation systems, air traffic management, urban management, as well as all the different types of utility networking such as power, telephone, water, and gas. A spatial network is a richly connected graph where nodes are labeled with a geographical (x, y) location of the specific entity to be modeled, and edges link pairs of entities. As an example, a spatial graph for a road transportation system would be modeled by having a node for each road intersection and an edge for each road segment connecting two intersections. Nodes would be labeled with Euclidean (x, y) coordinate of the intersection, while edges would be associated to various information depending upon the target application (e.g., length of the road segment, name of the street, type of street, etc.). Note that spatial networks are also suitable to model three dimensional spaces.

In order to fully capture the dynamic nature of our mobile user based application we modified the standard spatial network model into a finer representation. In our system, a spatial graph G is created by a two-step procedure as described in the following. Firstly, apply to each polyline $p = (< x_0, y_0 >, < x_1, y_1 > , \ldots, < x_n, y_n >)$ in the input map the following:

1. Create a class k_0 in G and label k_0 with "$< x_0, y_0 >$";
2. For each vertex $< x_j, y_j > (j = 1, 2, \ldots, n)$ of p:
   (a) Create a new class k_j and label it with "$< x_j, y_j >$";
   (b) Create an edge e connecting k_j to k_{j-1} and label e with the Euclidean distance between $< x_j, y_j >$ and $< x_{j-1}, y_{j-1} >$.

Once the graph G is created apply the following recursive procedure to G.

1. While there is an edge e of G, connecting the classes k_1 and k_2, such that $e.label > MAX\_EDGE\_LENGTH$ (where MAX_EDGE_LENGTH is a user's defined threshold) do the following:
   (a) Remove e from G;
   (b) Compute $< x_m, y_m >$ as the mid-point coordinates of the two locations identified by (the label of) k_1 and k_2;
   (c) Create a new class k_m with label "$< x_m, y_m >$";

(d) Create a new edge e_1 connecting k_1 to k_m and label it with $e.label \div 2$;
(e) Create a new edge e_2 connecting k_m to k_2 and label it with $e.label \div 2$;

This procedure creates a set of spatial graphs: one for each polyline in the input map. Such set of graphs needs then to be combined into a single large fully connected spatial graph where road intersections are properly represented. Figure 2 depicts two polylines crossing each other: circles in the figure represent nodes in the spatial graph, notice that nodes exist for each vertex in a road as well as for the intersection.

In essence, our spatial graph has the following properties:

1. A node for each route vertex (node a in Figure 2);
2. A node for each road intersection (node b in Figure 2);
3. Intermediate nodes are created for long straight road segments in order to make sure that each segment is shorter than a certain threshold (i.e., MAX_EDGE_LENGTH) defined by the user (node c in Figure 2).
4. Edges connect adjacent nodes if a road segment exists connecting the two corresponding geographical locations.

A spatial graph in our system is basically an approximation of the coordinate space in the input map for all the points belonging to a route in the map. The approximation ratio is controlled by the parameter MAX_EDGE_LENGTH which is set by the user. The value of MAX_EDGE_LENGTH depends upon the current application and the accuracy of the related equipment (e.g., GPS tolerance).

At each refresh point, a mobile user's location is mapped to a particular class of the spatial graph. Henceforth, user's movement history is described by the sequence of classes *visited* by the user.

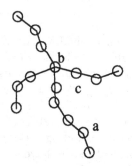

Fig. 2. Merging Map-Graphs

2.2 Modeling Interest Profiles of Mobile Users

A simple way of modeling an interest in information pertaining to a geographic region is to use a rectangular bounding box. While convenient, information requests modeled as rectangular bounding boxes can be inaccurate, especially for that of a group of users clustered together. We address the problem of the inefficiencies in using rectangular bounding boxes for clustering entities by marking off spatial coverage using grids. The entire playing area is divided into a rectangular grid and the geographic area of interest of each mobile user lies in its neighborhood along some path.

Once the spatial graph is created it can be used at run-time to dynamically compute distance among mobile users. At each refresh point, coordinates of moving entities are mapped into classes of the spatial graph G. That is, each entity location is mapped to the closest point described by a class of G. A neighborhood of a mobile user is the set of classes in the spatial graph that are within a threshold distance from the user's location at a specific time. In Figure 3 three objects (represented by a circle) and their respective neighborhoods (represented by the set of small squares) are depicted. Neighborhoods can easily be computed on the spatial graph by a linear time graph visit algorithm [16].

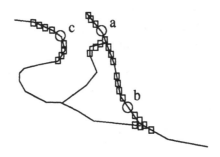

Fig. 3. Neighborhoods of Moving Entities

Vicinity among users is then defined based on the number of overlapping classes in their respective neighborhoods, the more classes overlap between the two neighborhoods the (spatially) closer the two users are. The adopted graph based representation discussed above allows an easy and efficient computation of the vicinity dimension which is then used by the profile clustering algorithm. The adopted spatial graph representation is suitable to compute distance between all possible pairs of mobile users by exploiting a conventional graph *shortest path* algorithm [16]. However, empirical results have shown us that due to the large number of moving objects to be modeled, computing the actual distance between all possible pairs would be too costly for a dynamic real-time application.

3   Dynamic Profile Clustering

A real-time mobile information dissemination environment generally consists of the mobile users with their specialized information requests and the many information streams created by the users and stationary data collection stations updating the situation in the field. Profile clustering is the process of incrementally clustering user information requests into a collection of group requests each covering all of its member users' requests.

Profile clustering is a conceptual clustering problem that has been studied extensively by the artificial intelligence community as a means to extract hidden inference information [5], and by the information retrieval community as a method to improve the quality of query results [2]. Clustering systems algorithms (e.g., COBWEB [5], CLUSTER/2 [10], and AUTOCLASS [3]) share the common goal of discovering structure in data but differ in the objective function used to evaluate clustering quality and the control strategy used to search the space of clusterings.

Our algorithm to generate profile clustering to allow effective information dissemination differs from existing conceptual algorithms in that the goal of the clustering is more than only to optimize the accuracy of clustering. In addition to accuracy in the clustering, also important in incremental profile clustering are simplicity (small number of groups), and stability (infrequent reassignment of user requests to groups). We desire a simple clustering networks generally have a practical limit on the number of multicast channels that can be supported. At the same time, it is important for the profile clustering to not change frequently because a large number of users are potentially interrupted when the clustering changes, while re-materializing a changed clustering in the form of new multicast channels can be an expensive process. Our algorithm alleviates the problem of instability by sacrificing some of the optimality in the clustering through the use of a request group cover diameter tolerance. In addition, we allow the bounding of the number of groups that translates to the number of multicast channels that is limited in a real network.

3.1 Static Clustering Framework

The submission, update, or cancellation of user requests triggers the profile clustering process. Given the needs of profile clustering for intelligent mobile information sharing, the algorithm to reconfigure an existing profile clustering when a new request is submitted consists of 3 main steps:

1. **Try to find an existing group in the clustering that closely covers the request.** A profile group covers a request if the group request completely subsumes the request. A threshold, $COVER\_THRESHOLD$, is defined as a means to control the inaccuracy of clustering that can be tolerated. A request is assigned to a group that completely covers it only if the difference between the their coverages is below the threshold. For example, if the threshold equals 0, a request is assigned to an existing group only if they have exactly the same coverage. On the other hand, if the threshold is 1, a request can be assigned to any existing group in the profile clustering.

2. **Try to expand existing group that is the closest in coverage to the request.** A threshold, $EXPAND\_THRESHOLD$, is defined to control the extent of expansion in a group's coverage that is allowed. Specifically, the

group that is closest to the request in coverage cannot be expanded to include the request if the increase in the group's coverage from its original coverage is larger than the threshold. If the threshold is 0, no group is allowed to expand in coverage. On the other hand, groups can be expanded arbitrarily if the threshold is 1. The threshold introduces a tunable means of allowing inaccuracy in clustering as a tradeoff for profile clustering stability.

3. **If all else fails, generate a new profile group that covers the user request.** The new group will serve as a new locus of clustering to which new user requests are attracted to. Moreover, it may later be merged with other groups as its coverage migrates due to changes in its membership and the coverage of its members.

3.2   Adaptive Reclustering

Once the objects are clustered, we try to preserve the original clusters as far as possible - as objects move or require new information, the group they are in is modified to reflect these changes. However, over a period of time this may lead to improper clustering of entities. The entities in a particular group may drift far apart enough from each other such that the coverage area of the group becomes undesirably large. Also, two clusters may change their coverage such that their information coverage overlaps to an extent that it would be justified to merge the two clusters into a single cluster.

We tackle the dynamics of user profiles by using a two-phased approach towards maintaining group assignments. Periodically, the assignment of entities to clusters is "reviewed". This review consists of two phases: a splitting phase and a merging phase.

– **Splitting Phase.** In the splitting phase, each group and the entities in it are examined and if the entities in the group have moved significantly away from each other, the group is split into multiple smaller clusters. We tackle the task of deciding whether a group should be split into multiple clusters by reclustering the entities in the group within themselves. If the entities in the group should logically belong to a single cluster, the reclustering will result in a single cluster. If, on the other hand the entities in the group have moved far apart from each other since the last time the clusters were reviewed, so much so that their being in the same group cannot be justified, the result of the reclustering will be two or more smaller clusters consisting of subsets of the entities in the original cluster.

Reclustering the entities within a group is less expensive because the number of entities being reclustered is a lot smaller than the total number of entities in the system. Within a group, the first entity picked should be one on an extreme corner of the space defined by other entities (the one on the bottom left corner, say) If there is no single entity at the corner, an entity closest to one of the corners may be chosen as the first one to cluster. All other entities in the system should be ranked in order of increasing distance from the entity we start with. The iterative procedure cycles through the entities in this order.

– **Merging Phase.** In the merging phase, the clusters (subclusters generated after the first phase) are examined. If two or more clusters overlap to a sufficient degree that it is justifiable to merge them into a single cluster, they are merged into one. Considering the merger of one group into another is similar to that of merging an entity into a group described earlier. The ratio of cells in the spatial coverage of the group to be merged that are not covered by the group it is being merged into to the total number of cells covered by the group to be merged is computed. If this ratio is less than the parameter $GROUP\_MERGE\_THRESHOLD$, the clusters can be merged, based upon the spatial coverage attribute. In other words, as the threshold increases, the less overlap 2 groups have to have for them to be merged and the more relaxed the merging criteria is.

4   Experiment Results

The profile clustering algorithm contains a number of parameters that can be adjusted to tune the balance between the overhead, accuracy, and simplicity of the clustering generated. They include the reclustering frequency, the cluster merging threshold, and the group joining threshold. One of the most important parameters that can be used to adjust the quality of clustering is the group merging threshold that controls the allowable overlap between overlapping groups before they are combined into an aggregated group. In this section, we will show some simulation results and discuss the effects of changing one of the parameters, namely the cluster merging threshold, on mobile information dissemination.

Given fixed $COVER\_THRESHOLD$ of 0.1, $EXPAND\_THRESHOLD$ of 0.1, and reclustering period of 10 seconds, Figure 4 shows a plot of the average number of groups for different group merging thresholds The simulated user environment consists of 30 vehicles following random paths in the road grid shown in Figure 7. As expected, the number of groups decreases as the group merging threshold increases since groups are allowed to be merged only with a decreasing amount of overlap. More interestingly, since each group maps into a multicast channel in our information dissemination system, a limit on the number of allowed multicast channels in the networking infrastructure imposes a limit on the maximum group merging threshold. For example, if the network infrastructure limits the number of multicast channels to 12, then the group merging threshold should be set to at least 0.55 given that all the other parameters remain the same.

Figure 5 shows a plot of the total area covered by the groups for different group merging thresholds. In this case, groups includes the coverages of its members and all 5x5 unit grids along the map-graph connecting them. The total coverage size increases as the group merging threshold increases and group count decreases, since more clustering causes more space "between" users to be included in groups. A group merging threshold of 0.55 as dictated by a multicast channel limit of 12 results in a total group coverage of 870 units. Assuming that approximately the same amount of data are being generated by mobile users

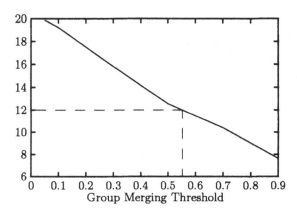

Fig. 4. Plot of Group Count against Group Merging Threshold

and data collection centers to describe the situation at each area, a plot showing the network bandwidth requirement will have the same shape as Figure 5 for varying group merging thresholds. This information can be used, with the multicast channel limit, to determine the appropriate group merging threshold to use given different network bandwidth availability and allocation.

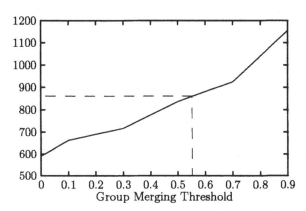

Fig. 5. Plot of Total Group Area against Group Merging Threshold Using Grid-Based Groups

Similar to Figure 5, Figure 6 also shows a plot of the total area covered by the groups for different group merging thresholds using the same clustering parameters. The difference is that in this case, the spatial coverage of a group is the rectangular bounding box of its members' coverage. A group merging threshold of 0.55 as dictated by a multicast channel limit of 12 results in a total group coverage of 7280 units, which is much higher than that achieved when group profiles are modeled as a collection of grids. Interesting, the total group coverage increases much faster as the merging criteria is loosened before leveling off. The two facts combine to indicating that bounding boxes is ineffective as a model for user groups, even for lightly clustered environments.

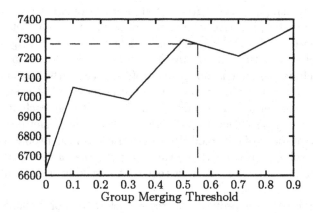

Fig. 6. Plot of Total Group Area against Group Merging Threshold Using Bounding Boxes

5   Mobile Information Dissemination System Prototype

We have implemented a prototype of the intelligent mobile information system described in this paper in a hybrid satellite-terrestrial networking environment [4] that combines high bandwidth satellite networking services (e.g., DirecPC) and terrestrial wireless networks d to support rapid deployment, user mobility, and wide-area support. It supports the dissemination and maintenance of extended situation awareness throughout such a network information infrastructure by means of adaptive reliable multicast-based information dissemination control from a clustering of mobile users.

The scalable group communication model of IP multicast forms a natural basis for large-scale data dissemination. However, IP multicast is unreliable, and a reliable multicast based reliable group data distribution protocol has to be used to effectively support dissemination of important information that demands absolute delivery reliability. We adopted the Multicast Dissemination Protocol (MDP) [21] for reliable multicast-based information dissemination. MDP is a receiver-reliable sender-oriented scheme that implements a NACK suppression algorithm in which receivers listen to the multicast channel for a NACK multi-casted by another receiver requested re-multicast of a data item if it recognizes the loss of the piece of data. The algorithm allows NACKs to be amortized since multiple receivers often miss the same data items unless the data loss happens at the "last mile" to a particular receiver. MDP is particularly suitable for situation awareness environments in which usres are resource-limited.

In addition to modeling user interest simply based on its current location, the profile of a mobile user includes the expected travel paths of the users in addition to the spatial interest coverage around their location. The expected rate and direction of change in users' geo-spatial interest domain serves as a basis for providing support for proactive information dissemination that anticipates users information needs and seamlessly fill each client's cache with neighborhood, or likely to be relevant, situation information. In particular, the x- and y- velocity vectors are included as dimensions in the profile space for the purpose of clustering. The intuitive idea would be to cluster entities moving in the same direction (more or less) together and ones moving rapidly apart from each other in different groups so that groups tend to preserve each other for longer periods of time. By taking into account of expected changes in user profiles, profile-oriented data dissemination achieves predictive information push that anticipates future user needs and minimizes latency of data request by making data available before they are explicitly requested.

Even though the user profile captures expected temporal changes in coverage of the interest domain, the actual future interest need may still differ. Plain dead-reckoning is a simple position update policy for maintaining accurate user location despite diversion in trajectory from plan. A mobile user updates its position, and hence its information request, when it deviates from the expected position over some pre-defined threshold. From the user's perspective, updating a request results in the possible reassignment of the request to a different group in the profile clustering. In addition, groups get merged, splitted, and deleted, during profile reclustering even if mobile users do not do anything. A user is notified of changes when the profile group, and hence the multicast information channel, it is assigned to changes. While seemingly complicated, all the above interaction between a user and the information dissemination infrastructure is transparently handled by the client software that updates the user's profile and provides a seamless transition between multicast channels when its channel assignment changes.

As a simulation and monitoring tool, we have implemented a dynamic clustering monitor software that displays the changes in the grouping of mobile objects over time. Figure 7 shows a screen dump of the tool monitoring a run of the experiment described in the last Section.

5.1 Applications

Intelligent mobile information sharing has many real-life military, civil, and commercial applications. Battlefield situation awareness and emergency response are two important information-centered applications that require the capability to effectively disseminate multimedia information to large groups of geographically distributed mobile users collaborating on a common mission and with interests in common situation domain. For example, after a natural disaster, emergency response teams and residents of affected areas need to share situation information

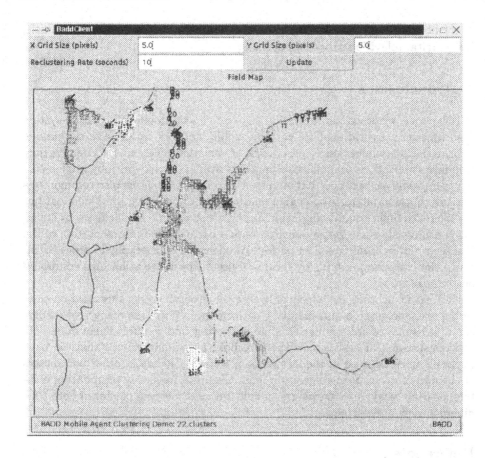

Fig. 7. Screen Dump of Dynamic Clustering Monitor Software

(e.g., road condition) as well as their up-to-date locations in order to coordinate evacuation; at the same time, they also receive a large variety of multimedia information from outside source (e.g., real-time video from CNN, satellite sensor data from NASA, weather maps and forecasts) to enhance decision making. Shared situation awareness, during real-time mission execution, will be achieved by a hierarchical propagation of information throughout the operational organization. To effectively adapt and react to rapidly evolving scenarios, units at all levels of command must perceive an extended awareness of the situation and often act autonomously while remaining globally consistent in the overall mission objective.

At the same time, future passenger vehicles will be equipped with networking facilities that allow their occupants to remain connected to the Internet while on the move. The value of advanced traveler information systems can be greatly increased by extending beyond providing simple traffic related information and guidance, to support scalable multimedia dissemination services to vehicles for informational (e.g., fleet management and communication) and entertainment (e.g., digital television services) purposes.

6 Conclusions

In this paper, we described the design of an intelligent mobile information system that runs on a hybrid satellite-wireless mobile network. It satisfies the unique information dissemination requirements of situation awareness and emergency response applications and effectively utilizes available bandwidth by implementing user profile aggregation that incrementally aggregates users into communities sharing common interests to enable multicast-based information dissemination.

We reach higher quality of disseminated information to mobile users by tracking continuously their spatial location onto a cartographic representation of the real scenario. Reasoning is then performed on such a cartography in order to improve the clustering criteria by means of users' spatial relationships computed on the real map.

We have discussed the effects of tuning the group merging threshold on clustering and information dissemination performance. There are many more different dimensions of optimizing dynamic clustering under different environments and constraints. In addition to the threshold for joining groups that we have explored and discussed in the last section, we can also adjust other parameters of the dynamic clustering algorithm (e.g., temporal frequency of updating clusters, spatial "grid" size, frequency of splitting and merging clustering, etc.) for different system environments.

References

1. G. Biswas and J. Weinberg and C. Li: ITERATE: A Conceptual Clustering Method for Knowledge Discovery in Databases. Innovative Applications of Artificial Intelligence in the Oil and Gas Industry. (1994)
2. M. Charikar and C. Chekuri and T. Feder and R. Motwani: Incremental Clustering and Dynamic Information Retrieval. Proceedings of the Conference on Theory of Computation. (1997)
3. P. Cheeseman and J. Kelly and M. Self and J. Stutz and W. Taylor and D. Freeman: AUTOCLASS: A Bayesian Classification System. Proceedings of the Fifth International Machine Learning Conference. (1988) 54-64
4. S. Dao and B. Perry: Information Dissemination in Hybrid Satellite/Terrestrial Networks. IEEE Data Engineering Bulletin. Vol. 19. No. 3 (1996) 12-19
5. D. Fisher: Knowledge Acquisition via Incremental Conceptual Clustering. Machine Learning. Vol. 2 (1987)
6. M. Hadzikadic and D. Yun: Concept Formation by Incremental Conceptual Clustering. Proceedings of the International Joint Conference in Artificial Intelligence. (1989) 831-836
7. R. Katz et al.: The Bay Area Research Wireless Access Network (BARWAN). Proceedings of the Spring COMPCON Conference. (1996)
8. M. Lebowitz: Experiments with Incremental Concept Formation: UNIMEM. Machine Learning. Vol. 2 (1987) 103-138

9. K. McKusick and P.Langley: Constraints on Tree Structure in Concept Formation. Proceedings of the International Joint Conference on Artificial Intelligence. (1991) 810-816
10. R.S. Michalski and R. Stepp: Automated Construction of Classifications: Conceptual Clustering versus Numerical Taxonomy. IEEE Transaction on Pattern Analysis and Machine Intelligence. Vol. 5 (1983) 219-243
11. A. J. Nevins: A Branch and Bound Incremental Conceptual Clusterer. Machine Learning. Vol. 18 (1995) 5-22
12. Y. Reich and S. Fenves: The Formation and Use of Abstract Concepts in Design. Concept Formation: Knowledge and Experience in Unsupervised Learning. (1991)
13. P. Utgoff: An Improved Algorithm for Incremental Induction of Decision Trees. Proceedings of the Eleventh International Conference on Machine Learning. (1994) 318-325
14. C.S. Wallace and D.L. Dowe: Intrinsic Classification by MML - the SNOB Program. Proceedings of the Seventh Australian Joint Conference ion Artificial Intelligence. (1994) 37-44
15. T.W. Yan and H. Garcia-Molina: SIFT - a Tool for Wide-Area Information Dissemination. Proceedings of the 1995 USENIX Technical Conference. (1995)
16. Mark P. Weiss: Algorithms, Data Structures, and Problem Solving with C++. Addison Wesley. (1996)
17. S. Shekhar and D.R. Liu: CCAM: A Connectivity-Clustered Access Method for Networks and Network Computations. IEEE Transactions of Knowledge and Data Engineering. Vol. 9. No. 1 (1997)
18. S. Shekhar and D. R. Liu and A. Fetterer: Genesis: An Approach to Data Dissemination in Advanced Traveler Information Systems. Bulletin of the Technical Committee on Data Engineering: Special Issue on Data Dissemination. Vol. 19. No. 3 (1996)
19. W. C. Collier and R. J. Weiland: Smart Cars, Smart Highways. IEEE Spectrum. (1994) 27-33
20. O. Wolfson and S. Chamberlain and S. Dao and L. Jiang: Location Management in Moving Object Databases. Proceedings of the Second International Workshop on Satellite-Based Information Service. (1997)
21. J. Macker and W. Dang: The Reliable Dissemination Protocol Framework. IETF Internet Draft. (1997)

On the Generation of Spatiotemporal Datasets

Yannis Theodoridis[1], Jefferson R.O. Silva[2], and Mario A. Nascimento[2]

[1] Computer Technology Institute
P.O. Box 1122, GR-26110 Patras, Hellas
yannis.theodoridis@cti.gr
[2] Institute of Computing, State University of Campinas
P.O. Box 6176, 13083-970 Campinas SP, Brazil
nascimento@computer.org

Abstract. An efficient benchmarking environment for spatiotemporal access methods should at least include modules for generating synthetic datasets, storing datasets (real datasets included), collecting and running access structures, and visualizing experimental results. Focusing on the dataset repository module, a collection of synthetic data that would simulate a variety of real life scenarios is required. Several algorithms have been implemented in the past to generate static spatial (point or rectangular) data, for instance, following a predefined distribution in the workspace. However, by introducing motion, and thus temporal evolution in spatial object definition, generating synthetic data tends to be a complex problem. In this paper, we discuss the parameters to be considered by a generator for such type of data, propose an algorithm, called "*Generate_Spatio_Temporal_Data*" (GSTD), which generates sets of moving point or rectangular data that follow an extended set of distributions. Some actual generated datasets are also presented. The GSTD source code and several illustrative examples are currently available to all researchers through the Internet.

1. Introduction

A field of ongoing research in the area of spatial databases and Geographical Information Systems (GIS) involves the accurate modeling of real geographical applications, i.e., applications that involve objects whose position, shape and size change over time. Real world examples include storage and manipulation of trajectories, fire or hurricane front monitor, simulators (e.g., flight simulators), weather forecast, etc.

Database Management Systems (DBMS) should be extended towards the efficient modeling and support of such novel applications. Towards this goal, recent research efforts have aimed at:

– modeling and querying time-evolving spatial objects (e.g., [19, 3, 24]),
– designing index structures and access methods (e.g., [13, 25]),
– implementing appropriate architectures and systems (e.g., [26]).

R.H. Güting, D. Papadias, F. Lochovsky (Eds.): SSD'99, LNCS 1651, pp. 147-164, 1999.
© Springer-Verlag Berlin Heidelberg 1999

In the recent literature, one can find work on formalization and modeling of spatiotemporal databases and a wide set of definitions about spatiotemporal objects. In the rest of the paper, we adopt the *discrete* definition for spatiotemporal objects that appears in [22]:

Definition: A *spatiotemporal object*, identified by its *o_id*, is a time-evolving spatial object, i.e., its evolution (or 'history') is represented by a set of instances (o_id, s_i, t_i), where s_i is the location of object o at instant t_i (s_i and t_i are called *spacestamp* and *timestamp*, respectively).

According to the above definition, a two-dimensional time-evolving point (region) is represented by a line (solid) in three-dimensional space. Figure 1 illustrates two examples: (a) a *moving point* and (b) a *moving region*, according to the terminology proposed by Erwig et al. in [3]. Although in the rest of the paper, we consider objects of dimensionality $d = 2$, the extension to higher dimensions is straightforward[1].

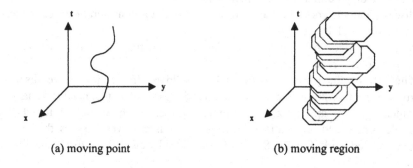

(a) moving point   (b) moving region

Fig. 1. Two-dimensional time-evolving spatial objects

One of the tasks that a SpatioTemporal Database Management System (STDBMS) should definitely support includes the efficient indexing and retrieval of spatiotemporal data. This task demands robust indexing techniques and fast access methods for a wide set of possible queries on spatiotemporal data. Either extensions of existing spatial access methods [27, 23, 13, 25] or new 'from-the-scratch' methods could be reasonable candidates. All proposals, however, should be evaluated under extensive experimentation on real and synthetic data. For instance, query processing and/or index building time (either real wall-clock time, or number of disk I/Os), space requirements and combinations thereof are all possible parameters against which one may want to evaluate a given index proposal.

Overall, there is a lack of consistent performance comparison among the proposed approaches, with respect to the space occupied, the construction time, and the response time in order to answer a variety of spatial, temporal, and spatiotemporal queries. Moreover, Zobel et al. [28] suggest that *"experiments of indexing techniques should be based on benchmarks such as standard sets of data and queries"*.

---

[1] Popular examples of spatial datasets with dimensionality $d > 2$ include, among others, virtual reality worlds ($d = 3$) and feature-based image databases (usually $d \leq 256$).

Following that guideline, the general architecture of a benchmarking environment for spatiotemporal access methods (STAMs) that is currently under design includes the following:

a) *a module that generates synthetic data and query sets*, which would cover a variety of real life examples,

b) *a repository of real datasets* (such as the extensively used TIGER files for - static - spatial data),

c) *a collection of STAMs* for experimentation purposes,

d) *a database of experimental results*, and

e) *a visualization tool* that could be able to visualize datasets and structures, for illustrative purposes.

Our study continues an attempt towards a *specification and classification scheme* for STAMs initiated in [22]. Within the above framework, in this paper we concentrate on module (a) and, in particular:

- discuss parameters that have to be taken into consideration for generating spatiotemporal datasets, and

- propose an algorithm that generates datasets simulating a variety of scenarios with respect to user requirements.

The rest of the paper is organized as follows: In Section 2 we discuss the motivation for this study. Section 3 discusses the parameters that need to be taken into consideration. An appropriate algorithm is presented in Section 4 together with example results and applications. Section 5 discusses several issues that arise and surveys related work. Finally, Section 6 concludes by also giving directions for future work.

2. Motivation

In the literature, several access methods have been proposed for spatial data without, however, taking the time aspect into consideration. Those methods are capable of manipulating geometric objects, such as points, rectangles, or even arbitrary shaped objects (e.g., polygons). An exhaustive survey is found in [4]. On the other hand, temporal access methods have been proposed to index valid and/or transaction time, where space is not considered at all. A large family of access methods has been proposed to support multiversion / temporal data, by keeping track of data evolution over time (e.g., assume a database consisting of medical records, or employees' salaries, or bank transactions, etc.). For a recent survey on temporal access methods see [17].

To the best of our knowledge, there is a very limited number of proposals that consider both spatial and temporal attributes of objects. In particular, *MR-trees* and *RT-trees* [27], *3D R-trees* [23], and *HR-trees* [13] are based on the R-tree family [7, 1, 9] while *Overlapping Linear Quadtrees* [25] are based on the Quadtree structure [18]. These approaches have the following characteristics:

- 3D R-trees treat time as another dimension using a 'state-of-the-art' spatial indexing method, namely the R-tree,

- MR-trees and HR-trees (respectively, Overlapping Linear Quadtrees) embed the concept of overlapping trees [12] into R-trees (Quadtrees) in order to represent successive states of the database, and
- RT-trees couple time intervals with spatial ranges in each node of the tree structure by adopting ideas from TSB trees [11].

The majority of proposed spatiotemporal access structures are based on the R-tree (one exception is [25]), as such we focus on such structures and a short survey of the R-tree based approaches follows.

Assuming *time* to be another dimension is a simple idea, since several tools for handling multidimensional data are already available [4]. The 3D R-tree implemented in [23] considers time as an extra dimension in the original two-dimensional space and transforms two-dimensional rectangles in three-dimensional boxes. Since the particular application considered in [23] (i.e., multimedia objects in an authoring environment) involves Minimum Bounding Rectangles (MBRs) that do not change their location through time, no dead space is introduced by their three-dimensional representation. However, if the above approach were used for moving objects, a lot of empty space would be introduced (Figure 2).

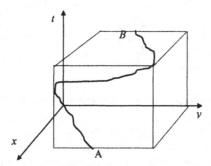

Fig. 2. The MBR of a moving object occupies a large portion of the data space

The approach followed by the RT-tree [27] only partially solves that problem. Time information is incorporated, by means of time intervals, inside the (two-dimensional) R-tree structure. Each entry, either in a leaf or a non-leaf RT-tree node, contains entries of the form (S, T, P), where S is the spatial information (MBR), T is the temporal information (interval), and P is a pointer to a subtree or the detailed description of the object. Let $T = (t_i, t_j)$, $i \leq j$, t_j be the current timestamp and t_{j+1} be the consecutive one. If an object does not change its spatial location from t_j to t_{j+1}, then its spatial information S remains the same, whereas the temporal information T is updated to T', by increasing the interval upper bound, i.e., $T' = (t_i, t_{j+1})$. However, as soon as an object changes its spatial location, a new entry with temporal information $T = (t_{j+1}, t_{j+1})$ is created and inserted into the RT-tree. This insertion strategy makes the structure mostly efficient for databases of low mobility; evidently, if we assume that the number of objects that change is large, then many entries are created and the RT-tree grows considerably. An additional criticism is based on the fact that R-tree node construction depends on spatial information S while T only plays a complementary role. Hence the RT-tree is not able to support temporal queries (e.g., "*find all objects that exist in the database within a given time interval*").

On the other hand, MR-trees and HR-trees are influenced by the work on overlapping B-trees [12]. Both methods support the following approach: different index instances are created for different transaction timestamps. However, in order to save disk space, common paths are maintained only once, since they are shared among the structures. The collection of structures can be viewed as an *acyclic graph*, rather than a collection of independent tree structures. The concept of overlapping tree structures is simple to understand and implement. Moreover, when the objects that have changed their location in space are relatively few, then this approach is very space efficient. However, if the number of moving objects from one time instant to another is large, this approach degenerates to independent tree structures, since no common paths are likely to be found. Figure 3 illustrates an example of overlapping trees for two different time instants t_0 and t_1. The dotted lines represent links to common paths / subpaths.

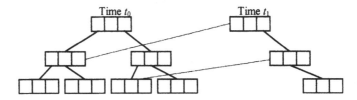

Fig. 3. Overlapping trees for two different time instants t_0 and t_1.

Among the aforementioned proposals, the 3D R-tree has been implemented and experimentally tested [23] using synthetic (uniform) datasets. The retrieval cost for several pure temporal, pure spatial and spatiotemporal operators was measured and appropriate guidelines were extracted. Recently, Nascimento et al. [14] have compared the HR-tree with the 3D R-tree and another structure, called 2+3 R-tree, using two R-trees and a rationale similar to the 2R approach presented in [10]. The basic conclusion is that the HR-tree is far more efficient in terms of query processing for *time point* queries while that is not true for *time interval* queries. Also, the HR-tree usually results to a rather large structure.

3. A Set of Operations and Parameters

Theodoridis et al. [22] have discussed a list of specifications to be considered when designing and evaluating efficient STAMs with respect to: (i) *data types and datasets* supported, (ii) issues on *index construction*, and (iii) issues on *query processing*. While the second and third ones mainly address the internal structure of a method and hence should be considered by STAM designers, the first group of specifications highly affect the design of an efficient benchmarking environment since they focus on

database characteristics for evaluation purposes. In particular, the specifications that
are addressed in [22] with respect to type (i) are the following:
- *Spec 1: on the data type(s) supported.* Appropriate STAMs could support either
  point or *non-point* spatial objects. In some cases, point objects could be considered
  as special cases of non-point objects but this depends on the underlying modeling.
- *Spec 2: on the time dimension(s) supported.* A second classification concerns the
  time dimension(s) supported, i.e., valid and/or transaction time. Since at least one
  time dimension should be supported, spatiotemporal databases are classified in
  valid-time, transaction-time, and *bitemporal* ones.
- *Spec 3: on the dataset mobility.* Three cases are addressed, with respect to the
  motion of objects and the cardinality of the dataset through time, namely *evolving*
  (i.e., moving objects of a fixed cardinality through time), *growing* (i.e., static
  objects of varying cardinality through time), and *full-dynamic* (i.e., moving objects
  of varying cardinality through time) databases.
- *Spec 4: on the timestamp features.* Whether future instances could refer to past
  timestamps or not leads to a distinction between *chronological* and *dynamic*
  databases, i.e., collections of objects' instances (o_id, s_i, t_i) that either have or not
  to obey the so-called *rule of consecutive timestamps*: $t_{i+1} > t_i$.

In the rest of the paper we study the case of *temporally degenerate databases* that
obey the *rule of consecutive timestamps*, i.e., for each object in the database, the
following inequality exists between the timestamp of the current instance t_i and that of
the next instance t_{i+1} to be inserted into the database: $t_{i+1} > t_i$. The term *degenerate*
refers to the characteristic that the valid time of object instances is identical to their
transaction time. That is, an object is valid as long as it exists in the database. The
problem that arises when no such constraint exists[2] is clarified through the following
example: Consider that two instances (o_id, s_i, t_i) and (o_id, s_j, t_j) of an object o have
been inserted into the database (without a loss of generality, we assume that $t_i < t_j$) and
no instance (o_id, s_x, t_x) exists, such that $t_i < t_x < t_j$. Hence $[t_i, t_j)$ is the valid (and
transaction) time of instance i. Let now assume that a new instance (o_id, s_k, t_k) is
inserted into the database, such that $t_i < t_k < t_j$. Due to that action, (a) the valid time of
instance i has to be changed from $[t_i, t_j)$ to $[t_i, t_k)$ and (b) the validity interval of the
new instance k has to be set to $[t_k, t_j)$. No straightforward support for those operations
exists in current STAMs and, therefore, we currently postpone the study of that case.
Note however, that this assumption does not hold in the area of bitemporal databases
[17]. Indeed in bitemporal access structures the rule is that, by definition, only
transaction time is monotonically increasing. However, as already mentioned, adding
spatial features to bitemporal data is still an open area for research.

3.1. User Requirements

Three (*Spec1* to *Spec3*) out of the above four specifications are orthogonal to each
other. On the other hand, only the *chronological* case of *Spec4* is supported in this
study, as declared earlier, and, as a result of that, we currently treat *transaction-* and

---

[2] Applicable to valid-time only since transaction-time always obeys that rule.

valid- time under a uniform platform. Hence, we distinguish among 12 different database families (e.g., a *point* plus *transaction-time* plus *evolving* plus *chronological* database) according to the following options:

- *Spec1*: point vs. region database,
- *Spec2*: transaction- (or valid-) vs. bitemporal database,
- *Spec3*: evolving vs. growing vs. full-dynamic database,
- *Spec4*: chronological database.

In order for the user of a benchmarking environment to generate a synthetic dataset, he/she should be able to (a) select one among the above database options and, then (b) tune the cardinality of the dataset and an appropriate set of parameters and distributions.

A fundamental issue on generating synthetic spatiotemporal datasets is the definition of a *complete set of parameters* that control the evolution of spatial objects. Towards this goal, we first address the following three operations:

- *duration of an object instance*, which involves change of timestamps between consecutive instances,
- *shift of an object*, which involves change of spatial location (in terms of center point shift), and
- *resizing of an object*, involves change of an object's size (only applicable to non-point objects).

In a more general case, the latter one could be regarded as *reshaping of an object*, as not only size but also shape could change. However, as the MBR is the most common approximation used by indices, we only consider that case, and thus shape changes are not an issue.

A description of each operation follows. In particular, the goal to be reached is the calculation of the consecutive instances (o_id, s_i, t_i) of an object o (recall the definition in Section 1) starting from an initial instance (o_id, s_1, t_1). We also assume that the spatial workspace of interest is the unit square $[0,1)^2$ and time varies from 0 to 1 (i.e., the unit interval). For illustration reasons, in Figure 4 we visualize four instances of a time-evolving two-dimensional region object o and the corresponding projections on the spatial plane and the temporal axis, respectively.

3.2. Parameters Involved

The *shift*, the *duration*, and the *resizing* of an object's instance are represented by the functions:

- **duration**(o_id,interval,curr_tstamp,new_tstamp)
- **shift**(o_id,Δc[],curr_sstamp_c,new_sstamp_c)
- **resizing**(o_id,Δext[],curr_sstamp_ext[],new_sstamp_ext[])

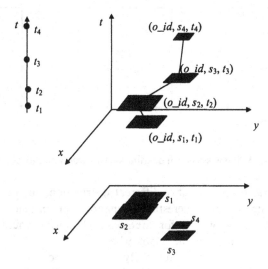

Fig. 4. Consecutive instances of a time-evolving object and the corresponding projections

which calculate the new timestamp and the new spacestamp's center and extent, called new_tstamp (a numeric value), new_sstamp_c (a 2-dimensional point), and new_sstamp_ext[] (an array of 2 intervals), respectively, of an object identified by its o_id, as the sums of the respective current values and the respective parameters, namely interval, Δc[], and Δext[].

In summary, Table 1 lists the parameters of interest and their corresponding domains. All parameters should follow a (user-defined) distribution, such as the ones we discuss in the following subsection.

Table 1: Parameters for generating time-evolving objects

Parameter	Type	Domain
Interval	number	$(0 \dots 1)$
Δc[]	2-dimensional vector	$(-1 \dots 1)^2$
Δext[]	2-dimensional vector	$(-1 \dots 1)^2$

3.3. Distributions

A benchmarking environment should support a wide set of well-established initial data distributions. Figure 5 illustrates three popular two-dimensional initial distributions, namely *uniform*, *gaussian*, and *skewed*. In addition to the initial spatial distributions, there exist several other parameters requiring some kind of statistical distribution, especially those mentioned above (interval, Δc[], and Δext[]).

(a) uniform   (b) gaussian   (c) skewed

Fig. 5. Basic statistical distributions in two-dimensional space

Through careful use of possibly different distributions for the above parameters one may simulate several interesting scenarios, for instance, using a random distribution for $\Delta c[i]$ as well as for `interval`, all objects would move equally fast (or slow) and uniformly on the map; whereas using a skewed distribution for `interval` one would obtain a relatively large number of slow objects moving randomly, and so on. Also, by properly adjusting the distributions for each $\Delta c[i]$, one may control the direction of the objects motion. For instance, by setting $\Delta c[i] =$ Uniform(0,1) $\forall i$, one would obtain a scenario where the set of objects eventually converge to the upper-right corner of the unit workspace, irrespectively from the initial distributions, but using the "*adjustment*" approach (see subsection 4.1). Similarly, if one likes the objects to move towards some specific direction (e.g., East), he/she can adjust Δc and put lower and upper bounds for the center's generated value, as will be discussed in detail in the following section.

Among the supported distributions, which are illustrated in Figure 5, the uniform distribution only requires *minimum* and *maximum* values while the other ones require extra parameters to be tuned by the user. In particular, the gaussian distribution needs *mean* and *variance* parameters as input and the skewed distribution needs a parameter to be declared, which controls the 'skewedness' of the distribution.

In the following section, we adopt the issues discussed earlier in order to present an algorithm that generates synthetic spatiotemporal datasets for benchmarking purposes.

4. The GSTD Algorithm

We propose an algorithm, called *Generate_Spatio_Temporal_Data* (GSTD), for generating time-evolving (i.e., moving) point or rectangular objects. For each object *o_id*, GSTD generates tuples of the format (o_id, t, p_l, p_u), where t is the instance's timestamp and p_l (p_u) is the lower (upper) coordinate point of the instance's spacestamp. The GSTD algorithm is listed in the Appendix.

4.1. Description of the Algorithm

GSTD gets several user-defined parameters as input:
− N and D correspond to the initial cardinality and density (i.e., the ratio of the sum of the areas of data rectangles over the workspace area) of the dataset,
− starting_id corresponds to the initial identification number of every object in the dataset,
− numsnapshots corresponds to the time resolution of the workspace,
− min_t and max_t correspond to the domain of the interval parameter,
− min_c[] and max_c[] correspond to the domain of the Δc[] parameter,
− min_ext[] and max_ext[] correspond to the domain of the Δext[] parameter,
and generates several tuples for each object, according to the following procedure: *"each object is initially active and, for each one, new instances are generated as long as timestamp t < 1; when all objects become inactive, the algorithm ends"*.

During the initialization phase (lines 01-04), all objects' instances are initialized, such that their center points are randomly distributed in the workspace, based on the distr_init distribution, and their extensions are either set to zero (in case of point datasets) or calculated according to extent(N,D) routine with respect to the input N and D parameters (in case of non-point datasets).[3]

During the main loop phase (lines 06-27), each new instance of an object is generated as a function of the existing one and the three parameters (interval, Δc[], and Δext[]). Then, *invalid instances* (i.e., those with coordinates located outside the predefined workspace) can be manipulated in three alternative ways as described below. In order for a new instance to be generated, the interval, Δc[], and Δext[] values are calculated by calling an RNG(distr, min, max) routine, i.e., a random number generator that generates random numbers between min and max following a predefined distr, which is a statistical distribution, such as the ones discussed in subsection 3.3.

Then, the output function checks whether the current instance of an object has a timestamp value greater than or equal to the value in next_snapshot. If so, the coordinates of the instance (given by the old_instance variable) before the current instance are printed, using the appropriate timestamp (which depends on the next_snapshot variable). In addition, the value of the next_snapshot variable is properly adjusted. Otherwise, the current instance is not output.

Obviously, it is possible that a coordinate may fall outside the workspace; GSTD manipulates *invalid* instances according to one among three alternative approaches:
− the *'radar'* approach, where coordinates remain unchanged, although falling beyond the workspace,
− the *'adjustment'* approach, where coordinates are adjusted (according to linear interpolation) to fit the workspace, and
− the *'toroid'* approach, where the workspace is assumed to be toroidal, as such once an object traverses one edge of the workspace, it enters back in the 'opposite' edge.

---

[3] In other words, an appropriate k = extent(N, D) value is set to achieve an initial density D of the dataset with respect to initial cardinality N.

In the first case, the output instance is appropriately flagged to denote its invalidity but the next generated instance is based on that. On the other hand, in the other two cases, it is the modified instance that is stored in the resulting data file and used for the generation of the next one. Notice that in the 'radar' approach, the number of objects present at each time instance may vary.

The three alternative approaches are illustrated in Figure 6 for the example of Figure 4. For simplicity, only the centers are illustrated; black (grey) locations represent valid (invalid) instances. In the example of Figure 6a, the 'radar' fails to detect s_3, hence it is not stored but the next location s_4 is based on that. Unlike 'radar', the other two approaches calculate a valid instance s_3' to be stored in the data file which, in turn, is used by GSTD for the generation of s_4. It is interesting to watch the behavior of s_4 in Figure 6c, where the calculated location finally stored (s_4') is actually identical to that in Figure 6a, as the effect of two consecutive calculations for s_3' and s_4'.

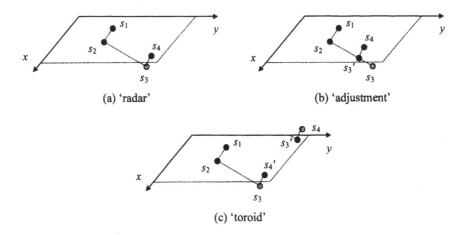

(a) 'radar'　　　　　　　　　　　　(b) 'adjustment'

(c) 'toroid'

Fig. 6. GSTD manipulation of invalid instances

4.2. Examples of Generated Datasets

As mentioned earlier, real world examples of (point or region) spatiotemporal datasets include trajectories of humans, animals, or vehicles, as, for instance, detected by a global positioning system (GPS), digital simulations of flights or battles, weather forecast and monitoring of fire or hurricane fronts. For example, detecting vehicle motion by GPS and storing the whole trajectory in a database is a typical every day life example. However, different motion scenarios correspond to different datasets, which an efficient structure should be evaluated on. *Random* versus *biased* direction, *fast* versus *slow* motion are some of the parameters that result to completely different applications.

In order to simulate some of those scenarios, in this subsection we present six example datasets consisting of point or rectangle objects generated by GSTD. For all files the following parameters were set: N = 1000, D = 0 or 0.5 (for points or rectangles, respectively), numsnapshots = 100. Illustrated snapshots correspond to *t* = 0, 0.25, 0.50, 0.75, and 1. Figure 7 presents the non-fixed input parameters and the generated snapshots for each file. Scenarios 1 and 2 follow the '*toroid*' and '*radar*' approach, respectively, to manipulate invalid instances, while scenarios 3 through 6 follow the '*adjustment*' approach.

Scenarios 1 and 2 illustrate points with initial gaussian distribution moving towards East and NorthEast, respectively. In the former case, where the *toroidal* world model was used, when the points traverse the right edge, they enter back in the left side of the map. Notice that to force the points moving to the East, $\Delta c[y] = 0$ and $\Delta c[x] > 0$. In the latter case, where the '*radar*' approach is simulated, the points move towards NorthEast and some of them fall beyond the upper-right corner (some quite early due to their speed), in fact some points move beyond the map. Notice that since $\Delta c[] > 0$ always, those points will never reappear in the map.

Scenario 3 illustrates the initially skewed distribution of points and the movement towards NorthEast. As the '*adjustment*' approach was used, the points concentrate around the upper-right corner. Scenario 4 includes rectangles initially located around the middle point of the workspace, which are moving and resizing randomly. The randomness of *shift* and *resizing* is guaranteed by the uniform distribution U(min,max) used for $\Delta c[]$ and $\Delta ext[]$, where $|min| = |max| > 0$.

Finally, scenarios 5 and 6 exploit the *speed* of a moving dataset as a function of the GSTD input parameters. By increasing (in absolute values) the min and max values of $\Delta c[]$, a user can achieve 'faster' objects while the same behavior could be achieved by decreasing the max_t value that affects interval. Thus, the speed of the dataset is considered to be a meta-information since it could be derived by the knowledge of the primitive parameters. Similarly, the *direction* of the dataset can be controlled, as presented in scenarios 1 through 3.

Alternatively, if the user's application makes necessary the conjunction of two (or more) scenarios, as for instance, a population of MBRs with only a small percentage of them moving towards some direction and the rest ones being static, two individual scenarios can be generated according to the above by properly setting the two starting_id input parameters and then merged, which is a straightforward task. Bottomline, by properly adjusting the parameters of Table 1, users can yield a wide spectrum of scenarios fitting their needs.

5. Discussion and Related Work

An alternative straightforward algorithm for generating N time-evolving objects would include the calculation of the spacestamp of each object at each snapshot, thus leading to an output consisting of $T = N \cdot$ numsnapshots tuples. Our approach outperforms that since it outputs a limited number T' of tuples ($T' \ll T$), i.e., the necessary ones in order to reproduce the dataset motion.

Scenario 1:
points moving
from center to
East ('*toroid*'
approach)

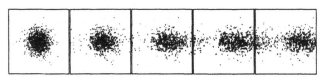

distr_init=G(0.5,0.1), interval=G(0,0.5),
Δc[x]=U(0,0.3), Δc[y]=U(0,0), Δext[x]=Δext[y]=U(0,0)

Scenario 2:
points moving
from center to
NorthEast
('*radar*'
approach)

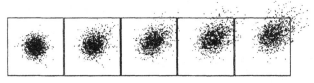

distr_init=G(0.5,0.1), interval=G(0,0.5),
Δc[x]=Δc[y]=U(0,0.4), Δext[x]=Δext[y]=U(0,0)

Scenario 3:
points moving
from SouthWest
to NorthEast

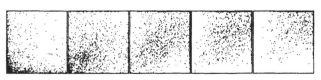

distr_init=S(1), interval=G(0,0.2),
Δc[x]=Δc[y]=U(0,0.3), Δext[x]=Δext[y]=U(0,0)

Scenario 4:
rectangles
moving (and
resizing)
randomly

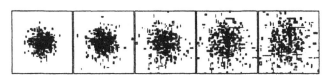

distr_init=G(0.5,0.1), interval=G(0,0.5),
Δc[x]=Δc[y]=U(-0.2,0.2), Δext[x]=Δext[y] =U(-0.01,0.01)

Scenario 5:
points moving
randomly (low
speed)

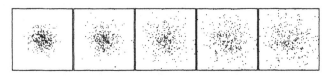

distr_init=G(0.5,0.1), interval=G(0,0.5),
Δc[x]=Δc[y]=U(-0.2,0.2), Δext[x]=Δext[y]=U(0,0)

Scenario 6:
points moving
randomly (high
speed)

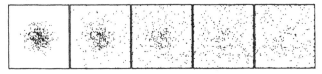

distr_init=G(0.5,0.1), interval=G(0,0.5),
Δc[x]=Δc[y]=U(-0.4,0.4), Δext[x]=Δext[y]=U(0,0)

Fig. 7. Example files generated by GSTD

However, a fundamental question arises: based on the knowledge of two instances (o_id, s_i, t_i) and (o_id, s_{i+1}, t_{i+1}) that correspond to consecutive timestamps, which is the location of an object at a time t_j, such that $t_i < t_j < t_{i+1}$? As an example, recall the instances of the object o illustrated in Figure 4. The status of its spacestamp between e.g., t_i and t_{i+1} is a fuzzy issue. Among others, two alternatives may be followed:
– *projection*: the spacestamp is considered to be static and equal to the one at time t_i,
– *linear interpolation*: the spacestamp is considered to be moving with respect to a start- (at time t_i) and an end- (at time t_{i+1}) position[4].

Both alternatives find applications in real world; cadastral systems, on the one hand, versus navigational systems, on the other hand, are popular examples. Figure 8 illustrates the two alternative scenarios for the example of Figure 4.

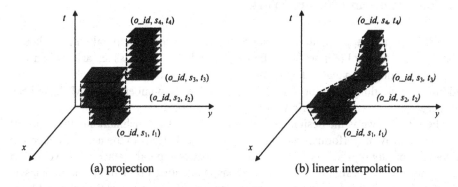

(a) projection   (b) linear interpolation

Fig. 8. Alternative scenarios for the location of an object between two timestamps

In any case, detecting the status of object o at a time instance during (t_1, t_2) is an open issue (e.g., uncertainty may need to be captured [16]). We argue that the proposed GSTD algorithm is independent of that issue. Actually, it generates a series of instances regardless of such an issue. Unlike the data generator, it is a visualization tool or a STAM construction algorithm that needs to support a specific scenario. Since in this study we are interested in spatiotemporal databases that follow the rule of consecutive timestamps, the knowledge of both the current and the new instances of an object, as supported by GSTD (see Appendix, line 13), are sufficient to deal with any alternative.

The need for independent platforms for benchmarking purposes or, in general, experiment management has been already addressed in the past [21, 8]. Such a need arises when a researcher aims to make a fair performance study or experimentation without the dilemma of building his/her own datasets for this purpose. Although extended related work is found in traditional database benchmarks and data generators (e.g., [2, 5]), in the field of spatial databases it is very limited [20, 6, 15]. Moreover,

---

[4] Linear interpolation assumes that a linear function represents boundary points' motion, i.e., intermediate locations are linear to the start- and end- points. Higher-order polynomials are hardly modeled.

when motion is introduced to support spatiotemporal databases, to our knowledge, no related work exists.

Relaxing this constraint, the most relevant to our work is the 'A La Carte' benchmark [6]. It is a WWW-based tool consisting of a rectangle generator that builds datasets based on user defined parameters (cardinality, coverage, coordinates' distributions) and an experimentation module that runs experiments on either user built or stored sample datasets (including parts of the Sequoia 2000 storage benchmark [21]). The module is actually a spatial join performance evaluator that supports several spatial join strategies.

6. Conclusion and Future Work

STDBMS require appropriate indexing techniques on spatiotemporal data. Although conceptually the problem seems to be easy to solve, several issues arise when one attempts to adopt a spatial indexing method to organize time-evolving objects by just adding an extra dimension for time. Therefore, a limited number of STAMs have been proposed in the literature as briefly surveyed in Section 2.

The effort towards the design and implementation of a benchmarking environment in order to provide performance comparison of STAMs leads to the need of collecting a variety of appropriate synthetic and real spatiotemporal datasets. However, in accordance to the design of efficient methods, generating efficient synthetic datasets is not a straightforward extension of generating spatial data, such as the ones that have been thoroughly used for experimental purposes in the spatial database literature. At a first step, several specifications that identify the type of the dataset have to be addressed and, at a second step, a set of parameters and corresponding distributions have to be tuned by the user. More specifically, we have discussed three operations, namely *duration of an object instance*, *shift* and *resizing of an object* (the latter one applicable to non-point objects) and derived a set of three parameters, namely interval, Δc, and Δext, which control the evolution of a spatial object through time in satisfactory terms.

Based on those parameters, we have designed and implemented the GSTD algorithm that generates sets of moving points or rectangles according to users' requirements, thus providing a tool that simulates a variety of possible scenarios. Some of those scenarios have been illustrated and discussed in Section 4. GSTD also includes alternative methodologies to support *invalid* instances, i.e., those with coordinates falling outside the workspace.

This study continues the work initiated in [22] towards a full and interactive support tool for designing, implementing, and evaluating access methods for the purposes of STDBMS. We are currently working on a WWW environment to make GSTD available to all researchers through the Internet[5].

---

[5] Mirror sites: http://www.dblab.ece.ntua.gr/~theodor/GSTD/ and http://www. dcc.unicamp.br/~mario/GSTD/.

We are also investigating some additional functionality on GSTD. For example, users may want to specify a movement flow to a specific point p in the workspace. Although, given the target p, it is not a complicate task, it is a specific implementation to a specific scenario. We currently study the parameterization of such specific scenarios by permitting GSTD input parameters to be (user-defined) functions rather than fixed values. Such an extension will enhance GSTD flexibility to simulate a variety of real applications.

Acknowledgements

Yannis Theodoridis was partially supported by the EC funded TMR project "CHOROCHRONOS: A Research Network for Spatiotemporal Database Systems". Jefferson R. O. Silva was supported by FAPESP (process number 97/11205-8) and is currently with the Brazilian Research and Development Center for Telecommunications (jeff@cpqd.com.br). Mario A. Nascimento was partially supported by CNPq (process number 300208/97-9) and by the Pronex/FINEP funded project "SAI: Advanced Information Systems" (process number 76.97.1022.00), he is also with CNPTIA – Embrapa.

References

1. N. Beckmann, H.-P. Kriegel, R. Schneider, B. Seeger, "The R*-tree: An Efficient and Robust Access Method for Points and Rectangles", *Proceedings of ACM SIGMOD Conference*, 1990.
2. D. Bitton, D. J. DeWitt, C. Turbyfill, "Benchmarking Database Systems: A Systematic Approach", *Proceedings of the 9th International Conference on Very Large Data Bases* (VLDB), 1983.
3. M. Erwig, R. H. Güting, M. Schneider, M. Vazirgiannis, "Abstract and Discrete Modeling of Spatio-Temporal Data Types", *Proceedings of the 6th ACM International Workshop on Geographical Information Systems* (ACM-GIS), 1998.
4. V. Gaede, O. Günther, "Multidimensional Access Methods", *ACM Computing Surveys*, 30(2): 170-231, June 1998.
5. J. Gray, P. Sundaresan, S. Englert, K. Backlawski, P. J. Weinberger, "Quickly Generating Billion-Record Synthetic Databases", *Proceedings of ACM SIGMOD Conference*, 1994.
6. O. Günther, V. Oria, P. Picouet, J.-M. Saglio, M. Scholl, "Benchmarking Spatial Joins A La Carte", *Proceedings of the 10th International Conference on Scientific and Statistical Database Management* (SSDBM), 1998.
7. A. Guttman, "R-trees: A Dynamic Index Structure for Spatial Searching", *Proceedings of ACM SIGMOD Conference*, 1984.
8. Y. Ioannidis, M. Livny, S. Gupta, N. Ponnekanti, "ZOO: A Desktop Experiment Management Environment", *Proceedings of the 22nd International Conference on Very Large Data Bases* (VLDB), 1996.
9. I. Kamel, C. Faloutsos, "Hilbert R-tree: An Improved R-tree Using Fractals", *Proceedings of the 20th International Conference on Very Large Data Bases* (VLDB), 1994.

10. A. Kumar, V. J. Tsotras, C. Faloutsos, "Designing Access Methods for Bi-temporal Databases", *IEEE Transactions on Knowledge and Data Engineering*, 10(1): 1-20, January - February 1998.

11. D. Lomet, B. Saltzberg, "Access Methods for Multiversion Data", *Proceedings of ACM SIGMOD Conference*, 1989.

12. Y. Manolopoulos, G. Kapetanakis, "Overlapping B+-trees for Temporal Data", *Proceedings of the 5th Jerusalem Conference on Information Technology* (JCIT), 1990.

13. M. A. Nascimento, J. R. O. Silva, "Towards Historical R-trees", *Proceedings of ACM Symposium on Applied Computing* (ACM-SAC), 1998.

14. M. A. Nascimento, J. R. O. Silva, Y. Theodoridis, "Access Structures for Moving Points", TimeCenter Technical Report TR–33, August 1998.

15. A. N. Papadopoulos, P. Rigaux, M. Scholl, "A Performance Evaluation of Spatial Join Processing Strategies", *Proceedings of the 6th International Symposium on Large Spatial Databases* (SSD), 1999.

16. D. Pfoser, C. S. Jensen, "Capturing the Uncertainty of Moving-Object Representations", *Proceedings of the 6th International Symposium on Large Spatial Databases* (SSD), 1999.

17. B. Salzberg, V. J. Tsotras, "A Comparison of Access Methods for Temporal Data", *ACM Computing Surveys*, 31(1), 1999.

18. H. Samet, "The Quadtree and Related Hierarchical Data Structures", *ACM Computing Surveys*, 16(2): 187-260, 1984.

19. A. P. Sistla, O. Wolfson, S. Chamberlain, S. Dao, "Modeling and Querying Moving Objects", *Proceedings of the 13th IEEE Conference on Data Engineering* (ICDE), 1997.

20. M. Stonebraker, J. Frew, J. Dozier, "The SEQUOIA 2000 Project", *Proceedings of the 3rd International Symposium on Large Spatial Databases* (SSD), 1993.

21. M. Stonebraker, J. Frew, K. Gardels, J. Meredith, "The SEQUOIA 2000 Storage Benchmark", *Proceedings of ACM SIGMOD Conference*, 1993.

22. Y. Theodoridis, T. Sellis, A. Papadopoulos, Y. Manolopoulos, "Specifications for Efficient Indexing in Spatiotemporal Databases", *Proceedings of the 10th International Conference on Scientific and Statistical Database Management* (SSDBM), 1998.

23. Y. Theodoridis, M. Vazirgiannis, T. Sellis, "Spatio-Temporal Indexing for Large Multimedia Applications" *Proceedings of the 3rd IEEE Conference on Multimedia Computing and Systems* (ICMCS), 1996.

24. N. Tryfona, "Modeling Phenomena in Spatiotemporal Applications: Desiderata and Solutions", *Proceedings of the 9th International Conference on Database and Expert Systems Applications* (DEXA), 1998.

25. T. Tzouramanis, M. Vassilakopoulos, Y. Manolopoulos, "Overlapping Linear Quadtrees: a Spatiotemporal Access Method", *Proceedings of the 6th ACM International Workshop on Geographical Information Systems* (ACM-GIS), 1998.

26. O. Wolfson, B. Xu, S. Chamberlain, L. Jiang, "Moving Objects Databases: Issues and Solutions", *Proceedings of the 10th International Conference on Scientific and Statistical Database Management* (SSDBM), 1998.

27. X. Xu, J. Han, W. Lu, "RT-tree: An Improved R-tree Index Structure for Spatiotemporal Databases", *Proceedings of the 4th International Symposium on Spatial Data Handling* (SDH), 1990.

28. J. Zobel, A. Moffat, K. Ramamohanarao, "Guidelines for Presentation and Comparison of Indexing Techniques", *ACM SIGMOD Record*, 25(3): 10-15, 1996.

Appendix: The GSTD Algorithm

Generate_Spatio_Temporal_Data algorithm

Input: values N, starting_id, numsnapshots, D, min_t, max_t

               arrays min_c[], max_c[], min_ext[], max_ext[]

               distributions distr_init, distr_t, distr_c, distr_ext

Output: instance (id, t, l_l_point, u_r_point), validity_flag

```
begin
01  for each id in range [starting_id .. N+starting_id] do
        //initialization phase
02      Set c[] = RNG(distr_init(), 0, 1), ext[] = extent(N, D)
03      Set t = 0, active = TRUE
04  end-for
05  Set step = 1 / numsnapshots
06  for each id in range [starting_id .. N+starting_id] do
        //loop phase
07      Set next_snapshot = step
08      while active do
            /* calculate delta-values and new instances */
09          Set interval = RNG(distr_t(), min_t, max_t)
10          Set Δc[] = RNG(distr_c(), min_c[], max_c[])
11          Set Δext[] = RNG(distr_ext(), min_ext[], max_ext[])
12          Set old_instance = instance
13          update_instance(instance)
            /* check instances and output */
14          if t > 1 then
15              active = FALSE
16              output(old_instance, current[i], next_snapshot)
17          else //check instance validity and output
18              Set validity_flag = valid(instance)
19              if validity_flag = FALSE and approach ≠ 'radar' then
20                  adjust_coords(instance, approach)
21              end-if
22              if t > next_snapshot then
23                  output(old_instance, current[i], next_snapshot)
24              end-if
25          end-if
26      end-while
27  end-for
end.
```

Spatial Data Mining and Classification

Efficient Polygon Amalgamation Methods for Spatial OLAP and Spatial Data Mining

Xiaofang Zhou[1], David Truffet[2], and Jiawei Han[3]

[1] Department of Computer Science and Electrical Engineering
University of Queensland, Brisbane QLD 4072 Australia
zxf@csee.uq.edu.au
[2] CSIRO Mathematical and Information Sciences
GPO Box 664, Canberra ACT 2601 Australia
david.truffet@cmis.csiro.au
[3] School of Computing Sciences
Simon Fraser University, Burnany, B. C., Canada V5A 1S6
han@cs.sfu.ca

Abstract. The polygon amalgamation operation computes the boundary of the union of a set of polygons. This is an important operation for spatial on-line analytical processing and spatial data mining, where polygons representing different spatial objects often need to be amalgamated by varying criteria when the user wants to aggregate or reclassify these objects. The processing cost of this operation can be very high for a large number of polygons. Based on the observation that not all polygons to be amalgamated contribute to the boundary, we investigate in this paper efficient polygon amalgamation methods by excluding those internal polygons without retrieving them from the database. Two novel algorithms, adjacency-based and occupancy-based, are proposed. While both algorithms can reduce the amalgamation cost significantly, the occupancy-based algorithm is particularly attractive because: 1) it retrieves a smaller amount of data than the adjacency-based algorithm; 2) it is based on a simple extension to a commonly used spatial indexing mechanism; and 3) it can handle fuzzy amalgamation.

Keywords: spatial databases, polygon amalgamation, on-line analytical processing (OLAP), spatial indexing.

1 Introduction

Following the success and popularity of on-line analytical processing (OLAP) and data mining in relational databases and data warehouses, an important direction in spatial database research is to develop spatial data warehousing, spatial OLAP and spatial data mining mechanisms in order to extract implicit knowledge, spatial relationships, and other interesting patterns not explicitly stored in spatial databases [7, 14]. Huge amounts of spatial data have been accumulated in the last two decades by government agencies and other organizations for various purposes such as land information management, asset and facility

R.H. Güting, D. Papadias, F. Lochovsky (Eds.): SSD'99, LNCS 1651, pp. 167–187, 1999.

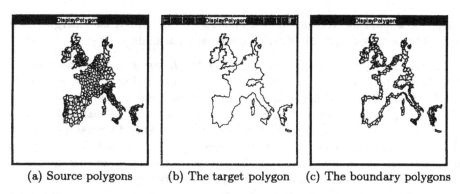

(a) Source polygons   (b) The target polygon   (c) The boundary polygons

Fig. 1. An example of polygon amalgamation.

management, resource management, and environment management. With the maturity of commercial spatial database management systems (SDBMS), it is a trend to migrate spatial data from proprietary file systems to an SDBMS. Thanks to various national spatial data initiatives and international standardization efforts, it is now both feasible and cost-effective to integrate spatial databases from different sources. Dramatic improvements have been made on the accessibility to extensive and comprehensive data sets in terms of geographical coverage, thematic layers and time. With huge amounts of integrated spatial data available, it is an imminent task to develop powerful and efficient methods for the analysis and understanding of spatial data and utilize them effectively.

For efficient OLAP and mining of spatial data, a spatial data warehouse needs to be built [15]. The cost of building a spatial data warehouse is intrinsically higher than building a relational data warehouse. Spatial operations are both I/O-intensive (for retrieving large amounts of spatial objects from database) and CPU-intensive (for performing complex spatial operations). Design of efficient spatial indexing structures and algorithms for processing various spatial operations and queries have been the focus themes in spatial database research [1, 3, 6, 10, 11, 12, 13, 20, 21, 23, 24, 26]. Satisfactory performance has been achieved for many spatial database operations. However, in the process of evolving SDMS towards spatial OLAP and spatial data mining, the performance of spatial data processing becomes the bottleneck again since these new applications analyze very large amounts of complex spatial data using costly spatial operations.

Among many spatial operations, we have found that special attention needs to be paid to one particular operation, the *polygon amalgamation* operation. Given a set of *source* polygons, this operation computes a new polygon (called the *target* polygon) which is the boundary of the union of the source polygons. Figure 1 (b) shows the target polygon from merging the source polygons shown in Figure 1 (a). While both intersection and union are basic operations on polygon data, the polygon amalgamation problem, unlike the polygon intersection problem, has received little attention so far in the context of spatial databases.

This operation, however, becomes a fundamental operation for emerging new applications such as spatial OLAP and spatial data mining.

Consider a typical scenario in spatial OLAP. A region is partitioned into a set of areas (represented as polygons), where each area is described by some non-spatial attributes (for example, area name, time, temperature and precipitation). Using the spatial data warehouse model proposed in [15, 17], a spatial data cube can be constructed with a spatial dimension (e.g., area) and several non-spatial dimensions (e.g., area name, time, temperature, precipitation). The measures used here can be non-spatial such as daily, weekly or monthly, or spatial such as combined areas according to certain concept hierarchy. Typical OLAP operations such as *drill-down* and *roll-up* can be applied to both spatial and non-spatial dimensions. A roll-up operation, such as generalizing temperature from degrees as recorded in the database into broader categories such as 'cold', 'mild' and 'hot', requires to merge areas according to their temperature degrees. Target polygons generated from such a generalization operation can be used for further operations (e.g., overlay with another spatial layer such as soil types). To spatial data mining, the polygon amalgamation operation is also important. For example, the user may want to group similar or closely related spatial objects into clusters, or to classify spatial objects according to certain feature classes (such as highly developed vs. poorly developed regions) [8, 17, 18]. Such mining will lead to combining polygons into large groups for high level description or inductive inference, using the polygon amalgamation operation.

The above discussion shows that an OLAP operation on a spatial dimension or a clustering operation on a group of spatial objects can result in new polygons at a high level of abstraction. Because of high processing cost associated with the polygon amalgamation operation, it is desirable to pre-compute target polygons and store them in the data cube in order to support fast on-line analysis. Obviously, it is a trade-off between the on-line processing cost for computing target polygons and the storage cost for materializing them. A similar problem in relational OLAP has been investigated by several researchers (e.g., [4, 16]). While materializing every view requires a huge amount of disk space, not materializing any view requires a great deal of on-the-fly and often redundant computation. Therefore, the cuboids (which are sub-cubes of a data cube) in a data cube are typically *partially* materialized. Even when a cuboid is chosen for materialization, it is still unrealistic to pre-compute *every* possible combination of source polygons when there are a large number of source polygons due to a prohibitive amount of storage required for newly generated polygons [15]. In other words, some polygon amalgamation tasks have to be performed on the fly. Moreover, certain types of multi-dimensional analysis require dynamic generation of hierarchies and dynamic computation of polygon amalgamation. In the previous example, different users may define different temperature ranges for 'cold', 'mild' and 'hot'. It might be necessary for some spatial data mining algorithms to try different classification (e.g., adding two more categories 'very cold' and 'warm' in order to find relationships between temperature and vegetation distribution). This type of analysis also demands dynamic polygon amalgamation.

Efficient polygon amalgamation is crucial for both building a spatial data warehouse and on-line processing. A straightforward method for polygon amalgamation is to retrieve *all* source polygons from database, and then merge them using some computational geometry algorithm such as described in [22]. Such a simplistic approach can be very time-consuming when the number of source polygons is large. We observe that there exist some *internal* polygons which do not contribute to the boundary of the target polygon. For example, if it is sufficient to use only the polygons shown in Figure 1 (c) to compute the target polygon in Figure 1 (b), a saving can be made by not to fetch and process other internal polygons. Savings from such an optimization can be significant as the number of polygons to be processed is reduced to be proportional to the perimeter of the target polygon, as opposed to its surface area. Obviously, the CPU cost of polygon amalgamation can be reduced by processing a smaller number of polygons. Whether the I/O cost can also be reduced depends on if those internal polygons can be identified *without* retrieving them from the database.

In this paper, we propose two novel methods for identifying internal polygons without retrieving them from the database. The first method uses the information about polygon adjacency, and the other takes an advantage of spatial indexing. Both algorithms are highly effective in reducing CPU and I/O costs. The latter, however, is particularly attractive for several reasons. Comparing with the adjacency-based algorithm, it takes less time in identifying internal polygons. More importantly, it is based on a simple extension to a popular spatial indexing mechanism that is supported by many SDBMSs. Thus, this algorithm can be incorporated easily and efficiently into the SDBMSs supporting that type of indexing mechanism. Another advantage comes when there are holes in target polygons. For spatial OLAP and spatial data mining applications, it is sometimes desirable to ignore holes smaller than certain threshold. These small holes might be insignificant to applications, or caused by imperfect data quality (for example, a small area with a high temperature surrounded by areas with low temperatures can be a 'noise'). We call it *fuzzy amalgamation* if the holes smaller than a specified threshold are to be ignored when merging polygons. Unlike the adjacency-based algorithm, the occupancy-based method can handle fuzzy amalgamation without incurring extra overhead for removing small holes.

The remaining of the paper is organized as follows. In Section 2, we give a basic amalgamation algorithm that processes all source polygons. The adjacency-based and occupancy-based algorithms are discussed in Section 3. A performance study of these algorithms is reported in Section 4. We conclude our discussion in Section 5.

2   A Simple Approach

In this section, we give a simple amalgamation algorithm that processes all source polygons. This algorithm provides a reference for evaluating the performance of other algorithms.

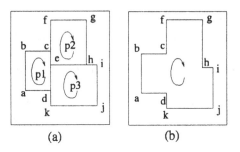

Fig. 2. Polygon representation

2.1 Polygon Representation

Let $\mathcal{D}$ be the data space, and $P = \{p_i | i = 1..n\}$ be a set of polygons inside $\mathcal{D}$ where p_i also denotes the identifier of polygon p_i. Polygon identifiers are unique in P. The boundary of a polygon may be disconnected. For example, the State of New South Wales (NSW) encloses the Australian Capital Territory (ACT), and ACT consists of two disconnected areas. We call a connected component of a polygon a *ring*. Thus, ACT is defined by two rings, and NSW is defined by three rings (one for its outer boundary and two for excluding ACT). In this paper we assume that a ring does not intersect with itself. We also assume that polygons in P do not overlap with each other. We use $t(S)$ to denote the target polygon amalgamated from a set $S \subseteq P$ of source polygons. A polygon p is a *boundary polygon* of S if it shares its boundary with that of $t(S)$. All boundary polygons of S are denoted as ∂S. The polygons in $(S - \partial S)$ (i.e., in S but not in ∂S) are *internal* polygons.

A point is represented by its coordinates (x, y). A line segment l is represented by its start and end points $(l.s, l.e)$. A polygon is represented as a sequence of points. For a polygon with k points $v_i \cdots v_k$, its boundary is defined by $k + 1$ connected line segments $(v_1, v_2), \cdots (v_{k-1}, v_k), (v_k, v_1)$. After a polygon is fetched from the database, it is unfolded into the form of line segments following, for example, the clockwise order (as in Figure 2). In other words, we view a polygon as a sequence of line segments in this paper. Thus, the polygons in Figure 2 (a) are represented as:

$$p_1 = \langle (a, b), (b, c), (c, d), (d, a) \rangle,$$
$$p_2 = \langle (e, f), (f, g), (g, h), (h, e) \rangle,$$
$$p_3 = \langle (e, i), (i, j), (j, k), (k, e) \rangle.$$

All the line segments in a polygon p are said to be in S if $p \in S$. We use $|S|$ and $\|S\|$ to denote the number of polygons and the number of line segments in S.

Among many possible relationships between two line segments l and l' on $\mathcal{D}$, we are interested in two relationships which are informally defined as:

1. *congruent*: l and l' are congruent if they are between the same pair of points (i.e., $l.s = l'.s$ and $l.e = l'.e$, or $l.s = l'.e$ and $l.e = l'.s$). In this case we also say these two line segments are *identical*.

2. *touching*: l touches l' if there exists one and only one common point v (termed the *joint point*) between l and l', such that v is an end point of l (i.e., $v = l.s$ or $v = l.e$) and v is not an end point of l' (i.e., $v \neq l'.e$ and $v \neq l'.s$). For example, (b, c) touches (e, f) at point c in Figure 2 (a).

The boundary of $t(S)$ consists of some existing line segments in S, and possibly some new line segments each of which is defined with at least one joint point. The polygon in Figure 2 (b) consists of a set of existing line segments in $S = \{p_1, p_2, p_3\}$ in Figure 2 (a), plus three new line segments (c, f), (h, i) and (k, d) where c, h, and d are joint points. For any algorithm to compute a target polygon, it is necessary to find at least the joint points that define the target polygon. In order to avoid costly operations of splitting lines at joint points, we apply a pre-processing step such that whenever a line segment l of polygon p touches l' of polygon p' at point v, l' in p' is replaced by two new line segments $(l'.s, v)$ and $(v, l'.e)$. Note that how lines split here depends only on data set P, regardless which subset of P is to be amalgamated. Thus, this operation can be done at the time of building spatial data cubes on P. After such a pre-processing step, the polygons in Figure 2 are represented in the database as

$$p_1 = \langle (a, b), (b, c), (c, e), (e, d), (d, a) \rangle,$$
$$p_2 = \langle (e, c), (c, f), (f, g), (g, h), (h, e) \rangle,$$
$$p_3 = \langle (e, h), (h, i), (i, j), (j, k), (k, d), (d, e) \rangle.$$

2.2   Removing Identical Line Segments

Under the above assumptions, it is clear that there are no identical line segments in a polygon. Further, a line segment l is on the boundary of $t(S)$ if and only if it has no identical line segments in S. Therefore, a straightforward algorithm to amalgamate polygons is to remove all identical line segments in S. Below is a sketch of such an algorithm:

Algorithm SIMPLE
Given a set of source polygons S, find $t(S)$.

1. (*Retrieve data*) fetch all the polygons in S into a set L of line segments, with the middle point of each line segment calculated.
2. (*Remove identical line segments*) sort line segments by their middle points (by x then y), and remove all line segments whose middle points appear more than once in L.
3. (*Finish*) return the remaining line segments in L as $t(S)$.

Since we assume that there are no overlapping polygons or self-intersecting rings in source data, in this algorithm we use the middle point to represent a line segment for identification of identical line segments. It is simpler and more efficient to process points than line segments. An additional advantage of representing a line segment by a point is that it becomes possible to apply a hash function to avoid sorting all line segments together [26]. That is, space $\mathcal{D}$ can be divided into cells, and a data bucket is associated with each cell. A line segment is mapped into the bucket whose corresponding cell contains the middle point of the line segment. After line segments are mapped into buckets, all identical line segments must be inside the same bucket. Thus, it is sufficient to sort the line segments bucket by bucket, instead of sorting them all together. This hashing-based method is particularly useful when the memory is not large enough for storing all line segments of S as it is now possible to apply those well-known methods developed in relational databases to handle similar problems (such as the hybrid join algorithm [5]).

3   Identifying Boundary Polygons

In this section we investigate two methods of identifying a subset $\overline{S} \subseteq S$ such that $t(\overline{S}) \equiv t(S)$. The performance of these algorithms will be discussed in Section 4.

3.1   Using Adjacency Information

Two identical line segments must come from two adjacent polygons. Strictly speaking, polygons can be adjacent to each other by edge or by point. There is no need to consider the latter because our interest here is to identify identical line segments. Using the data structures discussed in Section 2, one can simply define that two polygons are adjacent to each other if they have at least one pair of identical line segments. Moreover, the *adjacency table* of a set P of polygons is defined as a two column table ADJACENCY(p, p') where p, p' are identifiers of polygons in P. A tuple (p, p') is in table ADJACENCY if and only if polygon p is adjacent to polygon p'. The adjacency relation is reflective. For two adjacent polygons p and p', one can record the adjacency information redundantly (i.e., recording both (p, p') and (p', p)). Alternatively, a total order among polygon identifiers can be imposed (e.g., the alphabetical order if the identifiers are character strings) such that (p, p') in the adjacency table only if $p < p'$. As to be discussed in Section 4, this decision is an implementation issue, which has implications on the efficiency of the database queries to identify boundary polygons. For presentation simplicity, we assume the redundancy approach hereafter. Below is the adjacency table for the four polygons in Figure 3 (a) where $P = \{p_1, p_2, p_3, p_4\}$:

$$\{(p_1, p_2), (p_1, p_3), (p_1, p_4), (p_2, p_3), (p_2, p_4), (p_3, p_4), (p_2, \mathcal{D}), (p_3, \mathcal{D}), (p_4, \mathcal{D})\}$$

where $(p, \mathcal{D})$ means p has at least one line segment adjacent to no P polygons (in this case we say p is adjacent to a dummy polygon also labeled as $\mathcal{D}$). For $S \subseteq P$, by definition we have

$$\partial S = \{p | p \in S, \exists (p, p') \in \text{ADJACENCY}, p' \notin S\}$$

That is, a polygon is on the boundary of $t(S)$ if and only if it is inside S and it has at least one adjacent polygon which is not in S. Using the adjacency table, ∂S can be easily identified using SQL queries.

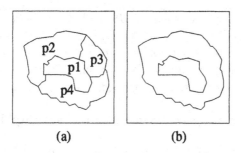

Fig. 3. An example of shadow ring

Unfortunately, it is not sufficient to use only ∂S to produce $t(S)$. When $\partial S \subset S$ (i.e., a true subset), two rings can be produced by removing all identical line segments in ∂S. For example, when merging the four polygons in Figure 3 (a) using $\partial S = \{p_2, p_3, p_4\}$, two rings are produced as shown in Figure 3 (b) where the inner ring is actually not part of $t(S)$. We call such an unwanted ring a *shadow ring*. The reason of producing shadow rings is simple: some line segments which would be otherwise found as identical cannot be recognized because their counterpart line segments are from polygons not in ∂S. In general, as a result of using only a subset of polygons, for each ring r which is part of $t(S)$ (called a *boundary ring*), there may exist a corresponding shadow ring r'. Note that r can either enclose or is enclosed by its shadow ring r' (the latter happens when r defines a hole of the target polygon). A shadow ring can be a boundary ring at the same time (e.g., when $S = \{p_2, p_3, p_4\}$ in Figure 3). The following example illustrates that it is not possible to tell whether a ring is a shadow ring by only looking at ∂S. While the inner ring in Figure 3 (b) is a shadow ring when $S = \{p_1, p_2, p_3, p_4\}$, it defines a hole in $t(S)$ when $S = \{p_2, p_3, p_4\}$. In order to identify possible shadow rings, we need to use a supplementary data set $\partial S^+ \subseteq S$ where ∂S^+ contains those S polygons adjacent to but not in ∂S polygons. That is,

$$\partial S^+ = \{p | p \in (S - \partial S), \exists p' \in \partial S, (p, p') \in \textbf{ADJACENCY}\}$$

We call the polygons in ∂S and ∂S^+ the *boundary polygons* and the *sub-boundary polygons* respectively. For a line segment $l \in \partial S$, if there exists a line segment $l' \in S$ and l' is identical to l, l' must be in either ∂S or ∂S^+. In other words, no ∂S line segments can form a shadow ring if ∂S polygons are processed together with ∂S^+ polygons. Of course, after removing identical line segments in $\partial S \cup \partial S^+$, all line segments from ∂S^+ need to be discarded.

Algorithm ADJACENCY
Given the adjacency table **ADJACENCY** for a set P of polygons and $S \subseteq P$, find $t(S)$.

1. (*Find ∂S*) $\partial S = \emptyset$; for each $p \in S$, add p to ∂S if $(p, p') \in \textbf{ADJACENCY}$ and $p' \notin S$.

2. (*Find ∂S^+*) $\partial S^+ = \emptyset$; for each $p \in \partial S$, add $p' \in (S - \partial S)$ to ∂S^+ if $(p, p') \in$ ADJACENCY.

3. (*Retrieve data*) retrieve all polygons in $\overline{S} = \partial S \cup \partial S^+$ into a set L of line segments, and mark the line segments from ∂S^+ as 'auxiliary'.

4. (*Remove line segments*) remove identical line segments in L (as in Algorithm SIMPLE).

5. (*Remove ∂S^+ line segments*) remove all the 'auxiliary' line segments from L.

6. (*Finish*) return the remaining line segments in L as $t(S)$.

Algorithm ADJACENCY computes $t(S)$ from $\overline{S} = \partial S \cup \partial S^+$, where ∂S and ∂S^+ are found using the adjacency table which contains no spatial data. That is, the internal polygons can be excluded from further processing without fetching and examining their spatial descriptions. It is not necessary to fetch all polygon identifiers for the database, as the adjacency table is a simple relational table and the first two steps in algorithm ADJACENCY can benefit from indices on both columns of the table. Note that the adjacency table only needs to be built once, independent of S. However, ∂S and ∂S^+ have to be built when S is given.

Like the *filter-and-refine* approach which is a standard approach in spatial data processing [3, 11, 26], algorithm ADJACENCY uses the adjacency table to perform a non-spatial filtering step to reduce the number of spatial objects to be processed. Because of the extra cost in constructing ∂S and ∂S^+, Algorithm ADJACENCY is efficient only when $|\overline{S}| << |S|$. As we will show in Section 4, even though $|\partial S^+|$ can be many times larger than $|\partial S|$, this filtering step is still very effective in improving the overall performance.

3.2   Using Occupancy Information

The adjacency-based approach above can be described as object-centric as it focuses on identifying boundary polygons. Now we propose a space-centric approach, which decomposes space $\mathcal{D}$ into small irregular regions and identifies a set of boundary regions.

Z-Values All SDBMSs support one or several spatial data access methods for fast retrieval of spatial objects. Spatial access methods have received extensive attention in the past from the spatial database research community (see [10] for a survey). Represented by the spatial indexing mechanisms based on the R-trees [13], R$^+$-trees [24] and the z-values [2], a spatial data access method establishes certain relationship between the data space and spatial objects or their approximations such as minimum bounding rectangles. The z-ordering technique is one of the most widely used spatial indexing mechanisms [2, 20, 23]. It approximates a given object's shape by recursively decomposing the embedding data space into smaller sub-spaces known as *Peano cells*. The z-ordering decomposition works as follows. The whole space $\mathcal{D}$ (represented as a rectangle) is divided into four smaller rectangles of the same size. The four quadrants are numbered as 1 to 4

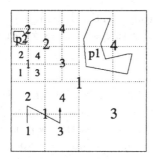

Fig. 4. Z-order and object approximation using z-values

following certain order (e.g., the z-order as shown in Figure 4). These quadrants can be further divided and numbered recursively. In such a way, $\mathcal{D}$ can be decomposed into a set of quadrants, or Peano cells, of varying sizes. Each Peano cell has a *z-value*, which can be determined elegantly by bit-interleaving [23]. The following is a simple description of one way to assign z-values (see Figure 4):

1. The z-value of the initial space, $\mathcal{D}$, is 1;
2. The z-values of the four quadrants of a Peano cell whose z-value is $z = z_1...z_n$, $1 \leq z_i \leq 4, 1 \leq i \leq n$, are $z1$, $z2$, $z3$ and $z4$ respectively following the z-order;

Thus, the z-values of the minimum Peano cells containing polygon p_1 and p_2 in Figure 4 are 14 and 1221 respectively. The z-values can have different length, reflecting different Peano cell sizes. The maximum number of decomposition level, also called resolution, determines the maximum length of z-values. In order to simplify processing, a number of '0's are often appended at the end of shorter z-values to make the length of all z-values identical (i.e., the maximum length).

Spatial objects can be approximated using a set of Peano cells. A polygon can be assigned with the z-value of the minimum Peano cell which fully encloses the polygon (so the z-values of polygons p_1 and p_2 in Figure 4 are 14 and 1221 respectively). The accuracy of approximation can be improved by assigning multiple z-values to a polygon (e.g., p_1 in Figure 4 can be approximated by three Peano cells whose z-values are 141, 142 and 143). The issue of approximating polygons by multiple z-values is a subject of several previous studies [2, 9, 19]. From a mathematical viewpoint, this decomposition is a transformation of a two- (or higher) dimensional object into a set of one-dimensional points (i.e., the z-values) which can be represented as numbers and therefore can be maintained by a ubiquitous one-dimensional access method such as the B$^+$-tree [1].

Let c and c' be two Peano cells. If c is nested inside c' and the z-value of c' has k non-zero digits, then c must have at least k non-zero digits, and they are digit-wise identical to the first k digits of the z-value of c. In other words, the containment relationship among Peano cells can be easily recognized by looking at their z-values. This property has a wide range of applications. For example, for selecting objects inside a given region, one can first find a set Z of z-values of the Peano cells covering the query region. Subsequently, the objects inside the query region can be quickly identified, using a B^+-tree index on z-values for example, by comparing their z-values with the z-values in Z.

Z-Values with Occupancy Let C be the set of all Peano cells with which S polygons overlap with. If $c \in C$ is not *completely* occupied by S polygons, we call c a *boundary cell*. An S polygon overlapping with a boundary cell is likely to be a boundary polygon. Now we look at how to use z-values to find boundary cells by extending the traditional z-value based spatial indices.

The spatial indices using z-values associate objects with Peano cells. That is, each index entry is of the form (z, p), stating that object p overlaps with Peano cell z. There is no information about what percentage of the cell is occupied by p, thus it is not sufficient to determine if a Peano cell is completely occupied by a set of objects. Therefore, we extend the index entry to the form of (z, p, α) where α is the *occupancy ratio*. Let $p \cap z$ be the polygon produced from clipping polygon p by the Peano cell z, then

$$\alpha = \frac{area(p \cap z)}{area(z)}$$

In other words, we record not only which polygons overlap with a Peano cell, but also the percentage of the area that each polygon occupies. The structure of the spatial indexing mechanisms based on traditional z-values, and the algorithms using such indices, need little modification to accommodate this additional piece of data, though some more efficient algorithms can be designed to take advantage of the occupancy information (e.g., the spatial join algorithm in [25]).

Identifying Boundary Peano Cells With the occupancy ratio for a polygon in a Peano cell, a boundary cell with respect to S can be identified simply by adding up the occupancy ratios of all S polygons overlapping with the cell. On the other side, we want to ensure that all S polygons which do not overlap with any boundary cells are internal polygons thus can be ignored by our amalgamation algorithm. While this is in general true, there is an exception when polygon p has a line segment l which coincides with the boundary of a Peano cell. One extreme case is a polygon that is of the same shape and size with a Peano cell. To solve this problem, we introduce zero-occupancy. That is, we use $(p, z, 0)$ if p is adjacent to but not inside cell z. In such a way, if p is a boundary cell but cannot be recognized by other cells, it can be picked up by cell z if z is a boundary cell. Now the problem of finding boundary polygons of S can be solved by finding all Peano cells which are not fully occupied by S polygons.

The algorithm for finding boundary cells can be complex because Peano cells are typically of different sizes, and some Peano cells may be nested inside others. Assume that c is nested inside c'. If c is an internal cell (i.e., fully occupied by S polygons), this fact needs to be propagated to its parent cell c' in order to find if c' is also fully occupied. This upwards propagation can be done using an algorithm of controlled-traversal similar to the one used in [20]. However, if a parent cell c'

is not fully occupied but one of its sub-cell c might be, it is difficult to translate the occupancy ratio from c' to c. This translation requires polygon clipping and re-calculation of the occupancy ratios in c for the polygons approximated in c'. In addition to a more complex algorithm to identify boundary cells, allowing nesting cells may lead to another disadvantage. That is, it is no longer possible to use simple SQL queries to find boundary cells; rather, all spatial index entries need to be pulled out and processed outside of the underlying DBMS.

On the other side, it is not efficient if there are too many polygons associated with a Peano cell. All source polygons associated with a boundary cell will be processed as possible boundary polygons. Thus, a number (termed *high watermark* (HWM)) is chosen such that a cell is to be further decomposed into quadrants when the number of polygons overlapping with the cell exceeding the high watermark. To avoid nesting cells, once a cell is decomposed all polygons approximated in that cell will be re-approximated at the lower level. In other words, while cells in the spatial index can be different sizes, there are no nesting cells.

Algorithm OCCUPANCY

Given a set $S \subseteq P$ of source polygons and a z-value-with-occupancy index $I = \{e_1, \ldots e_n\}$ for P where each e_i is of the form (z, p, α), find $t(S)$.

1. (*Identify boundary cells*) select $Z = \{e_i.z | e_i \in I, e_i.p \in S, \sum_{e \in I, e.z = e_i.z} e.\alpha < 100\%\}$.
2. (*Identify $\overline{S}$*) $\overline{S} = \emptyset$; add p to $\overline{S}$ if $(z, p, do\text{-}not\text{-}care) \in I$ and $z \in Z$.
3. (*Fetch $\overline{S}$ polygons and do line-clipping*) fetch polygons whose id in $\overline{S}$, and add the line segments which intersect with at least one cell in Z into L.
4. (*Remove duplicate line segments*) remove identical line segments in L (as in Algorithm SIMPLE).
5. (*Finish*) return L as $t(S)$;

Algorithm OCCUPANCY identifies boundary Peano cells and computes the target polygon using the occupancy information. Note that the first two steps in this algorithm can be implemented using a single SQL query with a group-by clause (i.e., "group by z having $sum(\alpha) < 100\%$"). Because a boundary cell tells not only which polygons to be fetched but also which parts of these polygons are to be used (that is, a line segment l of polygon p which overlaps with a boundary cell z contributes to the part of $t(S)$ in that cell only if l intersects with z). Based on this property, Algorithm OCCUPANCY can discard some line segments in step 2 and subsequently improve the performance of step 3.

In algorithm OCCUPANCY $\overline{S}$ is a superset of ∂S because an internal polygon can also overlap with a boundary cell. In general, it is likely to have more internal polygons in $\overline{S}$ if the Peano cells are large. This performance issue will be discussed in Section 4. Because of these internal polygons, on the other side, algorithm OCCUPANCY does not have the problem of producing shadow rings. One can see this from two aspects:

1. Any line segment which is not in any boundary cell is not part of the target polygon, thus can be discarded;

2. Any line segment inside a boundary cell either is part of the target polygon, or its counterpart from another polygon must also be fetched as this polygon also overlaps with the boundary cell.

Fuzzy Amalgamation Algorithm OCCUPANCY has a desirable advantage over the other two algorithms — the definition of boundary cells (thus boundary polygons) can be easily adjusted by the user. Instead of defining the internal cells as those with 100% aggregate occupancy, one can adjust to a lower *threshold* (for example, 95%). This is useful when the user wants to ignore data noises in polygon amalgamation (e.g., holes smaller than certain size, caused by either some abnormal or insignificant attribute values, or caused by inaccuracy in polygons definitions known as *sliver* polygons). If this threshold percentage is defined relatively to a Peano cell, one can simply replace 100% in algorithm OCCUPANCY with the threshold. In general, however, the user may want to use a threshold related to the data space $\mathcal{D}$. In this case one need to translate this threshold value to each Peano cell by considering actually the size of the cell. Let $d = length(z)$ be the number of non-zero digits of z-value z. The area of this cell is $1/2^d$ of that of $\mathcal{D}$. Thus, a threshold of t percent of the total data space is equivalent to $t \times 2^d$ percent in cell z. Algorithm OCCUPANCY can skip these insignificant holes easily and safely, with little extra overhead. This type of fuzzy amalgamation, often found as useful for spatial OLAP applications, cannot be achieved by the other two algorithms without forming those small polygons and calculating their sizes.

4   Performance Study

In this section we compare the three polygon amalgamation algorithms discussed in this paper. The primary performance index used in this section is the response time, which is measured as the elapsed time from when a predicate describing the source polygons is submitted to all the line segments of the target polygon are found. As pointed out in [10], the I/O cost-based measure such as the number of disk pages accessed is not necessarily a suitable performance indicator because the CPU cost can be equally important. Not measured are those once-off costs, including pre-processing of original polygons (to comply with the polygon data structures in Section 2), building the adjacency and occupancy tables and other necessary indices.

4.1   Cost Analysis

From a set $S \subseteq P$ of source polygons, an amalgamation algorithm takes three phases to produce the target polygon $t(S)$:

1. *the query phase* which identifies a subset $\overline{S} \subseteq S$ (where $\overline{S}$ is represented as a set of polygon ids);

2. *the fetch phase* which retrieves the polygons whose ids are in $\overline{S}$, and unfolds polygons from a sequence of points into an array L of line segments; and
3. *the merger phase* which computes the target polygon by removing duplicate line segments in L.

Let C_{query}, C_{fetch} and C_{merge} be the response times for the three phases respectively. The response time for an amalgamation algorithm, C_{total}, is the sum of these three components. That is:

$$C_{total} = C_{query} + C_{fetch} + C_{merge}$$

Two factors need to be considered for C_{query}. First, these three algorithms find an $\overline{S}$ with different numbers of polygons. A larger $|\overline{S}|$ may affect C_{query} as well as C_{fetch} (since more polygon ids and polygons need to be fetched). Second, these three algorithms use methods of different complexity to select $\overline{S}$. Algorithm SIMPLE simply uses $\overline{S} \equiv S$, thus it uses a straightforward selection query for this step. Algorithm ADJACENCY selects $\overline{S}$ in two steps: one to identify the boundary polygons and one to identify the sub-boundary polygons. Algorithms OCCUPANCY also uses two steps: identification of the boundary cells using a query with an aggregate function, followed by a set of queries to retrieve ids of the source polygons overlapping with the boundary cells. For both algorithms, it is possible to combine these two steps into one SQL query. However, such a complex query is far less efficient to execute than executing two separate queries for the two steps in an application program. For algorithm OCCUPANCY, both the final size of $\overline{S}$ and the cost for identifying $\overline{S}$ vary with HWM.

All the three algorithms retrieve the polygons whose ids are in $\overline{S}$. The cost for fetching objects (i.e., C_{fetch}) depends not only on how many polygons to be fetched but also on the sizes of these polygons. During this phase, algorithm ADJACENCY tags each line segment with its source (i.e., from a boundary or sub-boundary polygon). For algorithm OCCUPANCY, it clips line segments against boundary Peano cells such that only the line segments which intersect with at least one boundary Peano cell are kept for the next phase. With larger Peano cells (because of higher HWM), more internal polygons are included in $\overline{S}$ and less line segments can be dropped out by clipping. The middle point of each line segment is calculated in this phase for all the three algorithms.

The line segments are sorted by their middle points and the line segments that appear more than once are removed. Algorithm ADJACENCY needs to have an additional step to discard all line segments from the sub-boundary polygons. The remaining line segments form the target polygon. We do not include the time to order the line segments for the target polygon in C_{merge} as this cost is identical across all the three algorithms.

4.2   Databases and Parameters

Each polygon is stored as one object in the database (i.e., we do not consider object decomposition), whose schema is:

POLYGON(pid : INTEGER, boundary : POLYGON)

where data type POLYGON, which is essentially a sequence of points, can have different implementations according to different underlying DBMS. We use Oracle 8 in our tests (its object-relational features are not used, nor is the Spatial Data Cartridge, so POLYGON is simply implemented as a BLOB). Other attributes for a polygon are stored in a separate table and are linked to table POLYGON through pid.

A subset of the TIGER/LINE data (census blocks in California) is used for our performance testing (see http://www.census.gov/ftp/pub/geo/www/tiger/). A census block has an attribute county id, which is used for grouping source polygons in our experiments. A county consists of from 9 to 6,022 polygons. There are 21,648 polygons with 1,618,950 points in total. The number of points in a polygon ranges from 4 to 3,846, with an average of 75 points. We merge census blocks into counties, and adjacent counties into larger polygons. The preprocessing for splitting line segments at joint points and resolving data inconsistency problems are done using a GIS package ARC/INFO.

The adjacency table has the schema:

$$\text{ADJACENCY}(\text{pid} : \text{INTEGER}, \text{next\_to} : \text{INTEGER})$$

If polygon p and p' are adjacent to each other, we chose to store both (p, p') and (p', p) in table ADJACENCY. The number of tuples in the table is twice more than necessary, but the query to identify adjacent polygons is simpler and runs faster. The adjacency table for the data set we used has 137,978 rows. Both the boundary polygons and sub-boundary polygons for algorithm ADJACENCY are identified using this table, in two steps as mentioned before.

The occupancy information is recorded as

$$\text{OCCUPANCY}(\text{z} : \text{INTEGER}, \text{pid} : \text{INTEGER}, \text{occupancy} : \text{NUMBER})$$

Since all we need to find here is whether a cell is fully occupied by a set of polygons or not, we do not compute exact percentage of a polygon in a cell. Instead, we count the number of polygons in each cell and simply give each polygon an equal share of occupancy. For example, if there 8 polygons in a cell, the occupancy for each polygon in that cell is $1/8 = 0.125$, regardless their actual occupancy ratios. When the total occupancy for a given set of source polygons in the cell is 1, we know that the cell is not a boundary cell. (However, an accurate calculation of occupancy ratio is necessary in order to support fuzzy amalgamation.) The number of rows in table OCCUPANCY, as shown in Table 1, varies depending on HWM.

4.3 Experimental Results

Now we compare the performance of the three amalgamation algorithms empirically. For algorithm OCCUPANY, we also test with different HWMs. The algorithms are implemented using Microsoft Visual C++ and Oracle OCI interfaces. Both development and testing are done using a DELL notebook (Pentium II/266) with 128 MB memory. Indexes are created wherever necessary for all the

$HWM(bytes)$	Average num. of polygons per cell	Num. of rows in table OCCUPANCY
512	4.6	99638
1024	7.1	65405
2048	11.4	48168
4096	18.9	38336

Table 1. High water marks (HWM) for the occupancy table.

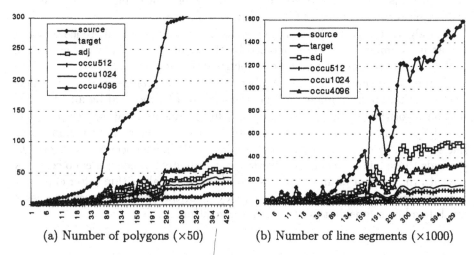

(a) Number of polygons (×50)   (b) Number of line segments (×1000)

Fig. 5. Number of spatial objects processed (x-axis = $|S|/50$)

tables used. Oracle's array fetch function is used whenever possible to improve the performance of data retrieval.

Data Size Reduction First, we look at the effectiveness for these algorithms in reducing the number of polygons to be processed. Figure 5 (a) shows the number of polygons actually fetched by different algorithms, where the x-axis in Figure 5 (and Figure 6) is the number of source polygons to be amalgamated (ranging from 17 to 21,528). Obviously, the maximum and minimal numbers of polygons to be fetched by any amalgamation algorithm are $|S|$ (i.e., all source polygons) and $|\partial S|$ (i.e., only the boundary polygons) respectively. These two numbers are labeled as 'source' and 'target' respectively in Figure 5.

Figure 5 (a) reveals three facts: (1) in comparison with algorithm SIMPLE which retrieves all the source polygons, both adjacency-based and occupancy-based algorithms fetch a much smaller number of polygons; (2) the performance of algorithms ADJACENCY and OCCUPANCY scales well when the number of source polygons increases (note that the number of polygons to be fetched depends not only on the number of source polygons but also the complexity of target polygon such as its shape and if there exist holes or not); and (3) the differences among the adjacency-based algorithm and the occupancy-based algorithms with different HWMs are significant (the differences among them may look deceptive in Figure 5 (a) due to much bigger differences between them and algorithm SIMPLE). The occupancy algorithm with HWM = 4096 performs consistently worse than the adjacency algorithm, which needs to retrieve about 50% more polygons than the occupancy algorithm with HWM=512. For the occupancy algorithm, when HWM increases, more polygons overlap with a Peano cell; thus, it is more likely to fetch internal polygons. The implication of different HWMs on response times will be examined later.

For algorithm ADJACENCY, $|\partial S^+|$ can be derived as the difference between the actual number of polygons fetched by the algorithm and $|\partial S|$. One can see that ∂S^+ contains about twice more polygons than ∂S.

Figure 5 (b) shows the total number of lines to be processed. Here the maximum and minimum numbers of line segments to be processed by any amalgamation algorithm are the total number of line segments in all source polygons (i.e., $||S||$) which is what algorithm SIMPLE has to process, and the number of line segments in the final target polygon (i.e., $||t(S)||$). The occupancy algorithm discards those line segments not overlapping with any boundary Peano cells. As a result, the number of line segments processed by the algorithm (after clipping) is much smaller than that by the adjacency algorithm, which in turn processes mush less line segments than the simple algorithm.

Response Times Figure 6 shows the response time for each phase as well as the total elapsed time. The query time is the elapsed time from when the predicate describing source polygons is submitted to when the polygons to be fetched are identified. Three factors may affect the query time. First, the size of the table used to produce $|\overline{S}|$ (i.e., the size of the adjacency and occupancy tables for algorithms ADJACENCY and OCCUPANCY respectively). Second, $|\overline{S}|$ itself. Third, the complexity of the query used for identifying candidate polygons. We avoid to use inefficient join query for the adjacency and occupancy algorithms (using two-passes as mentioned in section 4.1); thus the complexity of the queries used for these three algorithms are similar. Figure 6 (a) illustrates clearly that the response time for the query phase is primarily determined by the first factor. $|\overline{S}|$ is insignificant because of the use of array fetch. Algorithm SIMPLE is the fastest since it uses only a simple selection query on the base table that has a smaller number of rows than the adjacency or occupancy tables. The adjacency algorithm is the slowest, even when it results in smaller $|\overline{S}|$ in comparison with the occupancy algorithm with HWM=4096. For the occupancy algorithm, a smaller HWM results in a larger occupancy table (see Table 1) due to a higher probability for one polygon overlapping with many cells [19].

The time for fetching polygons from the database, as shown in Figure 6 (b), clearly dominates the whole polygon amalgamation process. It is our main motivation in this paper to reduce this time. We have achieved this goal by reducing this time by approximately 80% in comparison with algorithm SIMPLE.

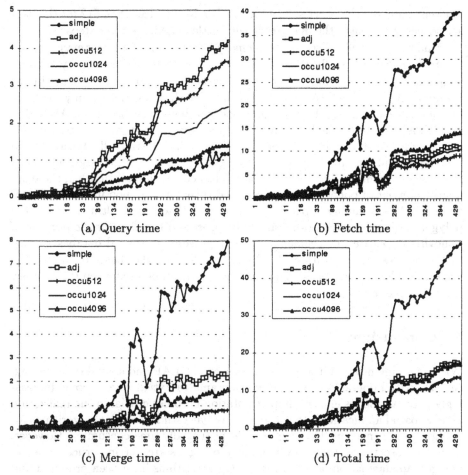

Fig. 6. Response time (seconds) (x-axis = $|S|/50$)

A clear winner is the occupancy algorithm with HWM=512, which is the most selective one as shown in Figure 5 (a).

The biggest time reduction in terms of percentage has been achieved for the merger phase. The cost of this phase, obviously, is determined by the number of lines to be processed (compare Figure 6 (c) with Figure 5 (b)). The occupancy algorithm, with all three HWMs, performs significantly better because of a smaller $|\overline{S}|$, and more importantly, because of discarding line segments by boundary cells at the end of the fetch step.

Figure 6 (d) shows the total response time. It is clear that the occupancy algorithm has achieved a remarkably better overall performance, in particular with an HWM which puts an average of 5 - 7 polygons to a Peano cell. That is, HWM = 512 or 1024 for the data set we use. A higher HWM degrades the performance because it becomes inefficient in filtering out internal polygons and line segments. On the other side, a too low HWM (i.e., less than 512 for the data set we used) increases the query cost with little benefit to other steps.

Finally, we briefly discuss the memory requirement for the amalgamation algorithms. Algorithm SIMPLE is the hungriest in terms of memory requirement among the three algorithms, as it needs to hold all line segments in the memory. Algorithm ADJACENCY consumes much less memory, only for holding line segments from the boundary and sub-boundary polygons. Algorithm OCCUPANCY needs the least amount of memory, as it stores only part of line segments for the polygons overlapping with boundary Peano cells. We assume in this paper that the memory is large enough for holding all line segments to be processed in memory. This assumption might become unrealistic when there are a large number of polygons to be processed (which is common in spatial OLAP and spatial data mining applications). However, based on the fact that only the polygons adjacent to each other are to be processed together for the purpose of removing duplicate line segments, those algorithms in spatial databases (such as the plane-sweep algorithm [3, 22]) or in relational databases to handle the similar problems (such as the hybrid hashing [5]) can be used when the memory is not big enough.

5   Conclusions

With emerging new applications such as spatial OLAP and spatial data mining, certain spatial operations such as polygon amalgamation have become increasingly popular and its efficient implementation becomes crucial in the realization of new spatial applications. In this paper, we have studied efficient algorithms for polygon amalgamation. This operation is intrinsically time-consuming. However, with the observation that only boundary polygons are playing crucial roles in polygon amalgamation, a set of interesting algorithms have been proposed and studied in this paper. Starting from improving a simplistic polygon amalgamation algorithm, we have proposed two methods, adjacency-based and occupancy-based, which exclude a large subset of polygons from being considered in the amalgamation algorithm without retrieving the spatial description of these polygons. The performances of these algorithms have been compared using real spatial data sets. With the support of a more sophisticated data storage structure, the occupancy-based method outperforms the adjacency-based method, whereas both methods are significantly more efficient than the algorithm which requires to fetch all objects to be merged.

The performance of the occupancy-based algorithm can be further improved by decomposing spatial objects. As implemented in some SDBMSs, a spatial object can be decomposed with the Peano cells approximating the object. In such a case, the occupancy-based algorithm only needs to fetch the parts of a spatial object that are inside a boundary cell, instead of the whole object. Such object decomposition can be done off-line when a spatial index is built. The online processing performance for the occupancy-based algorithm can be improved greatly as the amount of data to be retrieved is reduced and there is no need to do polygon clipping on-the-fly. Our work on this improvement will be reported in a separate paper. In the future we also plan to integrate our new polygon amalgamation algorithms with the research results in selective materialization for data cube construction [15] for supporting spatial OLAP and spatial data mining applications.

Acknowledgment: The first author conducted part of this research while in CSIRO. The work of the third author was partially supported by Natural Science and Engineering Research Council (NSERC) of Canada and Networks of Centres of Excellence (NCE) of Canada. The authors would like to thank Dr. Dave Abel, Dr. Volker Gaede and Professor Maria Orlowska for many helpful discussions.

References

[1] D. J. Abel. SIRO-DBMS: A database toolkit for geographical information systems. *Int. J. Geographical Information Systems*, 4(3):443 – 464, 1989.

[2] D. J. Abel and J. L. Smith. A data structure and algorithm based on a linear key for a rectangle retrieval problem. *Computer Vision, Graphics and Image Processing*, 24(1):1–13, 1983.

[3] T. Brinkhoff, H. P. Kriegel, and B. Seeger. Efficient processing of spatial joins using R-trees. In *Proc. ACM SIGMOD Int. Conf. on Management of Data*, pages 237–246, Washington, D. C., 1993.

[4] S. Chaudhuri and U. Dayal. An overview of data warehousing and olap technology. *SIGMOD Record*, 26(1), 1997.

[5] D. J. DeWitt and R. H. Gerber. Multiprocessor hash-based join algorithms. In *Proc. 1985 Int. Conf. on Very Large Data Bases*, pages 151–164, Austin, Texas, 1985.

[6] M. J. Egenhofer. Reasoning about binary topological relationships. In *LNCS 552: Proceedings of 2nd Int. Symp. on Spatial Databases (SSD'91)*, pages 143–160. Springer-Verlag, 1991.

[7] M. Ester, H.-P. Kriegel, and J. Sander. Spatial data mining: A database approach. In *LNCS 1262: Proceedings of the 5th Int. Symp. on Spatial Databases (SSD'97)*, pages 47–66, Berlin, Germany, 1997. Springer-Verlag.

[8] M. Ester, H.-P. Kriegel, J. Sander, and X. XU. Density-connected sets and their application for trend detection in spatial databases. In *Proc. 3rd Int. Conf. Knowledge Discovery and Data Mining (KDD'97)*, pages 10–15, Newport Beach, California, 1997.

[9] V. Gaede. Optimal redundancy in spatial database systems. In M. J. Egenhofer and J. R. Herring, editors, *LNCS 951: Proc. 4th Int. Symp. on Spatial Databases (SSD'95)*, pages 96–116, Portland, Maine, 1995. Springer-Verlag.

[10] V. Gaede and O. Günther. Multidimensional access methods. *ACM Computing Surveys*, 30(2):170–231, 1998.

[11] O. Günther. Efficient computation of spatial joins. In *Proceedings of 9th International Conference on Data Engineering*, pages 50–59, Vienna, Austria, 1993.

[12] R. H. Güting. An introduction to spatial database systems. *VLDB Journal*, 3(4):357–399, 1994.

[13] A. Guttman. R-trees: A dynamic index structure for spatial searching. In *Proc. ACM SIGMOD Int. Conf. on Management of Data*, pages 47–54, Boston, Massachusetts, 1984.

[14] J. Han, K. Koperski, and N. Stefanovic. GeoMiner: A system prototype for spatial data mining. In *Proc. ACM SIGMOD Int. Conf. on Management of Data*, pages 560–563, Tucson, Arizona, 1997.

[15] J. Han, N. Stefanovic, and K. Koperski. Selective materialization: An efficient method for spatial data cube construction. In *Proc. Pacific-Asia Conf. on Knowledge Discovery and Data Mining*, pages 144–158, Melbourne, Australia, 1998.

[16] V. Harinarayan, A. Rajaraman, and J. D. Ullman. Implementing data cubes efficiently. In *Proc. 1996 ACM SIGMOD Int. Conf. on Management of Data*, pages 206–216, Montreal, Canada, 1996.

[17] W. Lu, J. Han, and B. C. Ooi. Knowledge discovery in large spatial databases. In *Proc. Far East Workshop Geographic Information Systems*, pages 275–289, Singapore, 1993.

[18] R. Ng and J. Han. Efficient and effective clustering method for spatial data mining. In *Proceedings of 1994 International Conference on Very Large Data Bases*, pages 144–155, Santiago, Chile, 1994.

[19] J. Orenstein. Redundancy in spatial databases. In *Proc. ACM SIGMOD Int. Conf. on Management of Data*, pages 294–305, Portland, Oregon, 1989.

[20] J. Orenstein and F. A. Manola. Probe spatial data modeling and query processing in an image database application. *IEEE Trans. on Software Eng.*, 14(5):611–629, 1988.

[21] J. M. Patel and D. J. DeWitt. Partition based spatial-merge join. In *Proc. ACM SIGMOD Int. Conf. on Management of Data*, pages 259 – 270, Montreal, Canada, 1996.

[22] F. P. Preparata and M. I. Shamos. *Computational Geometry: an introduction*. Springer-Verlag, 1985.

[23] H. Samet. *Applications of Spatial Data Structures*. Addison-Wesley, 1990.

[24] T. Sellis, N. Roussopoulos, and C. Faloutsos. The R^+-tree: a dynamic index for multi-dimensional objects. In *Proc. 13th Int. Conf. on Very Large Data Bases*, pages 3–11, Brighton, England, 1987.

[25] D. Truffet and M. E. Orlowska. Two phase query processing with fuzzy approximations. In *Proc. 4th ACM International Workshop on Advances in Geographic Information Systems (ACM-GIS'96)*, Rockville, USA, 1996.

[26] X. Zhou, D. J. Abel, and D. Truffet. Data partitioning for parallel spatial join processing. In *LNCS 1262: Proc. 5th Int. Symp. on Spatial Databases (SSD'97)*, pages 178–196, Berlin, Germany, 1997. Springer-Verlag.

Efficiently Matching Proximity Relationships in Spatial Databases

Xuemin Lin[1]*, Xiaomei Zhou[1], and Chengfei Liu[2]

[1] School of Computer Science and Engineering
University of New South Wales, Sydney, NSW 2052, Australia
{lxue, xmei}@cse.unsw.edu.au
[2] School of Computing Sciences
University of Technology, Sydney, NSW 2009, Australia
liu@socs.uts.edu.au

Abstract. Spatial data mining recently emerges from a number of real applications, such as real-estate marketing, urban planning, weather forecasting, medical image analysis, road traffic accident analysis, etc. It demands for efficient solutions for many new, expensive, and complicated problems. In this paper, we investigate a proximity matching problem among *clusters* and *features*. The investigation involves proximity relationship measurement between clusters and features. We measure proximity in an average fashion to address possible nonuniform data distribution in a cluster. An efficient algorithm, for solving the problem, is proposed and evaluated. The algorithm applies a standard multi-step paradigm in combining with novel lower and upper proximity bounds. The algorithm is implemented in several different modes. Our experiment results do not only give a comparison among them but also illustrate the efficiency of the algorithm.

Keywords: Spatial query processing and data mining.

1 Introduction

Spatial data mining is to discover and understand non-trivial, implicit, and previously unknown knowledge in large spatial databases. It has a wide range of applications, such as demographic analysis, weather pattern analysis, urban planning, transportation management, etc. While processing of typical spatial queries (such as joins, nearest neighbouring, KNN, and map overlays) has been received a great deal of attention for years [2,3,4,28], spatial data mining, viewed as advanced spatial queries, demands for efficient solutions for many newly proposed, expensive, complicated, and sometimes ad-hoc spatial queries.

Inspired by a success in advanced spatial query processing techniques [2,3,4], [11,12,28], relational data mining [1,26,30], machine learning [9,10,22], computational geometry [27], and statistics analysis [17,29], many research results and system prototypes in spatial data mining have been recently reported [2,5,6,13],

---

* The work of this author is partially supported by a small ARC

R.H. Güting, D. Papadias, F. Lochovsky (Eds.): SSD'99, LNCS 1651, pp. 188–206, 1999.
© Springer-Verlag Berlin Heidelberg 1999

[15,18,19,21,24]. The existing research does not only tend to provide system solutions but also covers quite a number of special purpose solutions to ad-hoc mining tasks. These include efficiently computing spatial association rules [20], spatial data classification and generalization [13,15,21,24], spatial prediction and trend analysis [6], clustering and cluster analysis [5,7,18,25,32], mining in image and raster databases [8], etc.

Clustering has been proven one of the most useful tools to partition and categorize spatial data into *clusters* for the purpose of knowledge discovery. A number of efficient algorithms [5,7,25,31,32] have been proposed. Consider that the clustering technique might be too expensive to apply to approaching ad-hoc spatial data mining tasks. In [18,19], special purpose mining algorithms have been developed, as alternatives to clustering, for solving two ad-hoc problems. The first problem is to find the k closest *features* surrounding a set of points in two dimensional space. Such a set of points may be either a cluster obtained by a clustering algorithm or an existing spatial object (e.g. a residential area) in the database, while a feature is a polygon. The second problem in [18] is to compute the commonality among n sets of points (e.g. n residential areas), provided that their k closest features are pre-computed.

To complement the research in [18,19], in this paper we study the following ad-hoc *proximity matching* problem (PM) among n sets of points:

> Suppose that in two dimensional space, n sets of points and m sets of polygons are given. Regarding to a specific set C of points, find the sets of points such that each cluster C' meet the proximity matching condition - the "shortest distance" from C' to each set π of polygons is not greater than that between C and π. Further, if no sets of points meet the proximity matching condition then the "best" approximate solution is computed.

PM problem has a number of useful real applications. For instance, in real-estate spatial data, a set of points represent a residential area where each point represents a land parcel; a polygon corresponds to a vector representation of feature, such as a lake, golf course, school, motor way, etc. In this application, a house buyer or a real-estate developer may want to purchase a property in a well-known area C because of the proximity relationships to certain surround features but may not be able to do it due to either no property available in C or a budget limit. Therefore, the purchaser has to alternatively choose the available and affordable residential areas most similar to C with respect to these proximity relationships. Other applications include road traffic accident investigation, criminal analysis, etc.

PM will be formally defined in the next session. In PM, we assume that the "shortest distance" between a set of points and a set of polygon has not been pre-computed, nor stored in the database. Further, such a shortest distance will be defined in average sense to reflect non-uniform data distribution. These differentiate PM with KNN [2,13], the problem of searching commonalities among n sets of points [18], and incremental distance join problem [16].

A naive way to solve PM is to first precisely compute the distance information between each set of points and each set of polygons, and then to solve PM. However, in practice there may be many sets of points far from being part of solution to PM. Motivated by this, our algorithm adopts a standard multi-step technique [2,4,18,20] in combining with novel and powerful pruning conditions to filter out uninvolved features and sets of points. The algorithm has been implemented in several different modes for performance evaluation. Our experiments clearly demonstrates the efficiency of the algorithm.

The rest of the paper is organized as follows. In section 2, we present a precise definition of PM as well as brief an adopted spatial database architecture. Section 3 presents our algorithm for solving PM. Due to the length limitation, in this paper we sketch only the proofs of our theoretical results, and the interested readers may refer to our full paper [23] for the detailed proof. Section 4 reports our experiment results. In section 5, a discussion is presented regarding various modifications of our algorithm. This is followed by the conclusions and remarks.

2   Preliminary

In this section we precisely define the PM problem. A feature F is a *simple* and *closed* polygon [27] in the 2-dimensional space. A set C of points in the two dimensional space is called *cluster* for notation simplicity. Following [18], we assume that in PM a cluster is always *outside* [27] a feature. Note that this assumption may support many real applications. For instance, in real-estate data, a cluster represents a set of land parcel, and a feature represents a man-made or natural place of interest, such as lake, shopping center, school, park, entertainment center, etc. Such data can be found in many electronic maps in a digital library.

To efficiently access large spatial data (usually tera-bytes), in this paper we adopt an extended-relational and a SAND (spatial-and-non-spatial database) architecture [3]. That is, a spatial database consists of a set of spatial objects and a relational database describing non-spatial properties of these objects. For instance, a set of electronic data describing Sydney metropolitan area may be organized as follows.

- SUBURB (name, #houses, #units, average_price, ..., *g_des*),
- GOLF COURSE (name, #holes, ..., *g_des*),
- SCHOOL (name, type, ... , *g_des*),
- BEACH (name, type, ..., *g_des*).

In the above database schemata, the attribute *g_des* represents a spatial object, which is either a set of points or a polygon in PM. In order to achieve efficient access, in SAND the attribute *g_des* stores only a pointer in the relational table, pointing to the actual spatial object description. Below shows an example of PM:

Example 1. select s.*
> *from SUBURB s, SUBURB s1, GOLF COURSE g,*
> *BEACH w, SCHOOL sc*
> *where s1.name = 'Randwick' and s.name ≠ 'Randwick' and*
> *s.average_price ≤ 400,000 and g.#holes = 9 and sc.type = 'private'*
> **proximity-matching** *between (s.obj, s1.obj) regarding*
> *their shortest distances to g, sc, and w*   □

Example 1 is to find the suburbs with area average house price less than
$400,000, such that their individual shortest distances to the golf courses with
9 holes, to private schools, and to beaches are respectively smaller than those
between the suburb Randwick and the features. If such suburbs do not exist
then the suburbs most approximately meet the proximity matching conditions
will be reported.

Taking the above query as an example, we now formally define PM. In PM,
the input consists of:

- a cluster C_0 (e.g. the suburb Randwick in Example 1),
- a set S of clusters (e.g. the suburbs with average_price not greater than
  $400,000 in Example 1),
- a set $\Pi = \{\pi_j : 1 \leq j \leq m\}$ of groups of features (e.g. in Example 1, $m = 3$,
  π_1 is the set of golf courses with 9 holes, π_2 is the set private schools, and
  π_3 is the set of beaches).

Given a feature F and a point p outside F, the length of the actual (working
or driving) shortest path from p to F is too expensive to compute in the presence
of tens of thousands of different roads. In PM, we use the shortest Euclidean
distance from p to a point in the boundary of F, denoted by $d(p, F)$, to reflect the
geographic proximity relationship between p and F. We believe that on average,
the length of an actual shortest path can be reflected by $d(p, F)$. We call $d(p, F)$
the *distance* between p and F. Note that if F degenerates to a point p' then
$d(p, p')$ means the Euclidean distance between them; and F may also degenerate
to a line. Moreover, for the purpose of computing lower and upper proximity
bounds in Section 3, we need to extend the definition of $d(p, F)$ to cover the case
when p is inside or on the boundary of P; that is, $d(p, F) = 0$ if p is inside P.

A proximity value between a cluster C and a feature F can be defined in a
number of ways. We may define it by the shortest distance between the "boundary" of C and the boundary of F. However, as points in C admit an arbitrary
distribution, such a proximity value may not be the majority consensus from C;
this was shown in [18]. We use the following *average proximity* value to quanti-
tatively model the proximity relationship between F and C:

$$AP(C, F) = \frac{1}{|C|} \sum_{p \in C} d(p, F) \tag{1}$$

Consider that in PM a set π of features normally means a set of the same
kind of features. We define the distance between a set π of features and a cluster

C to be the smallest average proximity between C and a $F \in \pi$, and it is denoted by $D(C, \pi)$. That is,

$$D(C, \pi) = \min_{F \in \pi} \{AP(C, F)\} \qquad (2)$$

As mentioned earlier, if in PM no cluster meets the above requirements, then the proximity matching needs to find clusters that achieve the requirement most; in this case, we rank the importance of a set π_j of features by a positive value w_j. The more important a feature π_j is, the larger w_j is. The w_j can be assigned by either a user or the system default. Therefore, a set $\{w_j : 1 \leq j \leq m\}$ of positive values is also part of the input of PM. PM can now be modeled as to find the clusters C in S such that the following goal function is minimized.

$$PM_{C_0}(C, \Pi) = \sum_{j=1}^{m} w_j \, pm(D(C, \pi_j), D(C_0, \pi_j)) \qquad (3)$$

where,

$$pm(D(C, \pi_j), D(C_0, \pi_j)) = \begin{cases} 0 & \text{if } D(C, \pi_j) \leq D(C_0, \pi_j) \\ D(C, \pi_j) - D(C_0, \pi_j) & \text{otherwise} \end{cases} \qquad (4)$$

3   Algorithms for Solving PM

In this section, we present an efficient algorithm for solving PM. The algorithm is denoted by CPM, which stands for Computing the Proximity Matching.

An immediate way (brute-force) to solve PM is to 1) compute $AP(C, F)$ firstly for each pair of a cluster C and a feature F, 2) secondly compute $D(C, \pi_j)$ for every pair of a C and a π_j, 3) thirdly compute $PM_{C_0}(C, \Pi)$ for each cluster C in S, and 4) finally find the clusters C with the smallest values of $PM_{C_0}(C, \Pi)$. Note that $AP(C, F)$ can be easily computed in $O(|C||F|)$ according to the definition of $AP(C, F)$; and it is the dominant cost. Though the brute-force approach runs in quadratic time regarding the input size, there may be hundreds clusters and tens of thousands features involved in the computation. Moreover, each cluster (feature) may have a number of points (edges). This makes the brute-force approach computationally prohibitive in practice; and our experiment results in Section 4 confirm this.

An alternative way to approach PM is to adopt a multi-step paradigm [2,4], [18,20]. That is, we firstly apply a coarse and fast computation. Instead of computing the actual value of $AP(C, F)$ in quadratic time, we may compute a lower bound and an upper bound for $AP(C, F)$ in a constant time $O(1)$. By these bounds, for each cluster C in S we can rule out the features in a π_j, which are definitely not closest to C; and thus we do not have to precisely compute the average proximity values between these eliminated features and C. Secondly, we can deduce a lower bound and an upper bound for each $PM_{C_0}(C, \Pi)$ from the bounds of AP; and then filter out uninvolved clusters. This is the basic idea of our algorithm. In our algorithm CPM, we have not integrated our algorithm into a particular spatial index, such as R-trees, R^+-trees, etc, due to the following reasons:

– There may be no such a spatial index built.
– The PM problem may involve many features from different tables/electronic thematic maps; and thus, spatial index built for each thematic map may be different. This brings another difficulty to make use of spatial indices.
– A feature or a cluster, which is qualified in PM, may be only a part of a stored spatial object; for instance, user can be interested in only certain part of a residential area. This makes a possible existing index based on the stored spatial objects not applicable.
– The paper [18] indicates the existing spatial indexing techniques do not necessarily support well the computation of *aggregate* distances; the argument should be also applied to average distance computation.

The algorithm CPM consists of the following 5 steps:

Step 1: Read the relevant clusters into buffer.
Step 2: Read features batch by batch into buffer and determine their groups by validating the selection conditions against the relational tables.
Step 3: For each feature F in a π_j, compute lower and upper bounds of $AP(C, F)$ for a cluster C. Then determine whether or not F should be kept for the computation of $D(C, \pi_j)$.
Step 4: For each π_j and each cluster C, compute lower and upper bounds for $D(C, \pi_j)$; and then derive lower and upper bounds for (4). Filter out clusters which will not be part of the solution to PM.
Step 5: Apply the above brute-force method to the remaining clusters and their associated features to solve PM.

In the next several subsections we detail the algorithm step by step. Clearly, a success of the algorithm CPM largely relies on how good the lower and upper bounds of AP are. The goodness of lower and upper bounds means two things: 1) the bounds should be reasonably tight, and 2) the corresponding computation should be fast. We first present the lower and upper bounds.

Note that for presentation simplicity, the algorithms presented in the paper are restricted to the case when features and clusters qualified in PM are stored spatial objects in the database. However, they can be immediately extended to cover the case when a feature or a cluster is a part of a stored object.

3.1   Lower and Upper Bounds for Average Proximity

In this subsection, we recall first some useful notation. The *barycenter* (*centroid*) of a cluster C is denoted by $b(C)$. A *convex* [27] polygon encompassing a feature F is called a bounding convex polygon of F. The smallest bounding convex polygon of F is called the *convex hull* [27] of F and is denoted by P_F. An *isothetic* rectangle is orthogonal to the coordinate axis. The minimum bounding rectangle of F refers to the minimum isothetic bounding rectangle of F and is denoted by R_F. Similarly, we denote the convex hull of a cluster C by P_C and denote the minimum bounding rectangle of C by R_C. The bounds presented in the subsection are based on either minimum bounding rectangles or convex hulls.

Given a R_C and a R_F, an immediate idea is to use the shortest distance and the longest distance between R_C and R_F to respectively represent a lower bound and an upper bound of $AP(C, F)$. However, this immediate idea has two problems. The first problem is that when two rectangles intersect with each other (note that in this case C and F do not necessarily have an intersection), the shortest and longest distances between R_C and R_F are not well defined. The second problem is that the bounds may not be very tight even if the two rectangles do not intersect. These also happen similarly for convex hulls. Below, we present new and tighter bounds.

Our lower bound computation is based on the following Lemma.

Lemma 1. $\sum_{i=1}^{K} \sqrt{x_i^2 + y_i^2} \geq \sqrt{(\sum_{i=1}^{K} x_i)^2 + (\sum_{i=1}^{K} y_i)^2}$

Proof: It can be immediately verified that the inequality holds when $K = 2$. By mathematical induction, we can prove the Lemma. $\square$

From Lemma 1, Theorem 2 immediately follows.

Theorem 2. *Suppose that C is a cluster, F is a feature, and P is either the convex hull or the minimum bounding rectangle of F. Then, $AP(C, F) \geq d(b(C), P)$; in other words $d(b(C), P)$ is a lower bound of $AP(C, F)$.*

Figure 1 gives an example, and shows that our lower bound is tighter than the shortest distance between two rectangles.

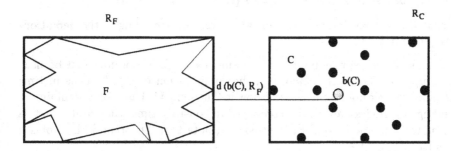

Fig. 1. A lower bound

The R_F of a feature F has four edges: the left boundary, right boundary, bottom boundary, and top boundary. Note that each boundary edge x is divided into several line segments (at least two). The two end points of such a line segment are either a) a pair of two adjacent intersection points between F and

R_F, or b) a vertex of R_F and an intersection point between F and R_F. We use $HR_{F,x}$ to denote the maximal segment length among these line segments in the boundary edge x, where $x \in \{l, r, b, t\}$ respectively represents either the left, or right, or bottom, or top boundary. For each boundary edge x, we use $VR_{F,x}$ to denote the maximal length of the perpendicular line segment from a vertex of an edge of F, which *faces* [27] x. Figure 2(a) illustrates these concepts.

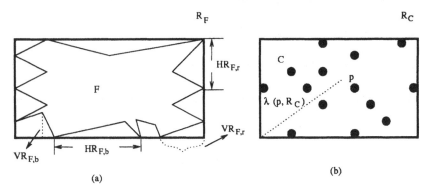

Fig. 2. Examples

Suppose that P' is either the convex hull or the minimum bounding rectangle of a cluster C, and p is a point inside P'. We use $\lambda(p, P')$ to denote the maximal distance between p and a point contained in P'. Clearly, $\lambda(p, P')$ is the maximal distance from p to one of the vertices of P'. Figure 2(b) illustrates this concept for the minimum bounding rectangle.

Below we present two upper bounds respectively for convex hulls, and minimum bounding rectangles.

Theorem 3. *Suppose that P_C is the convex hull of a cluster C, p is a point inside P_C, and F is a feature. Then $AP(C, F) \le d(p, F) + \lambda(p, P_C)$.*

Sketch of the Proof: The theorem can be verified according to the definitions of AP, d, and λ. □

It is clear the right hand side in the inequality of Theorem 3 can be used as an upper bound of AP. However, the computation of $d(p, F)$ runs in time $O(|F|)$. The lower bound presented below in Theorem 4 is based on the minimum bounding rectangles and can be computed in constant time, though it is not as tight as that in Theorem 3. First we should note that Theorem 3 also holds if we replace P_C by R_C.

Theorem 4. *Suppose that R_C of C is given, R_F of F is given, and p is a point contained in R_C. Then*

$$d(p, F) + \lambda(p, R_C) \le \min_{x \in \{l, r, b, u\}} \{\min\{HR_{F,x}, VR_{F,x}\} + d(p, x)\} + \lambda(p, R_C). \,(5)$$

Here x a boundary edge of R_F.

Proof: From the definition of HR and VR, the theorem can be immediately verified. $\square$

Theorem 4 together with Theorem 3 imply the right hand side in the inequality of Theorem 4 is another upper bound of AP. Further, it should be clear that this upper bound can be computed in constant time, provided that $HR_{F,x}$ and $VR_{F,x}$ are obtained and $\lambda(R_C, p)$ is computed for a given p.

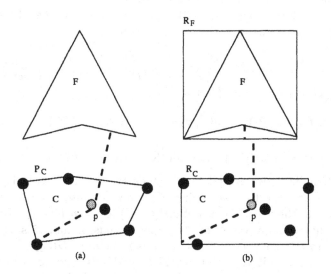

Fig. 3. Upper bounds

The total lengths of the thick dotted lines in Figure 3(a) and Figure 3(b) respectively show the upper bounds in Theorems 3 and 4. They also show that our bounds may be tighter than the longest distance between the convex hulls or between the minimum bounding rectangles. However, we cannot generally prove this because the tightness of the bounds depends on the choice of p. In our algorithm, we will choose the *centroid* of a cluster C in the upper bound computation since it has to be used to obtain a lower bound.

In the next several subsections, we present first the algorithm CPM based on the minimum bounding rectangles.

3.2   Read in Clusters

In Step 1, we first read in the clusters, specified in the query by a user, into buffer by execution of the data retrieval method [3]; for instance, regarding the query in Example 1 the suburbs with average house price less than \$400,000 are read into the database by querying the SUBURB table. Then, we compute R_C, $b(C)$, and $\lambda(b(C), R_C)$ for each cluster C. Clearly, this step takes linear time with respect to the total size of clusters.

3.3   Read In and Filter Out Features

This subsection presents Step 2 and Step 3. Consider that the number of the features to be processed may be very large, and each feature may have many edges. It may be impossible to keep all features in buffer all the time. Consequently, features should be read into buffer batch by batch. Once a batch of features are read in, they are first assigned group IDs against users specifications; for instance, in Example 1 three feature groups are retrieved.

After a feature is assigned a group ID, the algorithm CPM invokes the filtering process in Step 3. It is based on the following lemma.

Lemma 5. *Suppose that C is a cluster, and $\pi = \{F_1, F_2, ..., F_k\}$ is a group of features. For $1 \leq j \leq k$, let $LB_{AP}(C, F_j)$ and $UB_{AP}(C, F_j)$ be a lower bound and an upper bound of $AP(C, F_j)$. Then:*

$$\min_{1 \leq j \leq k}\{LB_{AP}(C, F_j)\} \leq D(C, \pi) \leq \min_{1 \leq j \leq k}\{UB_{AP}(C, F_j)\}.$$

Proof: The lemma immediately follows from the definition of D. □

For notation simplicity, we extend S to include the given C_0; that is, $S = \{C_i : 0 \leq i \leq n\}$. With respect to each pair of C_i and π_j, we use $LB^{i,j}$ to record the minimum value of the lower bounds of $AP(C_i, F)$ for each $F \in \pi_j$, and use $UB^{i,j}$ to record the minimum values of the upper bounds. Lemma 5 says that $LB^{i,j}$ and $UB^{i,j}$ are relatively a lower and an upper bound of $D(C_i, \pi_j)$.

In our algorithm, we initially set both $LB^{i,j}$ and $UB^{i,j}$ to ∞, and then gradually update them when a new feature in π_j is processed. We use a dynamic array $A_{i,j}$ to store the candidate features in π_j for computing $AP(C_i, \pi_j)$. Each element in $A_{i,j}$ stores the identifier FID of a feature F, the obtained lower bound of $AP(C_i, F)$, and the pointer g_des that points to the spatial description of the cluster.

Specifically, to process a $F \in \pi_j$, CPM firstly computes R_F and the values of VF and HF for R_F; this can done easily by scanning F only once. Secondly, for each C_i, we check if F should be included in $A_{i,j}$. According to Lemma 5, we add F to $A_{i,j}$ if the obtained lower bound of $AP(C_i, F)$ is less than the current value of $UB^{i,j}$. Further, we should update $LB^{i,j}$ and $UB^{i,j}$ each time after F is added to $A_{i,j}$; and then check $A_{i,j}$ to determine if some features in $A_{i,j}$ should be removed due to an update of $UB^{i,j}$. To prevent unnecessary scan of $A_{i,j}$ each time after $UB^{i,j}$ is updated, we also record the maximum value of the lower bounds of $AP(C_i, F)$ among the features F in the current $A_{i,j}$; it is recorded by $q^{i,j}$ ($q^{i,j}$ is initially zero).

For example, suppose that the features F_1, F_2, F_3, and F_4 are in π_1. Their lower and upper bounds of AP values against a cluster C_1 are $(4, 5)$ for F_1, $(6, 7)$ for F_2, $(3, 6)$ for F_3, and $(3, 3.5)$ for F_4; see Figure 4. Initially, $LB^{1,1} = UB^{1,1} = \infty$, and $q^{1,1} = 0$. Suppose that F_1 is first processed. We add F_1 to $A_{1,1}$ by recording its ID and the lower bound; and then $LB^{1,1} = 4$, $UB^{1,1} = 5$, and $q_{1,1} = 4$. Next, F_2 is processed. F_2 should not be included in $A_{1,1}$ because $6 > UB^{1,1}$. However, when F_3 is processed thirdly, F_3 should be added in $A_{1,1}$.

Consequently, $LB^{1,1} = 3$, $UB^{1,1} = 5$, $q^{1,1} = 4$. While processing F_4, we find that F_4 should be added to $A_{1,1}$. Accordingly, $LB^{3,3} = 3$, $UB^{1,1} = 3.5$, and $q^{1,1} = 4$. Then, since $q^{1,1} > UB_{1,1}$ we should check $A_{1,1}$ to delete F_1 from $A_{1,1}$.

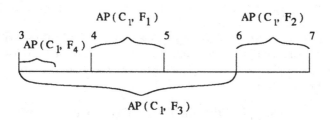

Fig. 4. Filter Out Features

More precisely, to process a feature $F \in \pi_j$, the above processes can be presented by the following pseudo codes:

```
compute R_F, VF_{R,x} and HF_{R,x}; /* x = {l, r, b, t} */
for each cluster C_i do {
    LB_AP(C_i, F) → a_1 (the lower bound); UB_AP(C_i, F) → a_2 (the upper bound);
    if a_1 < UB^{i,j} then {
        F → A_{i,j};
        min{a_1, LB^{i,j}} → LB^{i,j};
        min{a_2, UB^{i,j}} → UB^{i,j};
        max{q^{i,j}, a_1} → q^{i,j};
        if q^{i,j} ≥ UB^{i,j} then {
            remove features F from A_{i,j} if LB_AP(C_i, F) ≥ UB^{i,j};
            re-compute q^{i,j} from the remaining features in A_{i,j}; }
    }
}
```

Once a batch of features are processed by the above procedure, we do not keep them in buffer if no space is left for the next batch of features. In this situation, we will need to read in again the features not filtered out after Step 4, so that we can process Step 4; this is why we want to keep the pointer g_des for each feature object.

3.4   Filter Out Clusters

This subsection describes Step 4. After the computation in last subsection, we obtained $LB^{i,j}$ and $UB^{i,j}$ for each pair of C_i and π_j. By Lemma 5, we have:

$$LB^{i,j} \leq D(C_i, \pi_j) \leq UB^{i,j}. \qquad (6)$$

Note that from the definition, $pm(D(C_i, \pi_j), D(C_0, \pi_j)) = 0$ if $D(C_i, \pi_j) \leq D(C_0, \pi_j)$. This immediately implies that $pm(D(C_i, \pi_j), D(C_0, \pi_j)) = 0$ if one of the following two conditions applies:

1. $UB^{i,j} \leq LB^{0,j}$, or
2. $UB^{i,j} \leq D(C_0, \pi_j)$

Thus, intuitively $pm(D(C_i, \pi_j), D(C_0, \pi_j))$ is bounded by $UB^{i,j} - LB^{0,j}$. Further, it cannot be smaller than the minimum distance between the two closed intervals: $[LB^{i,j}, UB^{i,j}]$ and $[LB^{0,j}, UB^{0,j}]$. The intuition can be immediately verified and is stated in Lemma 6. Let

$$P^{i,j} = \begin{cases} UB^{i,j} - LB^{0,j} & \text{if } UB^{i,j} > LB^{0,j} \\ 0 & \text{otherwise} \end{cases}$$

and let

$$\alpha^{i,j} = \begin{cases} LB^{i,j} - UB^{0,j} & \text{If } LB^{i,j} > UB^{0,j} \\ 0 & \text{otherwise} \end{cases}$$

Lemma 6. $pm(D(C, \pi_j), D(c_i, \pi_j)) \leq P^{i,j}$, and $pm(D(C, \pi_j), D(c_i, \pi_j)) \geq \alpha^{i,j}$ for each pair of C_i and π_j,

Sketch of the Proof: Prove the theorem by applying (6) and the above two conditions. □

Lemma 6 gives us a lower and upper bound of $PM_{C_0}(C_i, \Pi)$ for each C_i:

$$\sum_{j=1}^{m} w_j \alpha^{i,j} \leq PM_{C_0}(C_i, \Pi) \leq \sum_{j=1}^{m} w_j P^{i,j}. \tag{7}$$

In Step 4, we firstly compute the lower and upper bounds for each cluster C_i ($i \neq 0$), as given in (7). Secondly, we compute the minimum value τ of the upper bounds among the clusters C_i in S but $i \neq 0$. Thirdly, we scan S to filter out clusters C_i if $\sum_{j=1}^{m} w_j \alpha^{i,j} > \tau$. This procedure runs in time $O(nm)$.

3.5   Precise Computation

This subsection describes Step 5. After pruning the clusters, the information of remaining features for solving PM is kept in each $A_{i,j}$. We need to read in these features again to perform the precise computation. Since two different $A_{i,j}$s may keep a same feature as a candidate, to efficiently read in required features we use a hashing method. A hash table H is created against feature ID - FID. We scan $A_{i,j}$ one by one to execute the following two steps.

Step 1: For each feature $F \in A_{i,j}$, hash its FID into an H entry and then determine if its spatial description is already in H.

Step 2: If the spatial description of F is not in H, then read it into buffer using g_des and store it in H. Then, compute $AP(C_i, F)$ for each remaining C_i.

After computing all AP values, we apply the remaining step of the brute-force method to the existing clusters to solve the problem. Note that before starting Step 5, we also check a special case - if $\alpha^{i,j} = P^{i,j}$ for all remaining clusters and features then CPM does not have to process Step 5 but outputs the remaining clusters as the solution.

3.6   Complexity of CPM

In this subsection, we analyze the complexity of CPM. Step 1 takes linear time with respect to the total sizes of clusters; that is $O(\sum_{i=0}^{n} |C_i|)$. Step 2 and Step 3 take time $O(n \times \sum_{j=1}^{m} |\pi_j|)$. Step 4 takes time $O(nm)$, while step 5 takes time $O(\sum_{\forall C_i, \forall F \in A_{i,j}} |C_i||F|)$ for the remaining clusters C_i.

Note that the brute-force algorithm runs in time $O(\sum_{\forall C, \forall F} |C||F|)$. It should be clear that in practical, the time complexity of the brute-force method is much higher than that of CPM. This is confirmed by our experiment results in Section 4.

3.7   Different Modes of CPM

The above mode of CPM uses only the minimum bounding rectangles; and it is denoted by CPM-R.

An alternative mode to CPM-R is to use a multiple-filtering technique [2,13,18]:

- first the minimum bounding rectangles are used in Steps 3 and 4, and then
- the convex hulls for features and clusters are adopted to repeat Steps 3 and 4 before processing Step 5.

It is denoted by CPM-RH. In CPM-RH, we employ the divide and conquer algorithm [27] to compute convex hulls for clusters. To compute a convex hull for a feature (simple polygon), we employ the last step in Graham's scan [27], which runs in linear time. We use the upper bound of AP in Theorem 3 to implement the procedure in Section 3.3 for convex hulls.

Another alternative mode to CPM-R is to use only the convex hulls instead of the minimum bounding rectangles. We denote this mode by CPM-H.

In next section, we will report our experiment results regarding the performances of the brute-force algorithm, CPM-R, CPM-RH, and CPM-H.

4   Experiment Results

The brute-force algorithm and the three different modes of CPM have been implemented by C++ on a Pentium I/200 with 128 MB of main memory, running Window-NT 4.0. In our experiments, we evaluated the algorithms for efficiency and scalability. Our performance evaluation is basically focused on Step 3 onwards, because the methods [3] of reading in clusters and features are not our contribution. Therefore, in our experiment we record only the CPU time but exclude I/O costs.

We developed a program to generate a benchmark. In the program, we first use the following random parameters to generate rectangles, such that a rectangle may intersect with at most another rectangle:

- M gives the number of rectangles, and
- wid_R controls the width of a rectangle R and h_R controls the height of R.

More specifically, we first generate M rectangles R with a random width wid_R and a random height h_R, where $1 \leq wid_R, h_R \leq 1000$. The generated rectangles are randomly distributed in the 2-dimensional space, and intersect with at most another rectangle. To generate n clusters, we randomly divide the whole region into n disjoint sub-regions and choose one rectangle from each region. We use a random parameter NC to control the average number of generated points in each rectangle among the chosen n rectangles. These give n clusters. The remaining $M - n$ rectangles correspond to features. We use another random parameter NF to control the average number of the vertices (points) generated in each remaining rectangle.

Note that if R intersects with R', we actually generate the points respectively in R and $R' - R$ if R' is not included in R. Further, in each rectangle R corresponding to a feature, we apply a Graham's scan-like algorithm [27] to produce a simple polygon connecting each vertex in R; the generated simple polygon is used as a feature. Therefore, in our benchmark we have two kinds of spatial objects - clusters and features.

In the experiments below, we adopt a common set of parameters: 1) a feature has 150 edges on average, 2) a cluster has 300 points on average, and 3) the features are grouped into 20 groups.

In our first experiment, we generate a database with 10 clusters and 1000 features. The experiment results are depicted in Figure 5, where the algorithm CPM-R, CPM-H, and CPM-RH are respective abbreviated to "R", "H", and "R-H".

From the first experiment, we can conclude that the brute-force is practically very slow. Another appearance is that H is slower than R and R-H due to the fact that in H, the computation of lower and upper bounds for each pair of cluster and feature does not take constant time. Intuitively, H should be significantly slower than R and R-H when the number of clusters and features increases; this has been confirmed by the second experiment.

The second experiment has been undertaken through two "dimensions":

- Fix the number of features to be 1000, while the number of clusters varies from 5 to 20. The results are depicted in Figure 6.
- Fix the number of clusters to be 200, while the number of features varies from 1000 to 10000. The results are depicted in Figure 7.

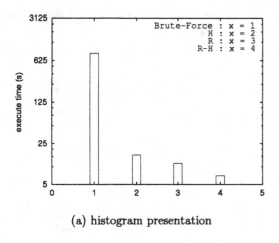

(a) histogram presentation

Algorithm	Average Run Time (s)
Brute-force	823.514
H	15.863
R	11.516
R-H	6.95

(b) table presentation

Fig. 5. A Comparison among four algorithms

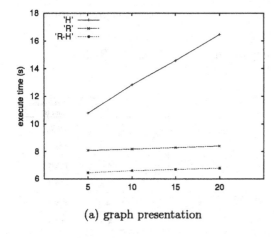

(a) graph presentation

#CL	Algorithm	Average Run Time (s)
5	H	10.772
	R	8.079
	R-H	6.447
10	H	12.827
	R	8.168
	R-H	6.606
15	H	14.575
	R	8.28
	R-H	6.706
20	H	16.458
	R	8.399
	R-H	6.8

(b) table presentation

Fig. 6. A Comparison among H, R, and R-H

Note that in the second experiment, we run R, H, and $R - H$ 40 times against each database. This experiment, together with the first experiment, also demonstrates that $R - H$ is faster than H on average.

In the third experiment, to test the scalability we vary the database size from 1000 features to 50000 features but fix the the number of clusters to be 200. For each database, we run both R and R-H 40 times against each database. Figure 8 illustrates our experiment results.

The three conducted experiments suggest that our algorithm is efficient and scalable. Secondly, we can see that though an application of convex hulls to our filtering procedures is more accurate than an application of minimum rectangles, but it is too expensive to use directly. The best use of convex hulls should follow an application of minimum bounding rectangles.

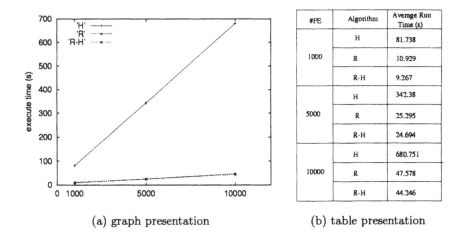

#FE	Algorithm	Average Run Time (s)
1000	H	81.738
	R	10.929
	R-H	9.267
5000	H	342.38
	R	25.295
	R-H	24.694
10000	H	680.751
	R	47.578
	R-H	44.246

(a) graph presentation   (b) table presentation

Fig. 7. Another Comparison among H, R, and R-H

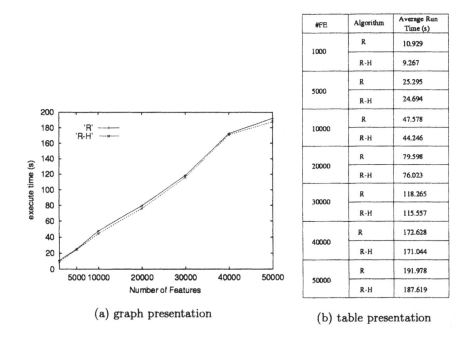

#FE	Algorithm	Average Run Time (s)
1000	R	10.929
	R-H	9.267
5000	R	25.295
	R-H	24.694
10000	R	47.578
	R-H	44.246
20000	R	79.598
	R-H	76.023
30000	R	118.265
	R-H	115.557
40000	R	172.628
	R-H	171.044
50000	R	191.978
	R-H	187.619

(a) graph presentation   (b) table presentation

Fig. 8. A comparison between R and R-H

5   Discussion

The problem PM and the algorithm CPM may be either generalized or constrained according to various applications. In this section, we present a discussion on these issues.

A slight modification of the algorithm CPM can be applied to the case where we specify the proximity matching conditions directly use the shortest distances instead of specifying a given cluster. For instance, in Example 1 we may directly specify that such suburbs are within 5 km away from a private school, 6 km away from a beach, and 2 km away from a golf course, instead of comparing to the suburb Randwick.

Another modification of the problem is to define $pm(D(C, \pi_j), D(C_0, \pi_j))$ as $|D(C, \pi_j) - D(C_0, \pi_j)|$. It can be shown [33] that our algorithms can be immediately modified accordingly to resolve this.

Our results and discussions, so far, are limited to the Euclidean distance. Note that the upper bounds presented in Section 3.1 are based on a triangular inequality in the Euclidean distance; that is $d(p_1, p_2) \leq d(p_1, p_3) + d(p_3, p_2)$. Since the triangular inequality is part of the definition of any metric distance, the upper bounds can be applied to any metric space. We should also note that the lower bound presented in Section 3.1 can be obtained in such a metric space that the metric distance γ follows the two constraints below:

- $\gamma(p_1, p_2) + \gamma(p_3, p_4) \geq \gamma(p_1 + p_3, p_2 + p_4)$, and
- $\gamma(c \times p_1, c \times p_2) = |c|\gamma(p_1, p_2)$ for any constant c.

Consequently, we can extend the problem PM and the algorithm CPM to any metric space where the above two constraints are satisfied. For instance, we can verify that the *Manhattan distance* [27] satisfied the two constraints; and thus our algorithm can be extended to 2-dimensional Manhattan distance space.

6   Conclusions

In this paper, we formalized a new problem (PM) in spatial data mining from real applications. We presented an efficient algorithm based on several novel pruning conditions, as well as various different modes of the algorithm. Our experiment results showed that the algorithm is very efficient and can support a number of real applications where data with huge volume are present.

Further, in Section 5, we showed that our work in this paper can be extended to many other metric spaces.

Note that the PM problem and the algorithm CPM are restricted to the case where a cluster is outside a feature. This restriction may be not generally applicable to some applications; we are now identifying such applications. Besides, we are currently working on the development of indexing techniques to support CPM. Further, a modification of PM to cover the applications where the distance from a cluster to a set of features is not necessarily restricted to one feature in the set seems more complicated; this is our another future study.

References

1. R. Agrawal and R. Srikant, Fast Algorithms for Mining Association Rules, *Proceedings of the 20th VLDB Conference*, 487-499, 1994.
2. M. Ankerst, B. Braunmuller, H.-P. Kriegel, T. Seidl, Improving Adaptable Similarity Query Processing by Using Approximations, *Proceedings of the 24th VLDB Conference*, 206-217, 1998.
3. W. G. Aref and H. Samet, Optimization Strategies for Spatial Query Processing, *Proceedings of the 17th VLDB Conference*, 81-90, 1991.
4. T. Brinkhoff, H. P. Kriegel, and R. Schneider, and B. Seeger, Multistep processing of spatial joins, *Proc. of ACM SIGMOD*, pp. 197-208, 1994.
5. M. Ester, H.-P. Kriegel, J. Sander and X. Xu, A density-based algorithm for discovering clusters in large spatial databases, *Proceedings of the Second International Conference on Data Mining KDD-96*, 226-231, 1996.
6. M. Este, H.-P. Kriegel, J. Sander, Spatial Data Mining: A Database Approach, *SSD'97*, LNCS 1262, 47-65, 1997.
7. V. Estivill-Castro and A.T. Murray, Discovering Associations in Spatial Data - An Efficient Medoid Based Approach, *Proceedings of the Second Pacific-Asia Conference on Knowledge Discovery*, LNAI 394, 110-121, 1998.
8. U. M. Fayyad, S. G. Djorgovski, and N. Weir, Automating the analysis and cataloging of sky surveys, *Advances in Knowledge Discovery and Data Mining*, AAAI/MIT Press, 1996.
9. U. Fayyad, G. Piatetsky-Shapiro, P. Smyth, and R. Uthurusamy, Eds. *Advances in Knowledge Discovery and Data Mining*, AAAI/MIT Press, Menlo Park, CA, 1996.
10. D. Fisher, Improving Inference through Conceptual Clustering, *Proceedings of 1987 AAAI Conferences*, 461-465, 1987.
11. R. H. Guting, An Introduction to Spatial Database Systems, *VLDB Journal*, 3(4), 357-400, 1994.
12. R. Guttman, A Dynamic Index Structure for Spatial Searching, *ACM-SIGMOD International Conference on Management of Data*, 47-57, 1984.
13. J. Han, Spatial Data Mining and Spatial Data Warehousing, *Invited Talk at SSD'97*, 1997.
14. J. Han, Y. Cai, and N. Cercone, Dynamic Generation and Refinement of Concept Hierarchies for Knowledge Discovery in Databases, *IEEE Trans. knowledge and Data Engineering*, 5, 29-40, 1993.
15. J. Han, K. Koperski, and N. Stefanovic, GeoMiner: A System Prototype for Spatial Data Mining, *Proceedings of 1997 ACM-SIGMOD International Conference on Management*, 553-556, 1997.
16. G. R. Hjaltason and H. Samet, Incremental Distance Join Algorithms for Spatial Databases, 237-248, *Proceedings of the 1998 ACM SIGMOD International Conference on Management of Data*, 1998.
17. L. Kaufman and P. J. Rousseeuw, *Finding Groups in Data: an Introduction to Cluster Analysis*, John Wiley & Sons, 1990.
18. E. M. Knorr and R. T. Ng, Finding Aggregate Proximity Relationships and Commonalities in Spatial Data Mining, *IEEE Transactions on Knowledge and Data Engineering*, 8(6), 884-897, 1996.
19. E. M. Knorr, R. T. Ng, and D. L. Shilvock, Finding Boundary Shape Matching Relationships in Spatial Data, *SSD'97*, LNCS 1262, 29-46, 1997.
20. K. Koperski and J. Han, Discovery of Spatial Association Rules in Geographic Information Databases, *Advances in Spatial Databases*, Proceeding of 4th Symposium (SSD'95), 47-66, 1995.

21. K. Koperski, J. Han, and J. Adhikary, Mining Knowledge in Geographic Data, to appear in *Communications of ACM*.
22. R. S. Michalski, J. M. Carbonnel, and T. M. Mitchell, editors, *Machine Learning: An Artificial Intelligence Approach*, Morgan Kaufman, 1983.
23. X. Lin, X. Zhou, and C. Liu, Efficient Computation of a Proximity Matching in Spatial Databases, *Tec. Report*, University of New South Wales, 1998.
24. W. Lu, J. Han, and B. C. Ooi, Knowledge Discovery in Large Spatial Databases, *Proceedings of Far East Workshop on Geographic Information Systems*, 275-289, 1993.
25. N. Ng and J. Han, Efficient and Effective Clustering Method for Spatial Data Mining, *Proceeding of 1994 VLDB*, 144-155, 1994.
26. J. S. Park, M.-S. Chen, and P. S. Yu, An Effective Hash-Based Algorithm for Mining Association Rules, *Proceedings of 1995 ACM SIGMOD*, 175-186, 1995.
27. F. Preparata and M. Shamos, *Computational Geometry: An Introduction*, Springer-Verlag, New York, 1985.
28. H. Samet, *The Design and Analysis of Spatial Data Structures*, Addison-Wesley, 1990.
29. G. Shaw and D. Wheeler, *Statistical Techniques in Geographical Analysis*, London, David Fulton, 1994.
30. H. Toivonen, Sampling Large Databases for Association Rules, *Proceedings of 22nd VLDB Conference*, 1996.
31. X. Xu, M. Ester, H.-P. Kriegel, Jorg Sander, A Distribution-Based Clustering Algorithm for Mining in Large Spatial Databases, *ICDE'98*, 324-331, 1998.
32. T. Zhang, R. Ramakrishnan and M. Livny, BIRCH: an efficient data clustering method for very large databases, *Proceeding of 1996 ACM-SIGMOD International Conference of Management of Data*, 103-114, 1996.
33. X. Zhou, *Efficiently Computing Proximity Relationships in Spatial Databases*, Master Thesis, University of New South Wales, under preparation, 1999.

3D Shape Histograms for Similarity Search and Classification in Spatial Databases

Mihael Ankerst, Gabi Kastenmüller, Hans-Peter Kriegel, Thomas Seidl

University of Munich, Institute for Computer Science
Oettingenstr. 67, 80538 Munich, Germany
http://www.dbs.informatik.uni-muenchen.de
{ankerst, kastenmu, kriegel, seidl}@dbs.informatik.uni-muenchen.de

Abstract. Classification is one of the basic tasks of data mining in modern database applications including molecular biology, astronomy, mechanical engineering, medical imaging or meteorology. The underlying models have to consider spatial properties such as shape or extension as well as thematic attributes. We introduce 3D shape histograms as an intuitive and powerful similarity model for 3D objects. Particular flexibility is provided by using quadratic form distance functions in order to account for errors of measurement, sampling, and numerical rounding that all may result in small displacements and rotations of shapes. For query processing, a general filter-refinement architecture is employed that efficiently supports similarity search based on quadratic forms. An experimental evaluation in the context of molecular biology demonstrates both, the high classification accuracy of more than 90% and the good performance of the approach.

Keywords. 3D Shape Similarity Search, Quadratic Form Distance Functions, Spatial Data Mining, Nearest Neighbor Classification

1 Introduction

Along with clustering, mining association rules, characterization and generalization, classification is one of the fundamental tasks in data mining [CHY 96]. Given a set of classes and a query object, the problem is to assign an appropriate class to the query object based on its attribute values. Many modern database applications including molecular biology, astronomy, mechanical engineering, medical imaging, meteorology and others are faced with this problem. When new objects are discovered through remote sensing, new tumors are detected from X-ray images, or new molecular 3D structures are determined by crystallography or NMR techniques, an important question is to which class the new object belongs. Further steps to deeper investigations may be guided by the class information: a prediction of primary and secondary effects of drugs could be tried, the multitude of mechanical parts could be reduced, etc.

As a basis for any classification technique, an appropriate model has to be provided. Classes represent collections of objects that have characteristic properties in common and thus are similar, whereas different classes contain objects that have more or less strong dissimilarities. In all of the mentioned applications, the geometric shape of the

R.H. Güting, D. Papadias, F. Lochovsky (Eds.): SSD'99, LNCS 1651, pp. 207-226, 1999.

objects is an important similarity criterion. Along with the geometry, also thematic attributes such as physical and chemical properties have an influence on the similarity of objects.

Data from real world applications inherently suffer from errors, beginning with errors of measurement, calibration, sampling errors, numerical rounding errors, displacements of reference frames, and small shifts as well as rotations of the entire object or even of local details of the shapes. Though no full invariance against rotations is generally required, if the objects are already provided in a standardized orientation, these errors have to be taken into account. In this paper, we introduce a flexible similarity model that considers these problems of local inaccuracies and may be adapted by the users to their specific requirements or individual preferences.

The paper is organized as follows: The remainder of this introduction surveys related work from molecular biology, data mining, and similarity search in spatial databases. In Section 2, we introduce the components of our similarity model: 3D shape histograms for object representation, and a flexible similarity distance function. Due to the large and rapidly increasing size of current databases, the performance of query processing is an important task and, therefore, we introduce an efficient multistep system architecture in Section 3. In Section 4, we present the experimental results concerning the effectiveness and efficiency of our technique in the context of molecular biology. Section 5 concludes the paper.

1.1 Classification in Molecular Databases

A major issue in biomolecular databases is to get a survey of the objects, and thus a basic task is classification: To which of the recognized classes in the database does a new molecule belong? In molecular biology, there are already classification schemata available. In many systems, classifying new objects when inserting them into the database requires supervision by experts that are very experienced and have a deep knowledge of the domain of molecular biology. What is desired is an efficient classification algorithm that may act as a fast filter for further investigation and that may be restricted e.g. to geometric aspects.

A sophisticated classification is available from the FSSP database (Families of Structurally Similar Proteins), generated by the Dali system [HS 94] [HS 98]. The similarity of two proteins is based on their secondary structure, that is substructures of the molecules such as alpha helices or beta sheets. The evaluation of a pair of proteins is very expensive, and query processing for a single molecule against the entire database currently takes an overnight run on a workstation.

Another classification schema is provided by CATH [OMJ+ 97], a hierarchical classification of protein domain structures, which clusters proteins at four major levels, class (C), architecture (A), topology (T) and homologous superfamily (H). The class label is derived from secondary structure content and cannot be assigned for all protein structures automatically. The architecture label, which describes the gross orientation of secondary structures, independent of connectivities, is currently assigned manually. The assignments of structures to topology families and homologous superfamilies are made by sequence and structure comparisons.

1.2 Nearest-Neighbor Classification

A lot of research has been performed in the area of classification algorithms; surveys are presented in [WK 91] [MST 94] [Mit 97]. All the methods require that a training set of objects is given for which both the attribute values and the correct classes are known a-priori. Based on this knowledge of previously classified objects, a classifier predicts the unknown class of a new object. The quality of a classifier is typically measured by the classification accuracy, i.e. by the percentage of objects for which the class label is correctly predicted.

Many methods of classification generate a description for the members of each class, for example by using bounding boxes, and assign a class to an object if the object matches the description of the class. Nearest neighbor classifiers, on the other hand, refrain from discovering a possibly complex description of the classes. As their name indicates, they retrieve the nearest neighbor p of a query object q and return the class label of p in order to predict the class label of q. Obviously, the definition of an appropriate distance function is crucial for the effectiveness of nearest neighbor classification. In a more general form, called k-nearest neighbor classification, k nearest neighbors of the query object q are used to determine the class of q. Thus, the effectiveness depends on the number k as well as on the weighting of the k neighbors. Both, appropriate similarity models as well as efficient algorithms for similarity search are required for successful nearest neighbor classification.

1.3 Geometry-Based Similarity Search

Considerable work on shape similarity search in spatial database systems has been performed in recent years. As a common technique, the spatial objects are transformed into high-dimensional feature vectors, and similarity is measured in terms of vicinity in the feature space. The points in the feature space are managed by a multi-dimensional index. Many of the approaches only deal with two-dimensional objects such as digital images or polygonal data and do not support 3D shapes.

Let us first survey previous 2D approaches from the literature. In [GM 93], a shape is represented by an ordered set of surface points, and fixed-sized subsets of this representation are extracted as shape features. This approach supports invariance with respect to translation, rotation and scaling, and is able to deal with partially occluded objects. The technique of [BKK 97] applies the Fourier transform in order to encode sections of polygonal outlines of 2D objects; even partial similarity is supported. Both methods exploit a linearization of polygon boundaries and, therefore, are hard to extend to 3D objects. In [Jag 91], shapes are approximated by rectangular coverings. The rectangles of a single object are sorted by size, and the largest ones are used for the similarity retrieval. The method of [KSF+ 96] is based on mathematical morphology and uses the max morphological distance and max granulometric distance of shapes. It has been applied to 2D tumor shapes in medical image databases. A 2D technique that is related to our 3D shape histograms is the Section Coding technique [Ber 97] [BK 97] [BKK 97a]. For each polygon, the circumscribing circle is decomposed into a given number of sectors, and for each sector, the area of the polygon inside of this sector divided by the total area of the polygon is determined. Similarity is defined in terms of the Euclidean distance of the resulting feature vectors. The similarity model in [AKS 98] handles 2D shapes in pixel images and provides a solution for the problem of small displacements.

The QBIC (Querying By Image Content) system [FBF+ 94] [HSE+ 95] contains a component for 2D shape retrieval where shapes are given as sets of points. The method is based on algebraic moment invariants and is also applicable to 3D objects [TC 91]. As an important advantage, the invariance of the feature vectors with respect to rigid transformations (translations and rotations) is inherently given. However, the adjustability of the method to specific applications is restricted. From the available moment invariants, appropriate ones have to be selected, and their weighting factors may be modified. Whereas the moment invariants are abstract quantities, the shape histograms presented in this paper are more intuitive and may be graphically visualized, thus providing an impression of the suitability for specific applications.

The Geometric Hashing paradigm for model-based 3D object recognition was introduced by [LW 88]. The objects are represented by sets of points; from these points, non-collinear triplets are selected to represent different orientations of a single object. For each of these orientations, every point of an object is stored in a hash table that maps 3D points to objects and their orientations. The query processing heuristic requires a certain threshold provided by the user. This threshold has a substantial impact on the effectiveness of the technique and, thus, an appropriate choice is crucial. If the threshold is too high, no answer is reported; if the threshold is too low, however, there is no guarantee and, moreover, no feedback whether the best matching object with respect to the underlying similarity model is returned. In contrast to that, the k-nearest neighbor algorithm used in our approach ensures that the k most similar objects are returned. There are no objects in the database which are more similar than the retrieved ones.

The approximation-based similarity model presented in [KSS 97] and [KS 98] addresses the retrieval of similar 3D surface segments. These surface segments occur in the context of molecular docking prediction where they represent potential docking sites. Since the segments are designed to model local portions of 3D surfaces but not to model the entire contour of a 3D solid, this technique is not applicable for searching 3D solids having similar global shapes.

1.4 Invariance Properties of Similarity Models

All the mentioned similarity models incorporate invariance against translation of the objects, some of them also include invariance against scaling which is not necessarily desired in the context of molecular or CAD databases. With respect to invariance against rotations, two approaches can be observed. Some of the similarity models inherently support rotational invariance, e.g. by means of the Fourier transform [BKK 97] or the algebraic moment invariants [TC 91]. Most of the techniques, however, include a preprocessing step that rotates the objects to a normalized orientation, e.g. by a Principal Axis Transform. If rotations should be considered nevertheless, the objects may be rotated artificially by certain angles as suggested in [Ber 97]. For some applications, eventually, rotational invariance may be not required, e.g. if mechanical parts in a CAD database are already stored in a standardized orientation.

An important kind of invariance has not yet be considered in previous work, the robustness of similarity models against errors of measurement, calibration, sampling errors, errors of classification of object components, numerical rounding errors, and small displacements such as shifts or slight rotations of geometric details. In our model, these problems are addressed and may be controlled by the user by specifying and adapting a similarity matrix for histogram bins.

2 A 3D Shape Similarity Model

In this section, we introduce our 3D shape similarity model by defining the two major ingredients: First, the shape histograms as an intuitive and discrete representation of complex spatial objects. Second, an adaptable similarity distance function for the shape histograms that may take small shifts and rotations into account by using quadratic forms.

2.1 Shape Histograms

The definition of an appropriate distance function is crucial for the effectiveness of any nearest neighbor classifier. A common approach for similarity models is based on the paradigm of feature vectors. A *feature transform* maps a complex object onto a feature vector in a multidimensional space. The similarity of two objects is then defined as the vicinity of their feature vectors in the feature space.

We follow this approach by introducing 3D shape histograms as intuitive feature vectors. In general, histograms are based on a partitioning of the space in which the objects reside, i.e. a complete and disjoint decomposition into cells which correspond to the bins of the histograms. The space may be geometric (2D, 3D), thematic (e.g. physical or chemical properties), or temporal (modeling the behavior of objects).

We suggest three techniques for decomposing the space: A shell model, a sector model, and a spiderweb model as the combination of the former two (cf. Figure 1). In a preprocessing step, a 3D solid is moved to the origin. Thus the models are aligned to the center of mass of the solid.

4 shell bins   12 sector bins   48 combined bins

Figure 1. Shells and sectors as basic space decompositions for shape histograms. In each of the 2D examples, a single bin is marked

Shell Model. The 3D is decomposed into concentric shells around the center point. This representation is particularly independent from a rotation of the objects, i.e. any rotation of an object around the center point of the model results in the same histogram. The radii of the shells are determined from the extensions of the objects in the database. The outermost shell is left unbound in order to cover objects that exceed the size of the largest known object.

Sector Model. The 3D is decomposed into sectors that emerge from the center point of the model. This approach is closely related to the 2D section coding method [BKK 97a]. However, the definition and computation of 3D sector histograms is more sophisticated, and we define the sectors as follows: Distribute the desired number of points uniformly on the surface of a sphere. For this purpose, we use the vertices of regular polyhedrons and their recursive refinements. Once the points are distributed, the Voronoi diagram of the points immediately defines an appropriate decomposition of the space. Since the points are regularly distributed on the sphere, the Voronoi cells meet at the center point of the model. For the computation of sector-based shape histograms, we need not to materialize the complex Voronoi diagram but simply apply a nearest neighbor search in 3D since typical number of sectors are not very large.

Combined Model. The combined model represents more detailed information than pure shell models and pure sector models. A simple combination of two fine-grained 3D decompositions results in a high dimensionality. However, since the resolution of the space decomposition is a parameter in any case, the number of dimensions may easily be adapted to the particular application.

In Figure 2, we illustrate various shape histograms for the example protein, 1SER-B, which is depicted on the left of the figure. In the middle, the various space decompositions are indicated schematically, and on the right, the corresponding shape histograms are depicted. The top histogram is purely based on shell bins, and the bottom histogram is defined by 122 sector bins. The histograms in the middle follow the combined model, they are defined by 20 shell bins and 6 sector bins, and by 6 shell bins and 20 sector bins, respectively. In this example, all the different histograms have approximately the same dimension of around 120. Note that the histograms are not built from volume elements but from uniformly distributed surface points taken from the molecular surfaces.

2.2 Shortcomings of the Euclidean Distance

In order to quantify the dissimilarity of objects, an appropriate distance function of feature vectors has to be provided. An obvious solution is to employ the classic Euclidean distance function which is well-defined for feature spaces of arbitrary dimension. In a squared representation, the Euclidean distance of two N-dimensional vectors p and q is defined as:

$$d^2_{euclid}(p, q) = \sum_{i=1}^{N} (p_i - q_i)^2 = (p - q) \cdot (p - q)^T.$$

However, the Euclidean distance exhibits severe limitations with respect to similarity measurement. In particular, the individual components of the feature vectors which correspond to the dimensions of the feature space are assumed to be independent from each other, and no relationships of the components such as substitutability and compensability may be regarded. The following example demonstrates these shortcomings in more detail.

Let us consider the three objects a, b, and c from Figure 3. From a visual inspection, we assess the objects a and b to be more similar than a and c or b and c since the two characteristic peaks are located more close together in the objects a and b than in object c. However, the peaks of a and b do not overlap the same sectors and, therefore, are mapped to distinct histogram bins. The Euclidean distance neglects any relationship of

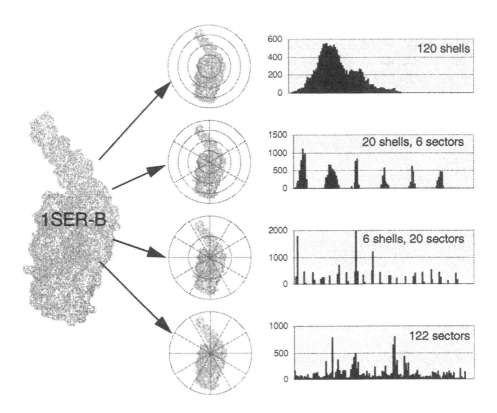

Figure 2. Several 3-D shape histograms of the example protein 1SER-B. From top to bottom, the number of shells decreases and the number of sectors increases

the vector components and does not reflect the close similarity of a and b in comparison to c. Thus, the three objects count for being equally similar, because their feature vectors have the same distance in pairs.

2.3 Quadratic Form Distance Functions

An approach to overcome these limitations has been investigated for color histograms in the QBIC project (Query by Image Content) at IBM Almaden [FBF+ 94] [HSE+ 95]. The authors suggest to use quadratic form distance functions which have also been successfully applied to several multimedia database applications [Sei 97] [SK 97] [KSS 97] [AKS 98] [KS 98]. A quadratic form distance function is defined in terms of a similarity matrix A as follows where the components a_{ij} of the matrix A represent the similarity of the components i and j in the underlying vector space.

$$d_A^2(x, y) = (x - y) \cdot A \cdot (x - y)^T = \sum_{i=1}^{N} \sum_{j=1}^{N} a_{ij}(x_i - y_i)(x_j - y_j).$$

From this definition, it becomes clear that the standard Euclidean distance is a special case of the quadratic form distance which is achieved by using the identity matrix Id

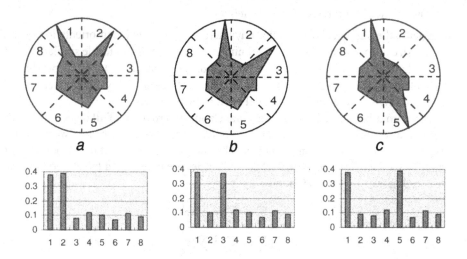

Figure 3. Shortcomings of the Euclidean distance. The Euclidean distance of the shape histograms does not reflect the similarity that is due to the proximity of neighboring sectors

as similarity matrix. Analogously, we obtain a weighted Euclidean distance function that has the weights $(w_1, w_2, ..., w_n)$ by using the diagonal matrix $\text{diag}(w_1, w_2, ..., w_n)$ as similarity matrix. In both cases, the non-diagonal components are set to zero which exactly corresponds to the fact that no cross-similarities of the dimensions are assumed.

The Euclidean distance of two vectors p and q is totally determined, there is no parameter which may be tuned. The weighted Euclidean distance is a little more flexible because it controls the effect of each vector component onto the overall distance by specifying individual weights for the dimensions. A new level of flexibility is supported by the general quadratic form distance function. On top of specifying the effect of individual dimensions onto the overall distance, cross-dependencies of the dimensions may be handled.

By using a quadratic form distance function as an adaptable similarity function, the problems of the Euclidean distance may be overcome. The neighborhood of bins in general and of shells or sectors in particular may be represented as similarity weights in the similarity matrix A. The individual similarity weights depend on the distances of the corresponding bins. Let us denote by $d(i,j)$ the distance of the cells that correspond to the bins i and j. For shells, we define the bin distance to be the difference of the corresponding shell radii, and for sectors, we use the angle between the sector centers as bin distances. When provided with an appropriate bin distance function, we compute the corresponding similarity weights by an adapted formula from [HSE+ 95] as follows:

$$a_{ij} = e^{-\sigma \cdot d(i,j)}.$$

The parameter σ controls the global shape of the similarity matrix. The higher σ, the more similar is the resulting matrix to the identity matrix. In any case, a high value of σ yields the matrix to be diagonally dominant. We observed good results for σ between 1.0 and 10.

2.4 Invariance Properties of the Models

In general, the 3D objects are located anywhere in the 3D, and their orientation as well as their size can vary arbitrarily. For defining meaningful and applicable similarity models, we have to provide invariance for translations, scaling and rotation, depending on the application. We can ensure these invariances in three ways, by a preprocessed normalization step, by the similarity model itself or by both steps.

In a normalization step, we perform translation and rotation of all objects. After the translation which maps the center of mass of each object onto the origin, we perform a Principal Axes Transform on the object. The computation for a set of 3D points starts with the 3×3 -covariance matrix where the entries are determined by an iteration over the coordinates (x, y, z) of all points:

$$\begin{bmatrix} \sum x^2 & \sum xy & \sum xz \\ \sum xy & \sum y^2 & \sum yz \\ \sum xz & \sum yz & \sum z^2 \end{bmatrix} .$$

The eigenvectors of this covariance matrix represent the principal axes of the original 3D point set, and the eigenvalues indicate the variance of the points in the respective direction. As a result of the Principal Axes Transform, all the covariances of the transformed coordinates vanish. Although this method in general leads to a unique orientation of the objects, this does not hold for the exceptional case of an object with at least two variances having the same value. In our experiments using the protein database, we almost never observed such cases and, therefore, assume a unique orientation of the objects.

The similarity models themselves have inherent invariance properties. Obviously, the sector model is invariant against scaling, whereas the shell model trivially has rotational invariance. Often, no full invariance is desired, instead just small displacement, shifts or rotations of geometric details occur in the data, for example caused by errors of measurement, sampling or numerical rounding errors. This variation of invariance precision which is highly application- and user-dependent is supported by the user-defined similarity matrix modeling the appropriate similarity weight for each pair of bins.

2.5 Extensibility of Histogram Models

What we have discussed so far is a very flexible and intuitive similarity model for 3D objects. However, the distance function of the similarity model is based just on the spatial attributes of the objects. Frequently, on top of the geometric information, a lot of thematic information is used to describe spatial objects. Particularly in protein databases, the chemical structure and physical properties are important. Examples include atom types, residue types, partial charge, hydrophobicity, electrostatic potential among others. A general approach to manage thematic information along with spatial properties is provided by combined histograms. Figure 4 demonstrates the basic principle. Assume we are given a spatial histogram structure as presented above, and additionally a thematic histogram structure to be given. A combined histogram structure is immediately obtained as the Cartesian product of the original structures.

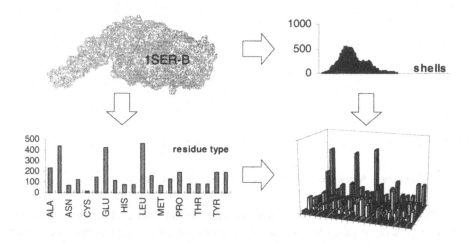

Figure 4. Example for a combined thematic and shape histogram for a molecule

Obviously, this product based approach leads to a tradeoff between a more powerful modeling versus a very high dimensionality. An investigation of the efficiency and effectiveness as well as the development of new techniques that meet the requirements of ultra high dimensional spaces is part of our future research plans.

3 Efficient Query Processing

Due to the enormous and still increasing size of modern databases that contain tens and hundreds of thousands of molecules, mechanical parts, or medical images, the task of efficient query processing becomes more and more important. In the case of quadratic form distance functions, the evaluation time of a single database object increases quadratically with the dimension. We measured 0.23 milliseconds in the average for 21D histograms, 6.2 milliseconds for 256D and 1,656 milliseconds in 4,096D space (cf. Figure 5). Thus, linearly scanning the overall database is prohibitive. In order to achieve a good performance, our system architecture follows the paradigm of multistep query processing: An index-based filter step produces a set of candidates, and a subsequent refinement step performs the expensive exact evaluation of the candidates [Sei 97] [AKS 98].

3.1 Optimal Multistep k-Nearest Neighbor Search

Whereas the refinement step in a multistep query processor has to ensure the correctness, i.e. no false hits may be reported as final answers, the filter step is primarily responsible for the completeness, i.e. no actual result may be missing from the final answers and, therefore, from the set of candidates. Figure 6 illustrates the architecture of our multistep similarity query processor that fulfills this property [SK 98]. Moreover, as an advantage over the related method of [KSF+ 96], our algorithm is proven to be optimal, i.e. it produces only the minimum number of candidates. Thus, expensive evaluations of unnecessary candidates are avoided, and we observed improvement factors of up to 120 for the number of candidates and 48 for the overall runtime.

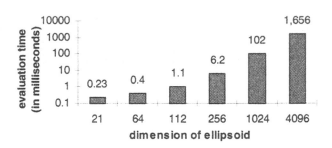

Figure 5. Average evaluation time for single ellipsoid distances

Based on a multidimensional index structure, the filter step performs an incremental ranking that reports the objects ordered by their increasing filter distance to the query object using an algorithm derived from [HS 95]. The number of accessed index pages is minimum as proven in [BBKK 97], and the termination is controlled by the refinement step in order to guarantee the minimum number of candidates [SK 98]. Only for the exact evaluation in the refinement step, the exact object representation is retrieved from the object server.

In order to guarantee no false dismissals caused by the filter step, the filter distance function d_f has to be a lower bound of the exact object distance function d_o that is evaluated in the refinement step. That is, for all database objects p and all query objects q, the following inequality has to hold:

$$d_f(p, q) \le d_o(p, q).$$

3.2 Reduction of Dimensionality for Quadratic Forms

A common approach to manage objects in high-dimensional spaces is to apply techniques to reduce the dimensionality. The objects in the reduced space are then typically managed by any multidimensional index structure [GG 98]. The typical use of common linear reduction techniques such as the Principal Components Analysis (PCA) or Karhunen-Loève Transform (KLT), the Discrete Fourier or Cosine Transform (DFT, DCT), the Similarity Matrix Decomposition [HSE+ 95] or the Feature Subselection [FBF+ 94] includes a clipping of the high-dimensional vectors such that the Euclidean distance in the reduced space is always a lower bound of the Euclidean distance in the high-dimensional space.

The question arises whether these approved techniques are applicable to general quadratic form distance functions. Fortunately, the answer is positive; an algorithm to reduce the similarity matrix from a high-dimensional space down to a low-dimensional space according to a given reduction technique was developed in the context of multimedia databases for color histograms [SK 97] and shapes in 2D images [AKS 98]. The method guarantees three important properties: First, the reduced distance function is a lower bound of the given high-dimensional distance function. Obviously, this criterion had to be a design goal in order to meet the requirements of multistep similarity query processing. Second, the reduced distance function again is a quadratic form and, there-

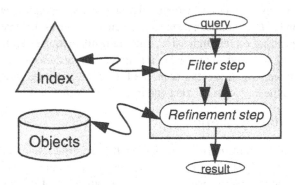

Figure 6. Multistep similarity query processing

fore, the complexity of the query model is not increased while decreasing the dimension of the space. Third, the reduced distance function is the greatest of all lower-bounding distance functions in the reduced space. As an important implication of this property, the selectivity in the filter step is optimal: In the reduced space, no lower-bounding distance function is able to produce a smaller set of candidates than the resulting quadratic form.

3.3 Ellipsoid Queries on Multidimensional Index Structures

The task remains to efficiently support k-nearest neighbor search and incremental ranking for quadratic form distance functions in low-dimensional spaces. Due to the geometric shape of the query range, a quadratic form-based similarity query is called an *ellipsoid query* [Sei 97]. An efficient algorithm for ellipsoid query processing on multidimensional index structures was developed in the context of approximation-based similarity search for 3-D surface segments [KSS 97] [KS 98]. The method is designed for index structures that use a hierarchical directory based on rectilinear bounding boxes such as the R-tree [Gut 84], R+-tree [SRF 87], R*-tree [BKSS 90], X-tree [BKK 96] [BBB+ 97], and Quadtrees among others; surveys are provided in [Sam 90] [GG 98]. The technique is based on measuring the minimum quadratic form distance of a query point to the hyperrectangles in the directory. Recently, an improvement by using conservative approximations has been suggested [ABKS 98].

An important property of the method is its flexibility with respect to the similarity matrix. The matrix does not have to be available at index creation time and, therefore, may be considered as a query parameter. Thus, the users may specify and adapt the similarity weights in the matrix even at query time according to their individual preferences or to the specific requirements of the application. In any case, the same precomputed index may be used. This property is the major advantage compared to previous solutions that were developed in the context of color histogram indexing in the QBIC project [FBF+ 94] [HSE+ 95] where the index depends on a specific similarity matrix that has to be given in advance.

The cost model of [BBKK 97] provides a theoretical analysis of the performance deterioration for multidimensional index structures with increasing dimensionality. An investigation in [WSB 98] results in the recommendation to use an accelerated sequential scan, and the VA-File was developed following this paradigm. However, the analyses are based on the L_2 (Euclidean distance), L_1, and L_∞ norms that may be evaluated in linear time depending on the dimension, and the results require careful reviewing and experimental evaluation when applied to quadratic form distance functions. Even if the index is substituted by a sequential scan, the filter-refinement architecture will still be necessary due to the high cost of exact quadratic form evaluations.

4 Experimental Evaluation

We implemented the algorithms in C++ and ran the experiments on our HP C160 workstations under HP-UX 10.20. For single queries, we also implemented a HTML/Java interface that supports query specification and visualization of the results. The atomic coordinates of the 3D protein structures are taken from the Brookhaven Protein Data Bank (PDB) [BKW+ 77]. For the computation of shape histograms, we use a representation of the molecules by surface points as it is required for several interesting problems such as the molecular docking prediction [SK 95]. The reduced feature vectors for the filter step are managed by an X-tree [BKK 96] of dimension 10.

The similarity matrices are computed by an adapted formula from [HSE+ 95] where the similarity weights a_{ij} of bin i and j are defined as $a_{ij} = e^{-\sigma \cdot d(i,j)}$. The distance $d(i,j)$ is equal to the difference of the corresponding shell radii in the shell model and is given by the angle between the sector axes in the sector model. In the combined model, the shell distance $d_{shell}(i,j)$ and the sector distance $d_{sector}(i,j)$ of the bins i and j are composed by using the Euclidean distance formula $d_{comb}(i,j) = \sqrt{d_{shell}^2(i,j) + d_{sector}^2(i,j)}$. We experimented with several values of the parameter σ but did not observe significant changes in the accuracy, so we set the parameter σ equal to 10 for the following evaluations [Kas 98].

4.1 Basic Similarity Search

In order to illustrate the applicability of the similarity model, we demonstrate the retrieval of the members of a known family. As a typical example, we chose the seven Seryl-tRNA Synthetase molecules from our database that are classified by CATH [OMJ+ 97] to the same family. The diagram in Figure 7 presents the result using shape histograms for 6 shells and 20 sectors. The seven members of the Seryl family rank on the top seven positions among the 5,000 molecules of the database. In particular, the similarity distance noticeable increases for 2PFK-A, the first non-Seryl protein in the ranking order.

4.2 Classification by Shape Similarity

For the classification experiments, we restricted our database to the proteins that are also contained in the FSSP database [HS 94] and took care that for every class, at least two molecules are available. From this preprocessing, we obtained 3,422 proteins assigned to 281 classes. The classes contain between 2 and 185 molecules. In order to measure the classification accuracy, we performed *leave-one-out* experiments for various histogram models. For each molecule in the database, the nearest neighbor classification was determined after removing that element from the database. Technically, we always used the same database and selected the second nearest neighbor since the query object itself is reported to be its own nearest neighbor.

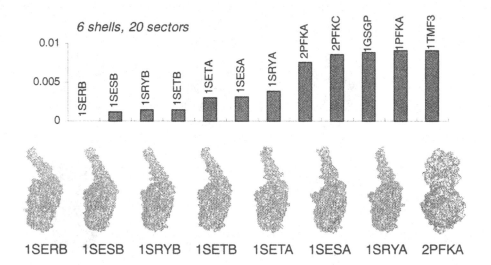

Figure 7. Similarity ranking for the Seryl-tRNA Synthetases 1SER-B. The diagram depicts the similarity distances of the 12 nearest neighbors to the query protein 1SER-B in ascending order. The illustration of the top eight molecules demonstrates the close similarity within the family, and the dissimilarity to the first non-Seryl protein in the ranking

Figure 8 demonstrates the results for histograms based on 12 shells, 20 sectors, and the combination of them. Obviously, the more fine-grained spiderweb model yields the best classification accuracy of 91.5 percent (top diagram), but even for the coarse sector histograms, a noticeable accuracy of 87.3 percent is achieved. These results compete with the accuracy of available protein classification systems such as CATH [OMJ 97] where also more than 90% of the class labels are predicted correctly. Whereas in CATH only four different class labels are used for the automatic classification, our experiments are based on a variety of 281 class labels.

The average overall runtime for a single query reflects the larger dimension of the combined model. It ranges from 0.05s for 12 shells over 0.2s for 20 sectors up to 1.42s for the combination. This runtime performance in the range of tens to thousands of milliseconds is a progress compared to established biomolecular systems for which query response times in the range of minutes and hours are reported [HS 98].

Figure 9 illustrates the effect of simply increasing the dimension of the model without combining orthogonal space partitionings. Again we observed the expected result that more information yields better accuracy. When increasing the histogram dimension by a factor of 10, the accuracy increases from 71.6 to 88.1 for the shell model, and from 87.3 to 91.6 for the sector model. For the task of classification, the increased granularity results in a better separation of the class members from the other objects. Obviously, the tradeoff for this gain is a larger space requirement and the increase of the runtime due to the high dimensionality. We plan to develop a cost model for obtaining the optimal number of bins in order to produce both accurate and fast results.

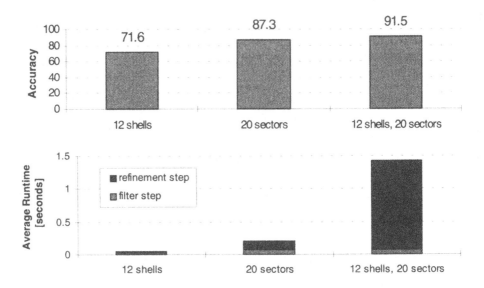

Figure 8. Classification accuracy (*top diagram*) and average runtime of query processing (*bottom diagram*) for histograms with 12 shells, 20 sectors, and their combination

In these experiments, we achieve the same accuracy for a fine-grained 122D sector model as we obtained from the 12 x 20 (240D) combined model. One may wonder why the combined model does not lead to the best accuracy. Although all proposed models yield good results in terms of accuracy and runtime, the sector model turns out to be most suitable for the tested data. One reason for this data-dependent result is that the decomposition of the 3D objects is computed for uniform sectors or equidistant shells. To reveal the properties of the space decompositions, we computed the standard deviation for each bin over all histograms. We present the observations by bar diagrams where the height of a bar represents the value of the standard deviation for the corresponding dimension.

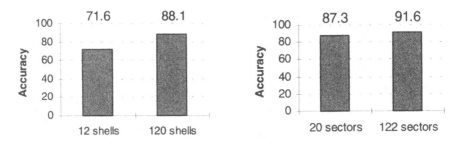

Figure 9. The accuracy increases with increasing granularity of the space partitioning for both, shell and sector histograms

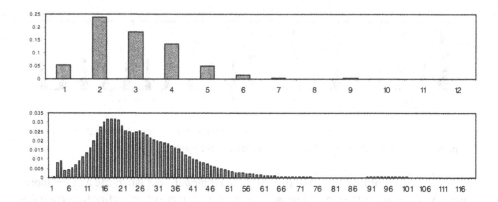

Figure 10. Standard deviations of the bins for the 12D (*top*) and 120D (*bottom*) shell models

Figure 10 demonstrates that for the shell model, the values of the standard deviations are distributed very unbalanced. For the 12D model, the highest standard deviation occurs in the shells 2 to 4 that contain the large majority of the surface points, and a low deviation is observed for the shells 7 to 12. For the 120D model, significant deviations occur only for the shells 3 to 60; there is only a low variance in the number of points for the other shells which are populated very sparse. Therefore, the corresponding histogram bins do not contribute to distinguish between different molecules but just increase the dimension and, as a consequence, the runtime becomes worse.

Figure 11 depicts the standard deviations for the two sector models, 20D and 122D. For every histogram bin, the standard deviation is high, and, therefore, all dimensions contribute to the distinction of different molecules. For the combined model, the standard deviations of the 240 bins are illustrated in Figure 12. These 240 bins result from the decomposition of the 3D space into 12 shells and 20 sectors. Therefore, the periodic pattern in the standard deviation reflects the previous observations for the shell and the sector models.

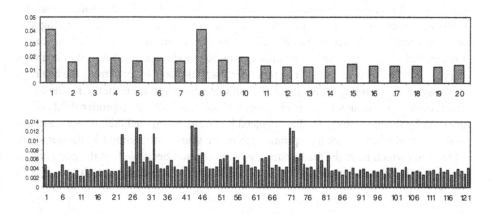

Figure 11. Standard deviations of the bins for the 20D (*top*) and 122D (*bottom*) sector models

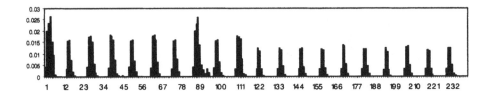

Figure 12. Standard deviations of the combined histogram bins. For each of the 20 sectors, the characteristic shape of the 12-shell histogram can be recognized

As a way to improve the properties of the shell model, we plan to use a more appropriate partitioning of the space. Instead of using equidistant radii to define the decomposition, the shell radii could be based on quantiles that are obtained from the distribution of the surface points in the space. As a consequence, this approach will also improve the effectiveness of the combined model.

5 Conclusions

In this paper, we presented a new intuitive and flexible model for shape similarity search of 3D solids. As a specific feature transform, 3D shapes are represented by using shape histograms for which several partitionings of the space are possible. This histogram model naturally is extensible to thematic attributes such as physical and chemical properties. In order to account for errors of measurement, sampling, numerical rounding etc., quadratic form distance functions are used that are able to take small displacements and rotations into account. For efficient query processing, a filter-refinement architecture is used that supports similarity query processing based on high-dimensional feature vectors and quadratic form distance functions. The experiments demonstrate both, the high classification accuracy of our similarity model, and the good performance of the underlying query processor.

The improvement of the space decomposition by using a quantile based method, the development of a cost model for determining the optimal number of bins, and the investigation of thematically extended histogram models are plans for our future work already mentioned so far. In addition, we will include a visualization of shape histograms as a Java applet in order to provide an explanation component for the classification system. This is an important issue since any notion of similarity is subjective in a high degree, and the users want to have as much feedback as possible concerning the behavior of the system depending on their queries and input parameters. Furthermore, the confidence of the users in an automatic classification increases with the reproducability of the decision by the user which can be enhanced by visualization methods. A more conceptual future work addresses the optimization of the space partitioning and the geometry of the cells which form the histogram bins. Both the number as well as the geometry of the cells affect the effectiveness and also the efficiency of similarity search and classification.

References

[ABKS 98] Ankerst M., Braunmüller B., Kriegel H.-P., Seidl T.: *Improving Adaptable Similarity Query Processing by Using Approximations*. Proc. 24th Int. Conf. on Very Large Databases (VLDB'98), New York, USA. Morgan Kaufmann (1998) 206-217

[AKS 98]   Ankerst M., Kriegel H.-P., Seidl T.: *A Multi-Step Approach for Shape Similarity Search in Image Databases*. IEEE Transactions on Knowledge and Data Engineering, Vol. 10, No. 6 (1998) 996-1004

[BBB+ 97] Berchtold S., Böhm C., Braunmüller B., Keim D., Kriegel H.-P.: *Fast Parallel Similarity Search in Multimedia Databases*. Proc. ACM SIGMOD Int. Conf. on Management of Data, Tucson, AZ. ACM Press (1997) 1-12, Best Paper Award

[BBKK 97]Berchtold S., Böhm C., Keim D., Kriegel H.-P.: *A Cost Model for Nearest Neighbor Search in High-Dimensional Data Spaces*. Proc. 16th ACM SIGACT-SIGMOD-SIGART Symp. on Principles of Database Systems (PODS), Tucson, AZ (1997) 78-86

[Ber 97]   Berchtold S.: *Geometry Based Search of Similar Mechanical Parts*. Ph.D. Thesis, Institute for Computer Science, University of Munich. Shaker Verlag, Aachen (1997) in German

[BKW+ 77]Bernstein F. C., Koetzle T. F., Williams G. J., Meyer E. F., Brice M. D., Rodgers J. R., Kennard O., Shimanovichi T., Tasumi M.: *The Protein Data Bank: a Computer-based Archival File for Macromolecular Structures*. Journal of Molecular Biology, Vol. 112 (1977) 535-542

[BKK 96] Berchtold S., Keim D., Kriegel H.-P.: *The X-tree: An Index Structure for High-Dimensional Data*. Proc. 22nd Int. Conf. on Very Large Data Bases (VLDB'96), Mumbai, India. Morgan Kaufmann (1996) 28-39

[BK 97]   Berchtold S., Kriegel H.-P.: *S3: Similarity Search in CAD Database Systems*. Proc. ACM SIGMOD Int. Conf. on Management of Data. ACM Press (1997) 564-567

[BKK 97]   Berchtold S., Keim D. A., Kriegel H.-P.: *Using Extended Feature Objects for Partial Similarity Retrieval*. VLDB Journal, Vol. 6, No. 4. Springer Verlag, Berlin Heidelberg New York (1997) 333-348

[BKK 97a] Berchtold S., Keim D.A., Kriegel H.-P.: *Section Coding: A Method for Similarity Search in CAD Databases*. Proc. German Conf. on Databases for Office Automation, Technology, and Science (BTW). Series Informatik Aktuell. Springer Verlag, Berlin Heidelberg New York (1997) 152-171; in German

[BKSS 90] Beckmann N., Kriegel H.-P., Schneider R., Seeger B.: *The R*-tree: An Efficient and Robust Access Method for Points and Rectangles*. Proc. ACM SIGMOD Int. Conf. on Management of Data, Atlantic City, NJ. ACM Press (1990) 322-331

[CHY 96]   Chen M.-S., Han J. and Yu P. S.: *Data Mining: An Overview from a Database Perspective*. IEEE Transactions on Knowledge and Data Engineering, Vol. 8, No. 6 (1996) 866-883

[FBF+ 94] Faloutsos C., Barber R., Flickner M., Hafner J., Niblack W., Petkovic D., Equitz W.: *Efficient and Effective Querying by Image Content*. Journal of Intelligent Information Systems, Vol. 3 (1994) 231-262

[GG 98]   Gaede V., Günther O.: *Multidimensional Access Methods*. ACM Computing Surveys, Vol. 30, No. 2 (1998) 170-231

[GM 93]   Gary J. E., Mehrotra R.: *Similar Shape Retrieval Using a Structural Feature Index*. Information Systems, Vol. 18, No. 7 (1993) 525-537

[Gut 84]   Guttman A.: *R-trees: A Dynamic Index Structure for Spatial Searching*. Proc. ACM SIGMOD Int. Conf. on Management of Data, Boston, MA. ACM Press (1984) 47-57

[HS 94]   Holm L., Sander C.: *The FSSP Database of Structurally Aligned Protein Fold Families*. Nucleic Acids Research, Vol. 22 (1994) 3600-3609

[HS 95]   Hjaltason G. R., Samet H.: *Ranking in Spatial Databases.* Proc. 4th Int. Symposium on Large Spatial Databases (SSD'95). Lecture Notes in Computer Science, Vol. 951. Springer Verlag, Berlin Heidelberg New York (1995) 83-95

[HS 98]   Holm L., Sander C.: *Touring Protein Fold Space with Dali/FSSP.* Nucleic Acids Research, Vol. 26 (1998) 316-319

[HSE+ 95] Hafner J., Sawhney H. S., Equitz W., Flickner M., Niblack W.: *Efficient Color Histogram Indexing for Quadratic Form Distance Functions.* IEEE Trans. on Pattern Analysis and Machine Intelligence, Vol. 17, No. 7. IEEE Press (1995) 729-736

[Jag 91]   Jagadish H. V.: *A Retrieval Technique for Similar Shapes.* Proc. ACM SIGMOD Int. Conf. on Management of Data. ACM Press (1991) 208-217

[Kas 98]   Kastenmüller G.: *Shape-oriented Similarity Search in 3D Protein Database Systems.* Diploma Thesis, Institute for Computer Science, University of Munich (1998) in German

[KS 98]   Kriegel H.-P., Seidl T.: *Approximation-Based Similarity Search for 3-D Surface Segments.* GeoInformatica Journal, Vol. 2, No. 2. Kluwer Academic Publishers (1998) 113-147

[KSF+ 96] Korn F., Sidiropoulos N., Faloutsos C., Siegel E., Protopapas Z.: *Fast Nearest Neighbor Search in Medical Image Databases.* Proc. 22nd VLDB Conference, Mumbai, India. Morgan Kaufmann (1996) 215-226

[KSS 97]   Kriegel H.-P., Schmidt T., Seidl T.: *3D Similarity Search by Shape Approximation.* Proc. Fifth Int. Symposium on Large Spatial Databases (SSD'97), Berlin, Germany. Lecture Notes in Computer Science, Vol. 1262. Springer Verlag, Berlin Heidelberg New York (1997) 11-28

[LW 88]   Lamdan Y., Wolfson H.J.: *Geometric Hashing: A General and Efficient Model-Based Recognition Scheme.* Proc. IEEE Int. Conf. on Computer Vision, Tampa, Florida, 1988 238-249

[Mit 97]   Mitchell T.M.: *Machine Learning.* McCraw-Hill, (1997)

[MST 94]   Michie D., Spiegelhalter D.J., Taylor C.C.: *Machine Learning, Neural and Statistical Classification.* Ellis Horwood (1994)

[OMJ+ 97] Orengo C.A., Michie A.D., Jones S., Jones D.T. Swindells M.B., Thornton, J.M.: *CATH – A Hierarchic Classification of Protein Domain Structures.* Structure, Vol. 5, No. 8 (1997) 1093-1108

[Sam 90]   Samet H.: *The Design and Analysis of Spatial Data Structures.* Addison Wesley (1990)

[Sei 97]   Seidl T.: *Adaptable Similarity Search in 3-D Spatial Database Systems.* Ph.D. Thesis, Institute for Computer Science, University of Munich (1997). Herbert Utz Verlag, Munich, http://utzverlag.com, ISBN: 3-89675-327-4

[SK 95]   Seidl T., Kriegel H.-P.: *A 3D Molecular Surface Representation Supporting Neighborhood Queries.* Proc. 4th Int. Symposium on Large Spatial Databases (SSD'95), Portland, Maine, USA. Lecture Notes in Computer Science, Vol. 951. Springer Verlag, Berlin Heidelberg New York (1995) 240-258

[SK 97]   Seidl T., Kriegel H.-P.: *Efficient User-Adaptable Similarity Search in Large Multimedia Databases.* Proc. 23rd Int. Conf. on Very Large Databases (VLDB'97), Athens, Greece. Morgan Kaufmann (1997) 506-515

[SK 98]   Seidl T., Kriegel H.-P.: *Optimal Multi-Step k-Nearest Neighbor Search.* Proc. ACM SIGMOD Int. Conf. on Management of Data, Seattle, Washington (1998) 154-165

[SRF 87]   Sellis T., Roussopoulos N., Faloutsos C.: *The R+-Tree: A Dynamic Index for Multi-Dimensional Objects.* Proc. 13th Int. Conf. on Very Large Databases, Brighton, England (1987) 507-518

[TC 91]   Taubin G., Cooper D. B.: *Recognition and Positioning of Rigid Objects Using Algebraic Moment Invariants.* in *Geometric Methods in Computer Vision,* Vol. 1570, SPIE (1991) 175-186

[WK 91]   Weiss S.M., Kulikowski C.A.: *Computer Systems that Learn: Classification and Prediction Methods from Statistics, Neural Nets, Machine Learning, and Expert Systems.* Morgan Kaufmann, San Francisco (1991)

[WSB 98]   Weber R., Schek H.-J., Blott S.: *A Quantitative Analysis and Performance Study for Similarity-Search Methods in High-Dimensional Spaces.* Proc. 24th Int. Conf. on Very Large Databases (VLDB`98), New York, USA. Morgan Kaufmann (1998) 194-205

Spatial Join

Multi-way Spatial Joins Using R-Trees: Methodology and Performance Evaluation

Ho-Hyun Park[1], Guang-Ho Cha[2]*, and Chin-Wan Chung[1]

[1] Department of Computer Science, KAIST, Taejon 305-701, Korea
{hhpark,chungcw}@islab.kaist.ac.kr
[2] IBM Almaden Research Center, San Hose, CA 95120, USA
ghcha@almaden.ibm.com

Abstract. We propose a new multi-way spatial join algorithm called *M-way R-tree join* which synchronously traverses M R-trees. The M-way R-tree join can be considered as a generalization of the *2-way R-tree join*. Although a generalization of the 2-way R-tree join has recently been studied, it did not properly take into account the optimization techniques of the original algorithm. Here, we extend these optimization techniques for M-way joins. Since the join ordering was considered to be important in the M-way join literature (e.g., relational join), we especially consider the ordering of the search space restriction and the plane sweep. Additionally, we introduce *indirect predicates* in the M-way join and propose a further optimization technique to improve the performance of the M-way R-tree join. Through experiments using real data, we show that our optimization techniques significantly improve the performance of the M-way spatial join.

1 Introduction

The spatial join is a common spatial query type which requires a high processing cost due to the high complexity and large volume of spatial data. Therefore, the spatial join is processed in two steps (the *filter step* and the *refinement step*) to reduce the overall processing cost [14,5]. Many 2-way spatial join methods have been published in the literature: the join using Z-order elements [14], the join using R-trees (called *R-tree join*) [3], the seeded tree join (STJ) [10], the spatial hash join (SHJ) [11], the partition based spatial merge join (PBSM) [20], the size separation spatial join (S^3J) [9], the scalable sweeping-based spatial join (SSSJ) [1] and the slot index spatial join (SISJ) [12]. However, there has been little research on the multi-way spatial join [16]. The M-way (M>2) spatial join combines M spatial relations using M-1 or more spatial predicates[1]. An example of a 3-way spatial join is "Find all buildings which are adjacent to roads that

---

* The work reported here was performed while Guang-Ho Cha was at Tongmyong University of Information Technology, Korea
[1] If the number of spatial predicates is less than M-1, the join necessarily includes cartesian products, in which case we regard the join not as one spatial join but as several spatial joins.

R.H. Güting, D. Papadias, F. Lochovsky (Eds.): SSD'99, LNCS 1651, pp. 229–250, 1999.
© Springer-Verlag Berlin Heidelberg 1999

intersect with boundaries of districts." An M-way spatial join can be modeled by a *query graph* whose nodes represent relations and edges represent spatial predicates.

One way to process M-way spatial joins is as a sequence of 2-way joins [12]. Another possible way, when all join attributes have spatial indexes and each join attribute is shared among the associated join predicates[2], is to combine the filter and refinement steps respectively as follows:

(1) Scan the relevant indexes synchronously for all join attributes to obtain a set of spatial object identifier tuples.
(2) Read objects for object identifier (oid) tuples obtained from Step (1), and perform an M-way spatial join using geometric computation algorithms.

Step (1) is called *combined filtering* and Step (2) *combined refinement* in [17]. Especially when the R-trees are used in Step (1), the combined filtering is called *M-way R-tree join* which is the scope of this paper. The M-way R-tree join is also called *synchronous traversal* (ST) in [16]. An advantage of the combined filtering is that it removes unnecessary refinement operations for some object pairs. For example, let Figure 1 be an MBR (Minimum Bounding Rectangle) combination of spatial objects for the above query. Let a, b and c be instances of the relations *buildings*, *roads* and *boundaries*, respectively. If it is processed by a sequence of 2-way joins and the evaluation order is determined to be $\langle a, b, c \rangle$ by a query optimizer, the refinement operation between a and b will be performed unnecessarily. However, the combined filtering can avoid this situation.

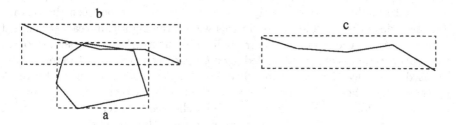

Fig. 1. An MBR combination in a 3-way join

The M-way R-tree join can be considered as a generalization of the *2-way R-tree join* of [3,7] and does not create intermediate results. Although a generalization of the 2-way R-tree join called *multi-level forward checking* (MFC) has recently been studied [15,16], it did not properly take into account the optimization techniques of the original 2-way R-tree join.

The main contributions of this paper are as follows: First, we generalize the 2-way R-tree join to consider the order of search space restrictions and plane sweeps because the join ordering was considered to be important in the M-way join

---

[2] In this case, only one spatial predicate per relation participates in the join.

literature (e.g., relational join) [8]. Second, we introduce *indirect predicates* in the M-way spatial join and propose a further optimization technique to improve the performance of the M-way R-tree join. Through experiments, we show that our optimization techniques significantly improve the performance of the M-way spatial join (especially the filter step) against MFC. Additionally, we find that the M-way R-tree join becomes CPU-bound as M increases.

The remainder of this paper is organized as follows: Section 2 provides some background by briefly explaining the 2-way R-tree join and the state-of-the-art M-way spatial joins using R-trees. In Section 3, we propose an algorithm of the M-way R-tree join, which considers the ordering of search space restrictions and plane sweeps, as a new generalization of the 2-way R-tree join, and further improve the performance of the M-way R-tree join using the concept of indirect predicates. In Section 4, we present some experiments for the performance analysis of our algorithms using the TIGER data [22]. Finally in Section 5, we conclude this paper and suggest some future studies.

2 Background

2.1 2-Way Spatial Joins Using R-Trees

Assuming that R-trees [4,2] exist for both join inputs, a join algorithm which synchronously traverses both R-trees using depth-first search was proposed [3]. The basic idea of the algorithm is as follows: First, it reads the root nodes of the R-trees and checks if the rectangles of entries of both nodes mutually intersect. Next, only for intersected entry pairs, it traverses the child node pairs by depth-first search and continuously checks the intersection between the rectangles of entries of both child nodes. In this way, if the algorithm reaches the leaf nodes, it outputs the intersected entry pairs and backtracks to the parent nodes. Two optimization techniques, called *search space restriction* and *plane sweep*, are used to reduce the CPU time. The search space restriction heuristic picks out the entries whose rectangles do not intersect with the rectangle enclosing the other node, before the intersection is actually checked between the rectangles of entries of both nodes. The plane sweep first sorts the rectangles of entries of both nodes for one axis, and then goes forward along the sweep line and checks the intersection for the other axis. The algorithm using the above techniques is shown below: (We skip the detailed algorithm for *SortedIntersectionTest* in Step (6) due to space limitation. Refer to [3] for details.)

```
RtreeJoin (Rtree_Node R, S)

(1)   FOR all E_i ∈ R DO
(2)       IF E_i.rect ∩ S.rect == ∅ THEN R = R-{E_i}; /* space restriction
          on R */
(3)   FOR all E_j ∈ S DO
(4)       IF E_j.rect ∩ R.rect == ∅ THEN S = S-{E_j}; /* space restriction
          on S */
(5)   Sort(R); Sort(S);
```

```
(6)     SortedIntersectionTest (R, S, Seq); /* plane sweep */
(7)     FOR i = 1 TO ‖Seq‖ DO
(8)         (E_R, E_S) = Seq[i];
(9)         IF R is a leaf page THEN /* S is also a leaf page */
(10)            output (E_R, E_S);
(11)        ELSE
(12)            ReadPage(E_R.ref); ReadPage(E_S.ref);
(13)            RtreeJoin (E_R.ref, E_S.ref);

END RtreeJoin
```

Additionally, the algorithm applied the *page pinning* technique for I/O optimization. The algorithm used only a local optimization policy to fetch the child node pairs. Later, a global optimization algorithm by breadth-first search was proposed [7]. In this paper, we call both of the join algorithms *2-way R-tree join* or simply *R-tree join*. When R-trees exist for both join inputs, it has been shown that the R-tree join is most efficient [10,20,12].

2.2   State-of-the-Art M-Way Spatial Joins Using R-Trees

In a recent study, two methods called *multi-level forward checking* (MFC) and *window reduction* (WR) were proposed to process structural queries for image similarity retrieval [15]. Later, they were applied to the multi-way spatial join [16]. MFC and WR were motivated by a close correspondence between multi-way spatial joins and constraint satisfaction problems (CSPs). A multi-way spatial join can be represented in terms of a binary CSP [16]:

- A set of n variables, $v_1, v_2, \ldots, v_n$, each corresponding to a dataset.
- For each variable v_i, a domain D_i which consists of the data in tree R.
- For each pair of variables (v_i, v_j), a binary constraint Q_{ij} corresponding to a spatial predicate.

If $Q_{ij}(E_{i,x}, E_{j,y}) = \text{TRUE}$, then the assignment $\{v_i = E_{i,x}, v_j = E_{j,y}\}$ is *consistent*. A solution is an n-tuple $\tau = \langle E_{1,w}, \ldots, E_{i,x}, \ldots, E_{j,y}, \ldots, E_{n,z} \rangle$ such that $\forall i, j, \{v_i = E_{i,x}, v_j = E_{j,y}\}$ is consistent. In the sequel, we use the terms variable/dataset/relation and constraint/predicate/join condition interchangeably.

1) Multi-level Forward Checking MFC is a kind of ST algorithms which synchronously traverses n R-trees as follows: It starts from the root nodes of n R-trees and checks all predicates for each n-combination (called *entry-tuple*) from the entries of the nodes. If an entry-tuple satisfies all the predicates, one of the following occurrs: If the *node-tuple* (an n-combination of the R-tree nodes) is in the intermediate level, the algorithm is recursively called for the child node-tuple pointed by the entry-tuple. Otherwise, i.e., if the node-tuple consists of leaf nodes, the algorithm outputs the entry-tuple and processes the next entry-tuple. If an entry-tuple does not satisfy at least one predicate, the entry-tuple is pruned. MFC was considered as a generalization of the 2-way R-tree join.

At each R-tree level, MFC applies *forward checking* (FC), which is known to be one of the most effective algorithms for solving CSP, to find the entry-tuples satisfying the predicates. FC maintains an $n * n * C$ *domain table* (n: number of variables, C: the maximum number of entries of an R-tree node) in main memory. $domain[i][j](0 \leq i, j < n)$ is a subset of an R-tree node N_j. FC works as follows [15]: First, $domain[0][j]$ is initialized to an R-tree node N_j for all j. When a variable v_0 is assigned a value u_k, $domain[1][j]$ is computed for each remaining v_j, by including only values $u_l \in domain[0][j]$ such that $Q_{0j}(u_k, u_l) =$ TRUE. In general, if u_k is the current value of v_i, $domain[i+1][j]$ is the subset of $domain[i][j]$ which is valid w.r.t. Q_{ij} and u_k. In this way, at each instantiation the domain of each *future variable* (un-instantiated variable) continuously shrinks. FC outputs a solution whenever the last variable is given a value. When the domain of the current variable is exhausted, the algorithm backtracks to the previous one.

For ordering of variable instantiations, MFC applies the *dynamic variable ordering* (DVO), which is also mainly used in CSPs. DVO dynamically reorders the future variables after each instantiation so that the variable with the minimum domain size becomes the next variable. Additionally, MFC adopts the *search space restriction* technique to improve performance. A slightly modified version of the space restriction algorithm used in [15] is shown below:

```
BOOLEAN SpaceRestriction_1 (Query_graph Q[][], Rtree_Nodes N[])

(1)   FOR i=0 TO n-1 DO
(2)       ReadPage (N[i]);
(3)       FOR all E_k ∈ N[i] DO
(4)           FOR j=0 TO n-1, i≠j DO
(5)               IF Q[i][j] == TRUE AND E_k.rect ∩ N[j].rect == ∅ THEN
(6)                   N[i] = N[i]-{E_k};
(7)                   BREAK;
(8)       IF N[i]==∅ THEN RETURN FALSE;
(9)   RETURN TRUE;

END SpaceRestriction_1
```

We do not adopt MFC for the following reasons: First, MFC does not apply the plane sweep technique, which is fairly efficient in the rectangle intersection problem [21,3], but uses FC-DVO which is just a special form of the nested loop. Second, during the space restriction, MFC does not consider the space restriction order among n R-tree nodes, i.e., which node should be checked first. In Section 3, we propose a new generalization of the 2-way R-tree join which considers both the space restriction ordering and the plane sweep technique.

2) Window Reduction WR maintains an $n * n$ *domain window* (instead of a 3D domain table) that encloses all potential values for each variable. When a variable is instantiated, a domain window for each future variable is shrunk to the intersection between the newly computed window according to the current variable instantiation and existing domain window. For the instantiation of the current variable, a window query is performed using the current domain window. In WR, the DVO technique was also applied to reorder the future variables, i.e., the future variable with the smallest domain window becomes the next variable to be examined. WR was considered as a special form of the indexed nested loop join. However it does not generate intermediate results. WR must essentially search the whole space in order to instantiate the first variable. To avoid the blind instantiation for the first variable, a hybrid technique called *join window reduction* (JWR) was proposed [15]. JWR applies the R-tree join for the first pair of variables and then WR for the rest of the variables.

In [16], a slightly different WR algorithm was proposed for the multi-way intersection join. In that algorithm, the instantiation order of variables is pre-determined according to an optimization method. As a query window, for acyclic queries (tree topology), the rectangle of the variable directly connected to the current variable among instantiated variables becomes the query window. For complete queries (clique topology), the common intersected rectangle of all instantiated variables becomes the query window. In our implementation and experiment, regardless of query types, among instantiated variables which are connected to the current variable, one whose value has the smallest rectangle was selected and the rectangle becomes a query window for the next variable instantiation.

3   New Methods for M-Way Spatial Joins Using R-Trees

3.1   A New Generalization of the 2-Way R-Tree Join

In this section, we propose a new M-way join algorithm which extends both the search space restriction and the plane sweep optimization techniques of the 2-way R-tree join. We emphasize the ordering of both optimization techniques, assuming only *intersect* (not disjoint) as a join predicate.

1) Search Space Restriction   Algorithm *SpaceRestriction_1* [15] does not consider ordering among M R-tree nodes. If no entry of an R-tree node passes over the space restriction, we do not have to check other nodes. Especially in an *incomplete join* (no join predicate between some variables), the possibility that no entry of an R-tree node may pass over the space restriction is high. In such a case, Algorithm *SpaceRestriction_1* may result in unnecessary reading of other nodes. Therefore, the space restriction order of the R-tree nodes becomes important. For example, Figure 2 shows an MBR intersection between intermediate nodes of the R-trees for a 4-way spatial join "X *intersect* Y and Y *intersect* Z and Z *intersect* W." Since no entry of node B simultaneously intersects with nodes A and C, the intermediate node-tuple ⟨A, B, C, D⟩ cannot pass over the space restriction and becomes a false hit. If the space restriction is performed first on node B, we do not have to check other nodes A, C and D and can save the I/O and CPU time.

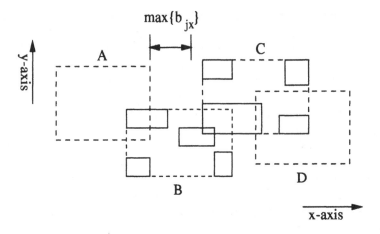

Fig. 2. Intersection between intermediate nodes of R-trees

We will explain the space restriction ordering (SRO) in the context of the query graph. For SRO, we use the following two metrics per node N_i, $0 \leq i \leq n\text{-}1$:

(1) *normalized common rectangle area* (NCRA): the area of the common intersection of N_i and its adjacent nodes divided by the area of the rectangle of N_i. Formally, for all N_j where $i = j$ or $Q_{ij} = \text{TRUE}$, $0 \leq j \leq n\text{-}1$, $\text{area}(\cap N_j.rect)/\text{area}(N_i.rect)$.

(2) *maximum inter-rectangle distance* (MIRD): the sum of squares of the maximum of distances per axis between the nodes adjacent to N_i. Formally, for all N_j and N_k where $Q_{ij} = \text{TRUE}$ and $Q_{ik} = \text{TRUE}$, $0 \leq j, k \leq n\text{-}1$, $j \neq k$, $\left(\max\{x\_dist(N_j, N_k)\} \right)^2 + \left(\max\{y\_dist(N_j, N_k)\} \right)^2$.

Using the above two metrics, we perform SRO on the basis of the following criteria:

(1) Choose a node with the minimum NCRA.
(2) If the minimum NCRA is zero for more than one node, choose a node such that MIRD is maximal.

In Figure 2, the common intersected rectangles for each node are A.rect ∩ B.rect, A.rect ∩ B.rect ∩ C.rect, B.rect ∩ C.rect ∩ D.rect and C.rect ∩ D.rect. Since nodes B and C have zero NCRA, these two nodes are selected by Criteria (1). Then, since MIRD (only between A and C in this case) of node B is longer than MIRD (between B and D) of node C, we perform the space restriction for node B first by Criteria (2).

In Metric (1), the reason we use the normalized area instead of the (absolute) common intersected rectangle area (CRA) is to choose a node which has large dead space (If CRAs are the same, a larger node is more likely to have more dead space). The dead space of the MBR of an intermediate node may be influenced by many factors such as the number of entries, the distribution of the rectangles of the entries, the density of the rectangles of the entries, and the MBR size of the node. If the other conditions are fixed, the smaller the number of entries of an intermediate node the more dead space the MBR of an intermediate node may have. Skewed distributions, low density and large MBRs may lead to large dead space. However, we cannot know the above characteristics except the MBR size unless we visit the node. Therefore, we choose only the MBR size among the above characteristics. We expect that NCRA behaves better than CRA especially in the complete query graph because CRAs of all nodes are always the same.

The time complexity of SRO is as follows: It takes $O(M)$ time to compute NCRA and MIRD per node. Therefore, it takes $O(M^2)$ time for all nodes. For sorting NCRA and MIRD, it takes $O(M \log_2 M)$ time. Therefore, the overall time complexity is $O(M^2)$. Algorithm *SpaceRestriction_2* is identical to Algorithm *SpaceRestriction_1* but considers ordering of nodes according to the above criteria.

2) Plane Sweep   In MFC, FC-DVO was used in a node-tuple join because it was known to be efficient in CSPs. However, the plane sweep algorithm was also known to be fairly efficient in the rectangle intersection problem [21]. Therefore, we use the plane sweep as the second optimization technique rather than FC-DVO. In the 2-way join, the plane sweep algorithm is applied only once. In the M-way join, however, the plane sweep algorithm must be applied multiple times because there are M variables and at least M-1 predicates. In this case, the ordering of plane sweeps among R-tree nodes becomes important.

Our plane sweep ordering (PSO) performs as follows: In PSO, we call the evaluated nodes *inner nodes* and the un-evaluated nodes *outer nodes*. In the following, the *cardinality* of a node is the number of entries in the node, and the *degree* of a node is the number of edges (i.e., the number of predicates) incident on the node. Before PSO starts, all R-tree nodes are initialized to outer nodes.

(1) Choose the first two connected nodes whose sum of the cardinalities / the maximal degree between the two nodes is minimal.
(2) Apply plane sweep between the selected two nodes and make the two nodes inner nodes.
(3) Choose an outer node which is adjacent to one or more inner nodes such that cardinality / degree is minimal.
(4) Choose an inner node which is adjacent to the selected outer node and whose cardinality is minimal.
(5) Apply plane sweep between the selected inner node and the selected outer node.
(6) Check additional predicates, if any, between the selected outer node and other inner nodes.
(7) Make the selected outer node an inner node.
(8) Stop if all nodes are inner nodes, otherwise go to Step (3).

In Step (1) and Step (3), the reason we divide the cardinality by the degree is because the more the number of predicates is, the smaller the intermediate result size may be. The time complexity of PSO excluding actual plane sweeps is as follows: It takes $O(M^2)$ time to choose the first two nodes. And, for the ordering of the rest variables, it also takes $O(M^2)$ time. Therefore, the overall time complexity of PSO is $O(M^2)$.

The direct application of PSO generates intermediate results M-2 times [19]. Since the number of solutions in a node-tuple join can be up to C^M (C: the maximum number of entries of an R-tree node) for the worst case, we need a main memory buffer which can store C^M tuples per R-tree level. For example, if the node size is 2048 bytes and the entry size is 20 bytes, C and C^M are about 100 and 100^5 respectively in a 5-way join. Although a much smaller buffer will be sufficient in general, this is a tremendous amount of main memory for the worst case. In order to solve this main memory problem, we can use pipelining.

For both the plane sweep and pipelining, we use M buffers ($Seq[]$ in Algorithm $MwayRtreeJoin\_1$) each of which holds intermediate entry-tuples. The buffer size is determined according to the main memory size. We apply plane sweep between the first two nodes selected by PSO. The intermediate entry-tuples produced by the first plane sweep are accumulated in $Seq[1]$. If $Seq[1]$ is full or all entries in the first two nodes are evaluated, we recursively call the plane sweep algorithm taking $Seq[1]$ and the next selected outer node as parameters. In general, the plane sweep between $Seq[m]$ and an outer node accumulates the intermediate result to $Seq[m+1]$. If plane sweep is called for the last outer node, the algorithm backtracks to the previous one. In PSO with pipelining, the actual ordering is determined once per node-tuple join. The M-way R-tree join algorithm using PSO with pipelining is shown below:

```
MwayRtreeJoin_1 (Query_graph Q[][], Rtree_Nodes N[])

(1)  IF NOT SpaceRestriction_2(Q[][], N[]) THEN RETURN;
(2)  PSO (Q[][], N[], outer_order[], inner_order[]);
(3)  i = outer_order[0]; j = inner_order[0];
(4)  Seq[0] = N[j];
(5)  FOR k=0 TO n-1, k≠j DO Sort (N[k]); /* sort all outer nodes */
(6)  PipelinedPlaneSweep (Q[][], N[], Seq[], i, j, 0);

     END MwayRtreeJoin_1

PipelinedPlaneSweep (Query_graph Q[][], Rtree_Nodes N[], Entry_Tuple_Buf
Seq[], int i, int j, int m)

(1)  Sort (Seq[m]);
(2)  SortedIntersectionTest_1 (N[i], Seq[m], j, Seq[m+1]); /* plane sweep +
     additional predicate checking until Seq[m+1] is full or all entries
     in N[i] and Seq[m] are evaluated */
(3)  IF m == n-2 THEN /* the last outer node is evaluated */
(4)      FOR all T_k ∈ Seq[m+1] DO
(5)          IF all N[l] are leaf nodes, 0≤l≤n-1 THEN
(6)              output T_k;
(7)          ELSE /* all tree heights are equal */
(8)              MwayRtreeJoin_1 (Q[][], T_k.ref[]); /* go downward */
```

```
 (9)   ELSE
(10)       i = outer_order[m+1]; j = inner_order[m+1];
(11)       PipelinedPlaneSweep (Q[][], N[], Seq[], i, j, m+1);
(12)   IF all entries in N[i] and Seq[m] are evaluated THEN
(13)       RETURN;
(14)   ELSE
(15)       empty Seq[m+1];
(16)       goto Step (2);

      END PipelinedPlaneSweep
```

In Algorithm *MwayRtreeJoin_1*, *SortedIntersectionTest_1* is the same as *Sorte-dIntersectionTest* in Algorithm *RtreeJoin* except for the following: First, one input is a sequence of entry-tuples. Second, additional predicate checks are done between the selected outer node and the non-selected inner nodes. Third, when $Seq[m + 1]$ is full, *SortedIntersectionTest_1* exits and the status of both loop counters[3] in the algorithm is saved for the next call. In our implementation of PSO, we did not use pipelining because all intermediate results fitted in main memory.

3.2   Consideration of Indirect Predicates

The maximum number of possible predicates in the M-way spatial join is M*(M-1)/2, i.e., all relation pairs have join conditions. We call such a join *complete*. If a join is not complete, i.e., the number of predicates is less than M*(M-1)/2, the join is *incomplete*.

As it was pointed out in [15], the M-way R-tree join may generate many false intersections in intermediate levels. As we can see in Figure 2, especially in an incomplete join, the possibility of a false intersection is high. In this case, if we can detect the false intersections before visiting the intermediate node-tuple, we can further reduce I/O and CPU time. For example, if we know in advance that no entry of node B can simultaneously intersect nodes A and C in Figure 2, we can avoid reading node B and checking the intersection between all entries of node B and other nodes (A and C) during space restriction. In this section, we propose a technique which detects a false intersection in intermediate levels of R-trees before visiting the node-tuple.

1) Indirect Predicates   In a query "X *intersect* Y and Y *intersect* Z and Z *intersect* W" like the one in Figure 2, it seems that there is no relationship between X and Z (or between Y and W, or between X and W). However, for a data tuple (i.e., a tuple of entries from leaf nodes) $\langle a, b, c, d \rangle$ which satisfies the query, $x\_dist(a, c) \leq b_x$ (or $x\_dist(b, d) \leq c_x$, or $x\_dist(a, d) \leq b_x + c_x$) must be satisfied on x-axis (b_x represents x-length for a data MBR b). The same condition holds on y-axis. Consequently, for the data tuple $\langle a, b, c, d \rangle$, $x\_dist(a, c) \leq \max\{b_{jx} \mid b_j \in dom(Y)\}$ (or $x\_dist(b, d) \leq \max\{c_{kx} \mid c_k \in dom(Z)\}$, or

---

[3] Two internal loops exist in *SortedIntersectionTest* [3].

$x\_dist(a, d) \leq \max\{b_{jx}\} + \max\{c_{kx}\})$ must be satisfied on x-axis ($\mathrm{dom}(Y)$ represents the domain (i.e., relation) of data MBRs for variable Y). The same condition holds on y-axis. We call the user predicates in the query such as "X *intersect* Y" and "Y *intersect* Z" the *direct predicates*, and the derived predicates such as $x\_dist(X, Z) \leq \max\{b_{jx}\}$ and $x\_dist(Y, W) \leq \max\{c_{kx}\}$ the *indirect predicates*. In R-trees, the x-length and y-length of MBRs of intermediate nodes may be longer than the max x-length and max y-length of the data MBRs in the domain. In Figure 2, if $x\_dist(A, C) > \max\{b_{jx}\}$ (or $x\_dist(B, D) > \max\{c_{kx}\}$, or $x\_dist(A, D) > \max\{b_{jx}\} + \max\{c_{kx}\}$), we do not have to visit the node-tuple $\langle A, B, C, D \rangle$ because the descendent node-tuples will never satisfy the query. Therefore, if we take advantage of the indirect predicates in intermediate levels of the M-way R-tree join, we can achieve more pruning effects. We call such pruning *indirect predicate filtering* (IPF). The max x-length and y-length can be obtained from the statistical information in the database schema.

2) Indirect Predicate Paths and Lengths   In Figure 2, we call the paths ABC, BCD and ABCD for indirect predicate pairs AC, BD and AD the *indirect predicate paths* (ipp), and the x-path lengths $\max\{b_{jx}\}$, $\max\{c_{kx}\}$, and $\max\{b_{jx}\} + \max\{c_{kx}\}$ the *indirect predicate x_path lengths* (x_ippl). The *indirect predicate y_path lengths* (y_ippl) are similarly defined. In Figure 2, since there is only one indirect predicate path for each indirect predicate pair, it is easy to compute indirect predicate paths and indirect predicate path lengths. However, there can be several indirect predicate paths for an indirect predicate pair in a general M-way join, and the x_path and y_path for the predicate pair can be different. Therefore, we need a systematic method to compute indirect predicate paths and their lengths.

We first draw a query graph whose nodes represent relations and edges represent direct predicates. Then, we assign weights to nodes. The weight of a node is the maximum x-length (x_max) and y-length (y_max) in the relation which the node represents. Since there can be multiple paths between a node pair, we compute the ipp and ippl by using the shortest path algorithm [6]. In order to get the shortest path between a node pair, we need edge weights but we have only node weights now. Therefore, we obtain the edge weights from node weights. The weight of an edge is obtained by summing weights of nodes on which the edge is incident. An example guery graph having both node weights and edge weights for a 5-way join is shown in Figure 3(a). We call this query graph *maximum weighted query graph*.

When there is no direct predicate between two nodes S and D in a maximum weighted query graph, the ipp and ippl between S and D can be obtained as follows: First, we calculate the shortest path and shortest path length per axis. Next, we subtract the weights of both S and D from the shortest path length and then divide the shortest path length by 2. This is because we want to get the sum of the weights of intermediated nodes in the shortest path, but the weights of S and D are included in the edge weights of the shortest path length, and the weights of the intermediated node are included twice. Therefore, the x_ippl between S and D can be calculated by Expression (1). The y_ippl is similarly defined.

$$x\_ippl(S, D) = (x\_shortest\_path\_length(S, D) - x\_max(S) - x\_max(D))/2 \tag{1}$$

The ipp's and ippl's for all indirect predicate pairs in Figure 3(a) are shown in Figure 3(b). In Figure 3, the x_ipp and y_ipp are different for indirect predicate pairs AD and AE.

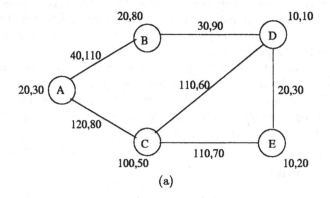

(a)

pairs	x_ipp	x_ippl	y_ipp	y_ippl
AD	ABD	20	ACD	50
BC	BDC	10	BDC	10
BE	BDE	10	BDE	10
AE	ABDE	30	ACE	50

(b)

Fig. 3. Maximum weighted query graph

The indirect predicates can be simultaneously checked with the additional predicates in *SortedIntersectionTest_1* of Algorithm *MwayRtreeJoin_1*. We call the algorithm doing the indirect predicate filtering *MwayRtreeJoin_2*.

3) Maximum Tagged R-Trees   Until now, we have used only one max x-length and y-length per relation. In this case, if there are several extremely large objects in a relation although other objects are not so large, the effect of indirect predicates can be considerably degenerated. One possible solution for this is to have the max x-length and y-length per R-tree node. A leaf node has the max x-length and y-length for MBRs of all entries in the node, and an intermediate node has the maximum value for the max x-lengths and max y-lengths of its child nodes. In the end, the root node has the max x-length and max y-length for the relation. The max x-length per R-tree node is recursively defined as in Expression (2). The max y-length is similarly defined.

$$x\_max(N) = \begin{cases} \max\{N_1.rect_x \ldots N_n.rect_x\} & \text{for leaf node} \\ \max\{x\_max(N_1.ref)\ldots x\_max(N_n.ref)\} & \text{for intermediate node} \end{cases} \quad (2)$$

where n is the number of entries in node N.

We call the max x-length and y-length per relation *domain max information* and those per R-tree node *node max information*. By using the node max information instead of the domain max information, we can have more prunning effects in indirect predicate filtering of the M-way R-tree join. Since only two max values are attached per R-tree node (one for x-length and the other for y-length), we can ignore the storage overhead due to the max lengths. And since the max lengths can be dynamically maintained with the R-tree insertion and deletion, we can always have exact max lengths per R-tree node. We call this R-tree *maximum tagged R-tree*.

We get only once the ipp's for each axis using the max information in root nodes of R-trees because calculating the shortest path for every node-tuple needs a large CPU time overhead[4]. However, we get the ippl's for every node-tuple based on the ipp's obtained from the root nodes. We call the algorithm using maximum tagged R-trees *MwayRtreeJoin_3*.

4 Experiments

To measure the performance of the M-way R-tree joins, we conducted some experiments using real data sets. The experiments were performed on a Sun Ultra II 170 MHz platform on which Solaris 2.5.1 was running with 384 MB of main memory. We implemented the three M-way R-tree join algorithms: *MwayRtree-Join_1* (MRJ1), *MwayRtreeJoin_2* (MRJ2) and *MwayRtreeJoin_3* (MRJ3). For performance comparisons, we also implemented the multi-level forward checking (MFC) algorithm with the dynamic variable ordering (DVO) and the join window reduction (JWR) algorithm which were proposed in [15,16]. Additionally, we implemented another MFC algorithm (MFC1) which uses our space restriction ordering (SRO) as well as FC-DVO to check the pure effect of SRO.

The real data in our experiments were extracted from the TIGER/Line data of US Bureau of the Census [22]. We used the road segment data of 10 counties of the California State in the TIGER data. The characteristics (statistical information) of the California TIGER data are summarized in Table 1. The original TIGER data of all counties were center-matched to join different county regions, i.e., the x and y coordinates of the original TIGER data were subtracted from those of the center point of each county. The center-matched data were divided by 10 for easy handling.

We implemented the insertion algorithm in [2] to build R*-trees for each county data. The node sizes of the R*-trees considered are 512, 2048 and 4096 bytes. The tree heights for all county data for each node size are 4, 3 and 3, respectively. The LRU buffers are 256 pages in every node size[5].

---

[4] The complexity of computing all pair's shortest paths is known to be $O(M^3)$ [6].

[5] We assume that an R*-tree node occupies one page.

Table 1. Characteristics of the California TIGER data

county	# of obj	domain area	max length	avg length	density
Alameda	49070	86222*44995	4662*3940	102*80	0.23
Contra Costa	40363	88025*33808	4676*5112	100*77	0.21
Fresno	58163	233238*151898	7190*4633	210*167	0.09
Kern	113407	257781*100758	8204*6497	212*169	0.26
Monterey	35417	175744*112068	9085*6194	234*192	0.20
Orange	91970	69999*55588	3658*6735	80*66	0.21
Riverside	91751	323725*65389	12113*10062	158*126	0.21
Sacramento	46516	75771*71218	6442*4103	111*86	0.24
San Diego	103420	151241*96476	8054*6828	122*104	0.22
Santa Barbara	64037	99301*58696	4541*6460	100*81	0.22

We selected the following 4 query types as input queries: complete, half, ring and chain. Example query graphs for each query type in a 5-way join are shown in Figure 4. The spatial predicate used for our experiments is *intersect*.

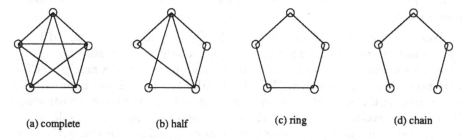

(a) complete   (b) half   (c) ring   (d) chain

Fig. 4. Example query graphs in a 5-way join

First, we measured the total response time (CPU time + I/O time) for various data sets and various query types, and a fixed node size of 2048 bytes. The total response time was measured by "the elapsed CPU time + the number of I/O * the unit I/O time." The unit I/O time was set to 10 ms which is a typical value for a random I/O [7,15]. For this experiment, we extracted the following three data sets from the TIGER data shown in Table 1. An M-way join for each data set was performed for the first M counties of the data set.

Data set 1:   Ora.   Sac.   S.B.   S.D.   Ala.   Kern   Riv.
Data set 2:   Ora.   Ala.   Sac.   S.D.   S.B.   Kern   Mon.
Data set 3:   C.C.   S.B.   Mon.   Ora.   Sac.   Ala.   Fre.

The total response time is shown in Table 2. The relative rates of the total response time compared to Algorithm *MwayRtreeJoin_1* (MRJ1) are shown in Figure 5 (only for the algorithms using the synchronous traversal (ST) technique). The numbers of solutions for each data set are also shown in Table 3.

First, we compared the relative performances among the ST algorithms such as MFCs and MRJs. In most cases, SRO considerably reduces the query response time (Compare MFC and MFC1 in Table 2 and Figure 5). FC-DVO has a better performance in complete and half queries while PSO has a better performance in the chain query. In the ring query, both have a similar performance or FC-DVO has a slightly better performance. (Compare MFC1 and MRJ1 in Table 2 and Figure 5.) The reason FC-DVO has a better performance in the complete and half queries is because FC-DVO prunes the entries of the future variables faster with many predicates while PSO does not prune the entries of the outer variables until they are actually evaluated. Since the chain and ring queries are more general in real life and more time consuming than other queries, we think that the optimization for these queries is more important. (According to Table 2, the differences of the query response time between MFC1 and MRJ1 in the complete and half queries are within 10 seconds, but the differences in the chain query reach about 1000 seconds.)

Sometimes, in data set 3, MFC1 does not work as well as MFC. This is due to the locality of LRU buffers and the CPU overhead of SRO. We observed that, in these cases, while MFC1 accessed fewer nodes, MFC performed a smaller or similar number of I/O's. However, in most cases, MFC1 performed a smaller number of I/O's.

Next, we measured the performance of indirect predicate filtering (IPF). In this measurement, we excluded the complete query type because no indirect predicates are in the complete query. In the half query, there is nearly no effect of indirect predicates (Compare MRJ1 and MRJ2 in Table 2 and Figure 5). We do not present the effect of the maximum tagged R-tree (MRJ3) in the half query because it is similar to that of MRJ2 in most cases. IPF has considerable impact on ring and chain queries. As the number of direct predicates decreases, the effect of indirect predicates increases. In summary, the three optimization techniques (SRO, PSO and IPF) improve efficiency. The maximum improvements compared to MFC are about 40%, 80%, 140% and 300%, respectively, for the complete, half, ring and chain queries.

A little later than the early version of this paper [19], other optimization techniques called *static variable ordering* (SVO) and *plane sweep and forward checking* (PSFC) were developed [13]. SVO orders the variables (or nodes) once according to the degrees before the algorithm starts. This static ordering is used both for the search space restriction and the forward checking. PSFC works as follows: The first variable is instantiated by a plane sweep, and a variant of the forward checking, called *sorted forward checking*, is used for the instantiations of remaining variables according to SVO. We believe that SRO is superior to SVO because it uses more sophisticated criteria. Actually, the experimental results in Table 2 and Figure 5 support our opinion. In complete and ring query graphs, the space restriction using SVO is the same as Algorithm *SpaceRestriction_1* used in MFC because the degrees of all nodes are the same. Since the experimental results show that MFC1 outperforms MFC in most cases, SRO will be superior to SVO. On the other hand, when there are many direct predicates, PSFC will

Table 2. Total response time for various data sets (node size: 2048, unit: sec)

		Data set 1					Data set 2					Data set 3				
	M	3	4	5	6	7	3	4	5	6	7	3	4	5	6	7
Complete	MFC	17	19	10	13	13	8	9	11	14	11	14	21	10	11	15
	MFC1	14	14	8	11	11	6	7	8	11	9	13	22	9	10	14
	MRJ1	15	17	9	11	12	6	8	9	12	9	13	23	9	10	14
	JWR	36	57	24	25	24	16	22	24	25	17	41	72	39	41	43
	M		4	5	6	7		4	5	6	7		4	5	6	7
Half	MFC		22	36	22	28		11	21	24	21		34	14	26	41
	MFC1		19	20	19	21		10	14	19	16		36	18	18	30
	MRJ1		21	20	20	22		11	13	19	17		46	14	17	33
	MRJ2		21	20	20	23		11	13	20	17		47	15	17	33
	JWR		78	33	56	285		74	29	29	72		223	194	58	223
	M		4	5	6	7		4	5	6	7		4	5	6	7
Ring	MFC		25	26	106	710		12	26	83	295		34	45	122	649
	MFC1		20	21	81	461		10	21	65	213		36	34	98	500
	MRJ1		22	21	77	470		11	22	68	196		37	36	119	623
	MRJ2		22	20	69	364		11	22	63	176		38	35	111	397
	MRJ3		21	19	63	301		11	22	60	152		40	34	101	333
	JWR		198	159	1048	226		41	97	172	217		172	171	209	183
	M	3	4	5	6	7	3	4	5	6	7	3	4	5	6	7
Chain	MFC	26	81	335	1469	3805	11	32	166	939	2105	21	251	191	1738	7851
	MFC1	24	69	244	969	2363	9	26	114	632	1428	19	223	162	1256	5396
	MRJ1	23	62	181	818	2026	8	22	92	538	1198	18	184	128	896	4324
	MRJ2	23	57	147	580	1341	8	20	76	396	859	18	196	127	779	3218
	MRJ3	23	55	131	454	954	8	20	87	342	719	18	201	126	655	2815
	JWR	40	212	455	1091	1557	17	44	106	135	428	58	277	2681	841	1276

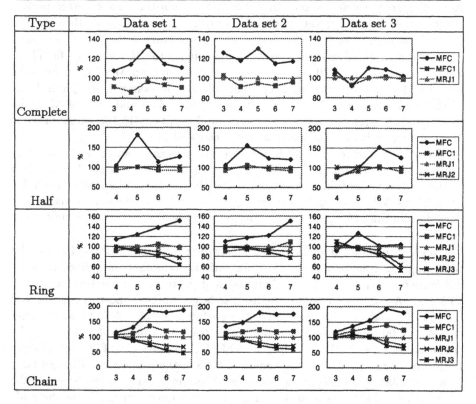

Fig. 5. Rates of total response time for various data sets (node size: 2048)

Table 3. Number of solutions for various data sets

M		3	4	5	6	7
Data Set 1	Complete	16,156	3,893	435	192	31
	Half		18,897	25,578	881	128
	Ring		11,590	7,098	12,298	4,220
	Chain	131,759	329,855	440,945	475,497	81,419
Data Set 2	Complete	4,733	1,719	435	192	77
	Half		15,327	2,371	825	318
	Ring		5,209	5,916	8,724	19,574
	Chain	23,155	61,446	56,295	254,505	177,627
Data Set 3	Complete	23,188	21,880	2,506	725	152
	Half		232,068	6,152	4,558	2,074
	Ring		161,611	72,346	51,600	42,238
	Chain	102,327	2,753,856	530,673	1,271,835	3,441,939

naturally outperform PSO because our experiments show that MFC outperforms PSO for numerous direct predicates.

Next, we compared the query response time between ST algorithms and JWR. As the variable instantiation order of JWR, we used the same as in PSO. According to the result shown in Table 2, ST algorithms have better performances in all Ms of complete and half queries and in most Ms of other queries. When M is high (6 or 7), JWR has a better performance than MFC for some data sets in ring and chain queries, which is similar to the result in [16]. There are some cases that JWR has a better performance than MFC for some data sets, but has a worse performance than MRJs. For example, see Table 2 for M=6,7 and data set 1, M=5 and data set 2, and M=6 and data set 3 in the chain query. Therefore, unlike the experimental results in [15,16], we can use our M-way R-tree join algorithms for a higher range of M.

Sometimes, the costs of JWR are abruptly increased (for example, M=7 in the half query of data set 1, M=6 in the ring query of data set 1, and M=5 in the chain query of data set 3). We think this is due to the evaluation order of variables. While real data sets are highly skewed, PSO does not consider the data distribution. However, the variable ordering worked properly in most other cases.

Next, we conducted an experiment for various node sizes. Table 4 shows the total response time of all algorithms for various node sizes and a fixed data set 2. Figure 6 illustrates the performance rates of the total response time compared to MRJ1. According to Figure 6, SRO has large effects in most cases. And the smaller the node size is, the better the performance of FC-DVO is. In other words, the larger the node size, the better the performance of PSO. (See the performance rate of MFC1 compared to MRJ1.) In particular, PSO has a better performance than FC-DVO for node size 4096 of the ring query although both have a similar performance for node size 2048. As the node size increases, the effect of IPF slightly decreases in ring and chain query types. When the node size is 4096, there is nearly no difference between the effect of indirect predicates using domain max information (MRJ2) and that using node max information

Table 4. Total response time for various node sizes (data set 2, unit: sec)

		512					2048					4096				
Complete	M	3	4	5	6	7	3	4	5	6	7	3	4	5	6	7
	MFC	29	32	34	47	31	8	9	11	14	11	5	6	7	10	9
	MFC1	20	25	24	58	26	6	7	8	11	9	4	5	6	9	8
	MRJ1	20	24	28	40	28	6	8	9	12	9	4	5	6	9	9
	JWR	39	49	47	50	34	16	22	24	25	17	14	20	22	23	16
Half	M		4	5	6	7		4	5	6	7		4	5	6	7
	MFC		38	77	77	65		11	21	24	21		8	17	20	21
	MFC1		32	50	63	54		10	14	19	16		7	11	15	15
	MRJ1		34	54	65	56		11	13	19	17		8	11	14	14
	MRJ2		33	51	68	59		11	13	20	17		7	11	15	15
	JWR		148	57	60	121		74	29	29	72		69	26	29	71
Ring	M		4	5	6	7		4	5	6	7		4	5	6	7
	MFC		42	87	281	791		12	26	83	295		9	21	78	323
	MFC1		32	66	236	622		10	21	65	213		8	17	62	222
	MRJ1		34	73	243	629		11	22	68	196		7	14	51	189
	MRJ2		34	70	222	537		11	22	63	176		7	14	47	155
	MRJ3		34	63	166	423		11	22	60	152		7	14	47	149
	JWR		81	172	273	371		41	97	172	217		43	108	200	246
Chain	M	3	4	5	6	7	3	4	5	6	7	3	4	5	6	7
	MFC	41	93	472	3521	6692	11	32	166	939	2105	8	27	147	774	1816
	MFC1	30	66	322	2383	4702	9	26	114	632	1428	7	23	107	528	1177
	MRJ1	30	69	303	2309	4647	8	22	92	538	1198	5	14	62	362	876
	MRJ2	29	65	252	1687	3467	8	20	76	396	859	5	13	48	259	600
	MRJ3	28	59	180	1027	2064	8	20	87	342	719	5	13	48	263	592
	JWR	40	85	166	292	604	17	44	106	185	428	15	45	113	219	446

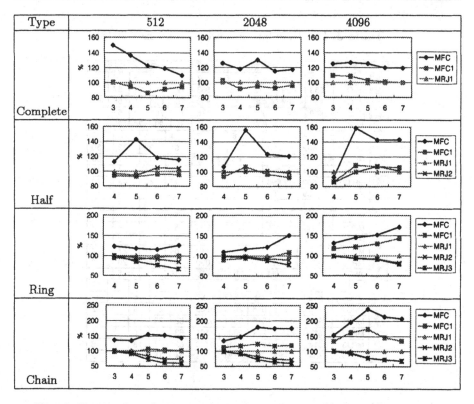

Fig. 6. Rates of total response time for various node sizes (data set 2)

(MRJ3). This is because the large node size leads to many entries per node and increases the node max information. However, still for a large node size (4096), the effect of IPF is large for the ring and chain queries.

The comparison between ST algorithms and JWR shows that ST algorithms perform better for large node sizes. This is due to the index probing overhead in JWR. Since there is no global ordering in multi-dimensional non-point objects, we should check all entries of a node during an R*-tree search. In addition, while ST algorithms have the best performance for all query types in node size 4096 compared to other node sizes, JWR has the best performance for the ring and chain queries in node size 2048.

Finally, we measured the I/O time (see Table 5 and Table 6). MRJs consume more I/O time than MFCs and JWR in high Ms. From Table 5 and Table 6, however, we found an important fact: the higher the value of M, the lower the rate of I/O time compared to the total response time. For ring and chain queries, the rate of I/O time considerably decreases as M increases. Therefore, the I/O time becomes less important and the M-way R-tree join becomes CPU-bound. The I/O rate also decreases along the node size. (According to Table 6, when the node size is 4096 and M is 7, the I/O rates in the ring and chain query types are less than or equal to 5%.)

Table 5. I/O time for data set 2 (node size: 2048)

		# of I/O					I/O rate (%)				
	M	3	4	5	6	7	3	4	5	6	7
Complete	MFC	553	629	712	787	563	72	69	64	57	52
	MFC1	439	466	495	512	448	70	66	61	55	51
	MRJ1	435	542	537	689	476	71	70	63	57	52
	JWR	575	631	609	657	541	35	29	25	26	31
	M		4	5	6	7		4	5	6	7
Half	MFC		692	799	835	617		61	39	35	30
	MFC1		564	601	765	561		57	43	41	35
	MRJ1		604	595	845	694		56	45	44	40
	MRJ2		602	595	841	649		57	45	43	38
	JWR		1429	602	739	1039		19	21	25	14
	M		4	5	6	7		4	5	6	7
Ring	MFC		720	858	1076	955		58	33	13	3
	MFC1		567	774	1127	1193		55	37	17	6
	MRJ1		720	1036	1776	1308		64	46	26	7
	MRJ2		716	1040	1714	1149		64	47	27	7
	MRJ3		716	1029	1671	1090		64	48	28	7
	JWR		652	714	777	763		16	7	5	4
	M	3	4	5	6	7	3	4	5	6	7
Chain	MFC	632	775	956	1347	1467	57	24	6	1	1
	MFC1	520	685	986	2014	2662	56	27	9	3	2
	MRJ1	514	845	1316	3351	6367	62	39	14	6	5
	MRJ2	514	829	1224	2800	4501	63	41	16	7	5
	MRJ3	514	821	1212	2510	4176	62	41	18	7	6
	JWR	584	654	704	784	794	34	15	7	4	2

In overall summary, we recommend the following based on the experimental results: First, always use SRO. Second, if there are many direct predicates as in

Table 6. I/O time for various node sizes (data set 2)

		# of I/O (512)				I/O rate (%)					# of I/O (4096)				I/O rate (%)			
	M	3	4	5	6	7	3	4	5	6	7	**M** 3	4	5	6	7	3	4	5	6	7
Complete	MFC	2563	2679	2655	3396	1796	87	82	78	72	59	298	331	365	420	383	59	56	50	41	40
	MFC1	1658	1782	1723	2372	1402	84	79	73	65	53	240	286	295	356	290	55	53	49	41	37
	MRJ1	1853	1902	2091	2687	1536	84	80	75	67	55	240	268	298	332	292	60	58	51	41	37
	JWR	2382	2591	2287	2449	1629	61	53	48	49	48	315	350	340	372	329	22	17	15	16	20
	M		4	5	6	7		4	5	6	7		4	5	6	7		4	5	6	7
Half	MFC		2931	3414	4216	2661		76	45	55	41		339	415	446	388		44	25	23	19
	MFC1		2309	2285	3429	2287		72	46	55	42		301	341	400	378		42	30	27	25
	MRJ1		2474	2427	3710	2564		73	45	57	46		315	349	420	391		38	33	31	27
	MRJ2		2451	2427	3953	2701		74	48	58	46		315	349	515	384		44	33	35	26
	JWR		5702	2342	2887	4109		39	41	48	34		689	333	413	527		10	13	14	7
	M		4	5	6	7		4	5	6	7		4	5	6	7		4	5	6	7
Ring	MFC		3106	4767	7918	8860		74	55	28	11		371	439	549	551		43	21	7	2
	MFC1		2326	3528	7318	7644		72	53	31	12		321	410	583	663		41	24	9	3
	MRJ1		2496	4332	8256	8258		74	59	34	13		326	411	678	750		49	28	13	4
	MRJ2		2498	4032	7954	8039		74	57	36	15		325	402	752	704		50	29	16	5
	MRJ3		2489	3773	7283	7229		73	60	39	17		325	402	753	701		49	30	16	5
	JWR		2673	2943	3214	3095		33	17	12	8		352	322	421	422		8	4	2	2
	M	3	4	5	6	7	3	4	5	6	7	3	4	5	6	7	3	4	5	6	7
Chain	MFC	3031	4740	7858	18627	29918	73	51	17	5	4	345	404	519	874	1106	42	15	4	1	1
	MFC1	2238	3249	6149	16064	23149	74	49	19	7	5	238	328	434	979	1338	40	15	4	2	1
	MRJ1	2269	3732	5855	20176	34121	75	54	19	9	7	236	327	519	1041	1298	54	24	8	3	1
	MRJ2	2255	3539	5659	17228	28152	77	55	22	10	8	236	327	504	1100	1172	53	26	10	4	2
	MRJ3	2250	3388	5384	14865	21579	79	58	30	14	10	236	327	499	1097	1160	53	25	10	4	2
	JWR	2378	2890	2900	3184	3334	59	32	17	11	6	313	351	362	418	433	21	8	3	2	1

the complete and half queries, use FC-DVO and no IPF. Third, if the number of direct predicates is small as in the ring and chain queries, use PSO and IPF. Fourth, if the node size is small and M is high, use JWR; otherwise, use ST algorithms.

5 Conclusions

In this paper, we study the generalization of the 2-way R-tree join. We proposed the following three optimization techniques: space restriction ordering (SRO), plane sweep ordering (PSO) and indirect predicate filtering (IPF). Through experiments using real data, we showed that our three optimization techniques have a great impact on improving the performance of synchronous traversal (ST) algorithms.

After completing the M-way R-tree join, an oid pair may appear several times in the resulting oid-tuples. If the oid-tuples are read in the combined refinement step without scheduling, it may access the same page several times and perform the same refinement operation several times. However, this can be solved by extending scheduling methods for oid pairs such as [23] to oid-tuples. In future studies, first, we will develop an efficient combined refinement algorithm for the M-way spatial join. Second, although we found that the I/O rate of the total response time decreases as M increases, the I/O rate is still high for a small M. Therefore, we will develop I/O optimization techniques for the M-way R-tree join. Last, we will combine the optimization techniques proposed in this paper with our rule-based optimization technique for spatial and non-spatial mixed queries called ESFAR (Early Separated Filter And Refinement) [17,18].

Acknowledgement

We would like to thank Dimitris Papadias and Nikos Mamoulis for their careful reading and valuable comments. We also thank the anonymous referees for their suggestions to improve the quality of this paper. This research was supported by the National Geographic Information Systems Technology Development Project and the Software Technology Enhancement Program 2000 of the Ministry of Science and Technology of Korea.

References

1. L. Arge, O. Procopiue and S. Ramaswary, "Scalable Sweeping-Based Spatial Join," Proc. of VLDB, 570-581, 1998.
2. N. Beckmann, H.-P. Kriegel, R. Schneider and B. Seeger, "The R*-tree: An Efficient and Robust Access Method for Points and Rectangles," Proc. of ACM SIGMOD, 322-331, 1990.
3. T. Brinkhoff, H.-P. Kriegel and B. Seeger, "Efficient Processing of Spatial Joins Using R-trees," Proc. of ACM SIGMOD, 237-246, 1993.
4. A. Guttman, "R-trees: A Dynamic Index Structure for Spatial Searching," Proc. of ACM SIGMOD, 47-57, 1984.
5. R. H. Güting, "An Introduction to Spatial Database Systems," VLDB Journal, Vol. 3, No. 4, 357-399, 1994.
6. E. Horowitz and S. Sahni, "Fundamentals of Computer Algorithms," Computer Science Press, 1978.
7. Y.-W. Huang, N. Jing and E. A. Rundensteiner, "Spatial Joins Using R-trees: Breadth-First Traversal with Global Optimizations," Proc. of VLDB, 396-405, 1997.
8. Y. E. Ioannidis and Y. C. Kang, "Left-deep vs. Bushy Trees: An Analysis of Strategy Spaces and Its Implications for Query Optimization," Proc. of ACM SIGMOD, 168-177, 1991.
9. N. Koudas and K. C. Sevsik, "Size Separation Spatial Join," Proc. of ACM SIGMOD, 324-355, 1997.
10. M. L. Lo and C. V. Ravishankar, "Spatial Joins Using Seeded Trees," Proc. of ACM SIGMOD, 209-220, 1994.
11. M. L. Lo and C. V. Ravishankar, "Spatial Hash-Joins," Proc. of ACM SIGMOD, 247-258, 1996.
12. N. Mamoulis and D. Papadias, "Integration of Spatial Join Algorithms for Processing Multiple Inputs," to appear in Proc. of ACM SIGMOD'99.
13. N. Mamoulis and D. Papadias, "Synchronous R-tree Traversal," Technical Report HKUST-CS99-03, 1999.
14. J. A. Orenstein, "Spatial Query Processing in an Object-Oriented Database System," Proc. of ACM SIGMOD, 326-336, 1986.
15. D. Papadias, N. Mamoulis and V. Delis, "Algorithms for Querying by Spatial Structure," Proc. of VLDB, 546-557, 1998.
16. D. Papadias, N. Mamoulis and Y. Theodoridis, "Processing and Optimization of Multi-way Spatial Joins Using R-trees," to appear in Proc. of ACM PODS'99.
17. H.-H. Park, C.-G. Lee, Y.-J. Lee and C.-W. Chung, "Separation of Filter and Refinement Steps in Spatial Query Optimization," KAIST, Technical Report, CS/TR-98-122, 1998. See also: http://islab.kaist.ac.kr/~hhpark/eng_tr_sfro.ps

18. H.-H. Park, C.-G. Lee, Y.-J. Lee and C.-W. Chung, "Early Separation of Filter and Refinement Steps in Spatial Query Optimization," Proc. of DASFAA, 161-168, 1999.

19. H.-H. Park, G.-H. Cha and C.-W. Chung, "Multi-way Spatial Joins Using R-trees: Methodology and Performance Evaluation," KAIST, Technical Report, CS/TR-99-135, 1999.

20. J. M. Patel and D. J. DeWitt, "Partition Based Spatial-Merge Join," Proc. of ACM SIGMOD, 259-270, 1996.

21. F. P. Preparata and M. I. Shamos, Computational Geometry: An Introduction, Springer-Verlag, 1985.

22. U.S. Bureau of the Census, Washington, DC., "TIGER/Line Files, 1995, Technical Documentation."

23. P. Valduriez, "Join Indices," ACM Transactions on Database Systems, Vol.12, No. 2, 218-246, 1987.

Algorithms for Joining R-Trees
and Linear Region Quadtrees

Antonio Corral*, Michael Vassilakopoulos**, and Yannis Manolopoulos

Data Engineering Lab
Department of Informatics
Aristotle University
54006 Thessaloniki, Greece
acorral@ualm.es, mvass@computer.org, manolopo@csd.auth.gr

Abstract. The family of R-trees is suitable for storing various kinds of multidimensional objects and is considered an excellent choice for indexing a spatial database. Region Quadtrees are suitable for storing 2-dimensional regional data and their linear variant is used in many Geographical Information Systems for this purpose. In this report, we present five algorithms suitable for processing join queries between these two successful, although very different, access methods. Two of the algorithms are based on heuristics that aim at minimizing I/O cost with a limited amount of main memory. We also present the results of experiments performed with real data that compare the I/O performance of these algorithms.

Index terms: Spatial databases, access methods, R-trees, linear quadtrees, query processing, joins.

1 Introduction

Several spatial access methods have been proposed in the literature for storing multi-dimensional objects (e.g. points, line segments, areas, volumes, and hyper-volumes). These methods are classified in one of the following two categories according to the principle guiding the hierarchical decomposition of data regions in each method.

- Data space hierarchy: a region containing data is split (when, for example, a maximum capacity is exceeded) to sub-regions which depend on these data only (for example, each of two sub-regions contains half of the data)
- Embedding space hierarchy: a region containing data is split (when a certain criterion holds) to sub-regions in a regular fashion (for example, a square region is always split in four quadrant sub-regions)

---

* Research performed under the European Union's TMR Chorochronos project, contract number ERBFMRX-CT96-0056 (DG12-BDCN), while on leave from the University of Almeria, Spain.
** Post-doctoral Scholar of the State Scholarship-Foundation of Greece.

R.H. Güting, D. Papadias, F. Lochovsky (Eds.): SSD'99, LNCS 1651, pp. 251–269, 1999.
© Springer-Verlag Berlin Heidelberg 1999

The book by Samet [20] and the recent survey by Gaede and Guenther [6] provide excellent information sources for the interested reader.

A representative of the first principle that has gained significant appreciation in the scientific and industrial community is the R-tree. There are a number of variations of this structure all of which organize multidimensional data objects by making use of the Minimum Bounding Rectangles (MBRs) of the objects. This is an expression of the "conservative approximation principle". This family of structures is considered an excellent choice for indexing various kinds of data (like points, polygons, 2-d objects, etc) in spatial databases and Geographical Information Systems.

A famous representative of the second principle is the Region Quadtree. This structure is suitable of storing and manipulating 2-dimensional regional data (or binary images). Moreover, many algorithms have been developed based on Quadtrees [20]. The most widely known secondary memory alternative of this structure is the Linear Region Quadtree [21]. Linear Quadtrees have been used for organizing regional data in Geographical Information Systems [19].

These totally different families of popular data structures can co-exist in a Spatial Information System. For example, in his tutorial in the same conference, Sharma [23] refers to spatial and multimedia extensions to the Oracle 8i server that are based on the implementation of a linear quadtree and a modified R*-tree. Each of these structures can be used for answering a number of very useful queries. However, the processing of queries that are based on both structures has not been studied in the literature. In this report, we present a number of algorithms that can be used for processing joins between the two structures.

For example, the R-tree data might be polygonal objects that represent swimming and sun-bathing sites and the Quadtree data a map, where black color represents a decrease of ozon and white color represents ozon safe areas. A user may ask which sites suffer from the ozon problem. The major problem for answering such a query is to make use of the space hierarchy properties of each of the structures, so that not to transfer in main memory irrelevant data, or not to transfer the same data many times. Three of the proposed algorithms are simple and suffer from such unnecessary transfers, when the buffering space provided is limited. We also propose another two more sophisticated algorithms that deal with this problem by making use of heuristics and achieve good performance with a limited amount of main memory.

The organization of the paper is as follows. In Section 2, we present in brief the families of R-trees and Linear Region Quadtrees. In Section 3, we review join processing in spatial databases. In Section 4, we present the algorithms that process R-Quad Joins. More specifically, in Subsections 4.1 to 4.3 we present the three simple algorithms, in Subsection 4.4 our heuristics and the buffering scheme used and in Subsections 4.5 and 4.6 the two sophisticated algorithms. In Section 5, we present our experimental setting and some results of experiments we performed with real data. These experiments compare the I/O performance of the different algorithms. In Section 6, we summarize the contribution of this work and discuss issues that require further research in the future.

2   The Two Structures

2.1   R-Trees

R-trees are hierarchical data structures based on B$^+$-trees. They are used for the dynamic organization of a set of k-dimensional geometric objects representing them by the minimum bounding k-dimensional rectangles (in this paper we focus on 2 dimensions). Each node of the R-tree corresponds to the minimum rectangle that bounds its children. The leaves of the tree contain pointers to the objects of the database, instead of pointers to children nodes. The nodes are implemented as disk pages.

It must be noted that the rectangles that surround different nodes may be overlapping. Besides, a rectangle can be included (in the geometrical sense) in many nodes, but can be associated to only one of them. This means that a spatial search may demand visiting of many nodes, before confirming the existence or not of a given rectangle.

The rules obeyed by the R-tree are as follows. Leaves reside on the same level. Each leaf contains pairs of the form (R, O), such that R is the minimum rectangle that contains spatially object O. Every other node contains pairs of the form (R, P), where P is a pointer to a child of the node and R is the minimum rectangle that contains spatially the rectangles contained in this child. An R-tree of class (m, M) has the characteristic that every node, except possibly for the root, contains between m and M pairs, where $m \leq \lceil M/2 \rceil$. The root contains at least two pairs, if it is not a leaf. Figure 1 depicts some rectangles on the right and the corresponding R-tree on the left. Dotted lines denote the bounding rectangles of the subtrees that are rooted in inner nodes.

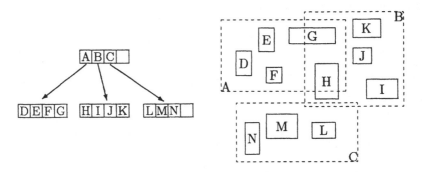

Fig. 1. An example of an R-tree

Many variations of R-trees have appeared. The most important of theses are packed R-trees [18], R$^+$trees [22] and R*-trees [2]. The R*-tree does not have the limitation for the number of pairs of each node and follows a node split technique that is more sophisticated than that of the simple R-tree. It is considered the most efficient variant of the R-tree family and, as far as searches are concerned, it can be used in exactly the same way as simple R-trees. This paper refers to simple R-trees or to R*-trees.

2.2 Region Quadtrees

The Region Quadtree is the most popular member in the family of quadtree-based access methods. It is used for the representation of binary images, that is $2^n \times 2^n$ binary arrays (for a positive integer n), where a 1 (0) entry stands for a black (white) picture element. More precisely, it is a degree four tree with height n, at most. Each node corresponds to a square array of pixels (the root corresponds to the whole image). If all of them have the same color (black or white) the node is a leaf of that color. Otherwise, the node is colored gray and has four children. Each of these children corresponds to one of the four square sub-arrays to which the array of that node is partitioned. We assume here, that the first (leftmost) child corresponds to the NW sub-array, the second to the NE sub-array, the third to the SW sub-array and the fourth (rightmost) child to the SE sub-array. For more details regarding Quadtrees see [20]. Figure 2 shows an 8×8 pixel array and the corresponding Quadtree. Note that black (white) squares represent black (white) leaves, while circles represent gray nodes.

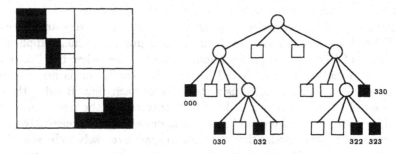

Fig. 2. An image and the corresponding Region Quadtree

Region Quadtrees, as presented above, can be implemented as main memory tree structures (each node being represented as a record that points to its children). Variations of Region Quadtrees have been developed for secondary memory. Linear Region Quadtrees are the ones used most extensively. A Linear Quadtree representation consists of a list of values, where there is one value for each black node of the pointer-based Quadtree. The value of a node is an address describing the position and size of the corresponding block in the image. These addresses can be stored in an efficient structure for secondary memory (such as a B-tree or one of its variations). There are also variations of this representation where white nodes are stored too, or variations which are suitable for multicolor images. Evidently, this representation is very space efficient, although it is not suited to many useful algorithms that are designed for pointer-based Quadtrees.

The most popular linear implementations are the FL (Fixed Length), the FD (Fixed length – Depth) and the VL (Variable length) linear implementations [21].

In the FL implementation, the address of a black Quadtree node is a code-word that consists of n base-5 digits. Codes 0, 1, 2 and 3 denote directions NW, NE, SW and SE, respectively, while code 4 denotes a do-not-care direction. If the black node resides on level i, where $n \geq i \geq 0$, then the first $n - i$ digits express the directions that constitute the path from the root to this node and the last i digits are all equal to 4. In the FD implementation, the address of a black Quadtree node has two parts: the first part is a code-word that consists of n base-4 digits. Codes 0, 1, 2 and 3 denote directions NW, NE, SW and SE, respectively. This code-word is formed in a similar way to the code-word of the FL-linear implementation with the difference that the last i digits are all equal to 0. The second part of the address has $\lceil log_2(n+1) \rceil$ bits and denotes the depth of the black node, or in other words, the number of digits of the first part that express the path to this node. In the VL implementation the address of a black Quadtree node is a code-word that consists of at most n base-5 digits. Code 0 is not used in addresses, while codes 1, 2, 3 and 4 denote one of the four directions each. If the black node resides on level i, where $n \geq i \geq 0$, then its address consists of $n - i$ digits expressing the directions that constitute the path from the root to this node. The depth of a node can be calculated by finding the smallest value equal to a power of 5 that gives 0 quotient when the address of this node is divided (using integer division) with this value.

In the rest of this paper we assume that Linear Quadtrees are represented with FD-codes stored in a B^+-tree (this choice is popular in many applications). The choice of FD linear representation, instead of the other two linear representations, is not accidental. The FD linear representation is made of base-4 digits and is thus easily handled using two bits for each digit. Besides, the sorted sequence of FD linear codes is a depth-first traversal of the tree. Since internal and white nodes are omitted, sibling black nodes are stored consecutively in the B^+-tree or, in general, nodes that are close in space are likely to be stored in the same or consecutive B^+-tree leaves. This property helps at reducing the I/O cost of join processing. Since in the same quadtree two black nodes that are ancestor and descendant cannot co-exist, two FD linear codes that coincide at all the directional digits cannot exist neither. This means that the directional part of the FD-codes is sufficient for building B^+-trees at all the levels. At the leaf-level, the depth of each black node should also be stored so that images are accurately represented. In Figure 2 you can see the directional code of each black node of the depicted tree.

3   Spatial Join Processing

In Spatial Databases and Geographical Information Systems there exists the need for processing a significant number of different spatial queries. For example, such queries are: nearest neighbor finding, similarity queries [16], window queries, content based queries [24], or spatial joins of various kinds. A spatial join consists in testing every possible pair of data elements belonging to two spatial data sets against a spatial predicate. This predicate might be *overlap, distance within, contain, intersect, etc.* In this paper we mainly focus on the *intersection spatial join* (the most widely used join type), or on spatial joins which are processed in the same way as the intersection join.

There have been developed various methods for processing spatial joins for spatial data using approximate geometry [13,14], two R-trees [3], PMR quad-trees [5], seeded trees when one [9] or none [10] of the data sets does not have a spatial index, spatial hashing [1,11,15], or sort merge join [8].

In this paper, we make the assumption that our spatial information system keeps non-regional data in R-trees or R*-trees and regional data in Linear Region Quadtrees, while users pose queries that involve both these two kinds of data. For example, the non-regional data might be cities and the regional data a map where black represents heavy clouds and white rather sunny areas. The user is very likely to ask which cities are covered with clouds.

Most spatial join processing methods are performed in two steps. The first step, which is called filtering, chooses pairs of data that are likely to satisfy the join predicate. The second step, which is called refinement, examines the predicate satisfaction for all these pairs of data. The algorithms presented in this paper, focus on the function of the filtering step and show how a number of pairs of the form (Quadtree block, MBR of object) can be produced (the two members of each pair produced intersect).

4   Join Algorithms

Before join processing, the correspondence of the spaces covered by the two structures must be established. A level-n Quadtree covers a quadrangle with $2^n \times 2^n$ pixels, while an R-tree covers a rectangle that equals the MBR of its root. Either by asking the user for input, or by normalizing the larger side of the R-tree rectangle in respect to 2^n, the correspondence of spaces may be determined. After this action, the coordinates used in the R-tree are always transformed to Quadtree pixel locations.

Joining of the two structures can be done with very simple ways, if it is ignored that both structures are kept in disk pages as multiway trees. These ways fall in two categories: either we scan the entries of the B^+-tree and perform window queries in the R-tree, or we scan the entries of the R-tree and perform window queries in the B^+-tree. More specifically, we designed and implemented the following three simple algorithms.

4.1 B^+ to R Join

- Descend the B^+-tree from the root to its leftmost leaf.
- Access sequentially (in increasing order) the FDs present in this leaf and for each FD perform a range search in the R-tree (reporting intersections of this FD and MBRs of leaves).

- By making use of the horizontal inter-leaf pointer, access the next B^+-tree leaf and repeat the previous step.

This algorithm may access a number of FDs (and the leaves in which they reside) that do not intersect with any data elements stored in the R-tree. Moreover, this algorithm is very probable to access a number of R-tree nodes several times.

4.2   R to B^+ Join with Sequential FD Access

- Traverse recursively the R-tree, accessing the MBRs in each node in order of appearence within the node.
- For the MBR of each leaf accessed, search in the B^+-tree for the FD of the NW corner of this MBR, or one of its ancestors.
- Access sequentially (in increasing order) the FDs of the B^+-tree until the FD of the SE corner of this MBR, or one of its ancestors is reached (reporting intersections of FDs and this MBR).

This algorithm may perform unnecessary accesses in both trees, while multiple accesses in B^+-tree leaves are very probable. The unnecessary accesses in the B^+-tree result from the sequential access of FDs. The following algorithm is a variation that deals with B^+-tree accessing differently.

4.3   R to B^+ Join with Maximal Block Decomposition

- Traverse recursively the R-tree, accessing the MBRs in each node in order of appearence within the node.
- For the MBR of each leaf accessed, decompose this MBR in maximal quad-tree blocks.
- For each quadblock, search in the B^+-tree for the FD of the NW corner of this quadblock, or one of its ancestors.
- Access sequentially (in increasing order) the FDs of the B^+-tree until the FD of the SE corner of this quadblock, or one of its ancestors is reached (reporting intersections of FDs and the respective MBR).

Although this algorithm saves many unnecessary FDs accessed, each search for a quadblock descends the tree. Nevertheless, the same intersection may be reported more than once. To eliminate duplicate results, a temporary list of intersections for the current leaf is maintained.

4.4   Heuristics and Buffering Scheme

In order to overcome the unnecessary and/or duplicate accesses of the previous algorithms, we propose a number of heuristics/rationales that focus on the opposite direction, that of increasing I/O performance.

- heuristic 1: Process small enough parts of the R-tree space so that the join processing of each part can be completed (in most cases) with the limited number of FD-codes that can be kept in main memory. At the presented form of the algorithm, each of these parts is a child of the root.
- heuristic 2: Process the children of an R-tree node in order that is close to the order in which the FD-codes (quadtree sub-blocks) are transferred in main memory. This order, called FD-order, is formed by sorting MBRs by the FD-code that corresponds to their NW corner.
- heuristic 3: While processing a part of the R-tree space, keep in memory only the FD-codes that may be needed at a later stage, drop all other FD-codes and fill up buffer with FD-codes that are needed but were not transferred in memory due to the buffer limit.
- heuristic 4: Use a buffer scheme for both trees that reduces the need to transfer in memory multiple times the same disk pages (explained in detail below).

A buffering scheme that obeys Heuristic 4 is presented graphically in Figure 3. In detail, this scheme is as follows.

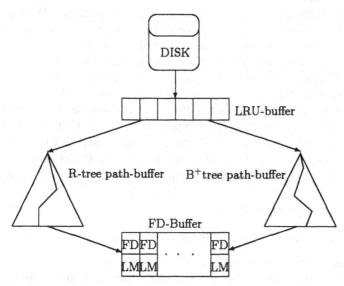

Fig. 3. The buffering scheme

- There is a path-buffer for R-tree node-pages (with number of pages equal to the height of the R-tree). However, the buffer pages of the R-tree buffer are larger than the actual R-tree disk pages, because for each entry (each MBR) an extra point is kept. This point is called START and expresses the pixel where processing of the relevant MBR has stopped (a special value, named MAX, specifies that processing of that MBR has been completed). This means that during transfers from disk to memory and the opposite an appropriate transformation of the page contents needs to be made.

- There is a path-buffer for B^+-tree node-pages (with number of pages equal to the height of the B^+-tree).
- It is assumed that the operating system keeps a large enough LRU-buffer for disk reads. The same assumption was made in [3]. This buffer is used for pages belonging in paths related to the current path that are likely to be accessed again in subsequent steps of the algorithm.
- The last buffer, called FD-buffer, is not a page buffer but one that holds the FDs needed for the processing of the current R-tree part. Each entry in this buffer contains also a level mark (LM), that is a number that expresses the level of the current R-tree path at which and below which the related FD might be needed for join processing. The size of this buffer is important for the I/O efficiency of the sophisticated algorithms.

Note that, the three simple algorithms described above can be easily made more effective by using a path-buffer and an LRU-buffer for each tree. As will be demonstrated in the experimentation section, by using adequately large LRU-buffers, the performance of the simple algorithms is comparable to that of the sophisticated ones.

The searching method used in the algorithms is as follows. When, for a point belonging in an R-tree MBR, the existence of a Linear Quadtree code that covers this point (its block contains this point) needs to be determined, we search the B^+-tree for the maximum FD-code M that is smaller than or equal to the FD-code P of the pixel related to this point. If $M = P$ and depth(M) = depth(P), then this specific black pixel exists in the Quadtree. If $M \leq P$, depth(M) < depth(P) and the directional codes of M and P coincide in the first depth(M) bits, then M is a parent of P (it represents a block that contains P). This searching method is used in lines 32 and 47 of the "One level FD-buffer join" and "Many levels FD-buffer join" algorithms, respectively. In the following, these two algorithms, which are designed according to the above heuristics, are presented.

4.5   One Level FD-Buffer Join

In very abstract terms, this algorithm works as follows:

- Process the children of the R-tree root in FD-order.
- Read as many FDs as possible for the current child and store them in FD-buffer.
- Call recursively the Join routine for this child.
- When the Join routine returns, empty the FD-buffer and repeat the previous two steps until the current child has been completely checked.
- Repeat for the next child of the root.

The Join routine for a node works as follows:

- If the node is a leaf, check intersections and return.
- If not (this is a non-leaf node), for each child of the node that has not been examined in relation to the FDs in FD-buffer, call the Join routine recursively.

In pseudo-code form this algorithm is as follows:

```
01  insert R-tree root in path-buffer;
02  for every MBR x in R-tree root
03      START(x) := NW-corner-of(x);
04  order MBRs in R-tree root according to FD of their START;
05  for every MBR x in R-tree root, in FD-order
06      while START(x) < MAX begin
07      |   read-FDs-in-buffer(x);
08      |   R-Quad-Join(node of x);
09      |   remove every FD from FD-buffer;
10      end;

11  Procedure R-Quad-Join(Z: R-tree node);
12  begin
13      if Z is not in path-buffer
14          insert Z in path-buffer;
15      if Z is internal then begin
16      |   for every MBR x in Z
17      |       START(x) := NW-corner-of(x);
18      |   order MBRs in Z according to FD of their START;
19      |   for every MBR x in Z, in FD-order
20      |       if START(x) < START(MBR of Z) begin
21      |       |   START(x) := first pixel of x after the last FD accessed,
        |       |   or MAX (if no such pixel exists);
22      |       |   if START(x) ≠ MAX or at least one FD in FD-buffer intersects x
23      |       |       R-Quad-Join(node of x);
24      |       end;
25      end
26      else
27          check and report possible intersection of MBR of Z and
            every FD in FD-buffer;
28  end;

29  Procedure read-FDs-in-buffer(Z: MBR);
30  begin
31      while START(Z) < MAX and FD-buffer not full begin
32      |   search in QuadTree for FD f covering START(Z) or
        |   for the next FD (in FD-order);
33      |   if no FD was accessed
34      |       f := MAX;
35      |   if f intersects Z
36      |       store f in FD-buffer;
37      |   START(Z) := first pixel of Z after f, or MAX (if no such pixel exists);
38      end;
39  end;
```

In Figure 4, a simple example demonstrating the simultaneous subdivision of space by a Quadtree for an 8×8 image (thin lines) and part of an R-tree (thick dashed lines) is depicted. The Quadtree contains two black quadrangles (the two dotted areas), Q1 and Q2. The MBR of the root of the R-tree is rectangle

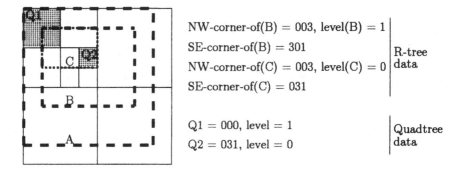

NW-corner-of(B) = 003, level(B) = 1	
SE-corner-of(B) = 301	R-tree
NW-corner-of(C) = 003, level(C) = 0	data
SE-corner-of(C) = 031	
Q1 = 000, level = 1	Quadtree
Q2 = 031, level = 0	data

Fig. 4. An example showing Quadtree (thin lines) and R-tree (thick dashed lines) subdivision of space.

A. However, in Figure 4, only one of the children of this root is depicted, the one with MBR B. Moreover, a child of B is depicted, the one with MBR C. We suppose that C is a leaf of the R-tree and it has no intersection with the other children of B (not shown in the figure). Consider a trivial situation, where the FD-buffer of the algorithm presented above can host one FD-code and the corresponding level mark (LM), only. Let's trace the most important steps of a partial run of this algorithm for this specific case.

1	START(B) = 003;	(l.3)
2	read-FDs-in-buffer(B);	(l.7)
3	search reaches Q1	(l.32)
4	FD-buffer: f = 000	(l.36)
5	START(B) = 012;	(l.37)
6	R-Quad-Join (B);	(l.8)
7	Since B is internal...	(l.15)
8	START(C) = 003;	(l.17)
9	C is the first MBR in B	(l.18)
10	Since START(C) < START(B)...	(l.20)
11	START(C) = 012;	(l.21)
12	R-Quad-Join(C);	(l.23)
13	Since C is a leaf...	(l.15)
14	report intersection of C and Q1;	(l.27)
15	R-Quad-Join(C) returns;	(l.28)
16	Other children of B are processed	(l.19)
17	R-Quad-Join(B) returns;	(l.28)
18	remove 000 from FD-buffer;	(l.9)
19	read-FDs-in-buffer(B);	(l.7)
20	search reaches Q2	(l.32)
21	FD-buffer: f = 031	(l.36)

22	START(B) := 032;	(l.37)
23	R-Quad-Join (B);	(l.8)
24	Since B is internal...	(l.15)
25	START(C) = 003;	(l.17)
26	C is the first MBR in B	(l.18)
27	Since START(C) < START(B)...	(l.20)
28	START(C) = MAX;	(l.21)
29	R-Quad-Join(C);	(l.23)
30	Since C is a leaf...	(l.15)
31	report intersection of C and Q2;	(l.27)
32	R-Quad-Join(C) returns;	(l.28)
33	Other children of B are processed	(l.19)
34	R-Quad-Join(B) returns;	(l.28)
35	remove 031 from FD-buffer;	(l.9)

The run of the algorithm continues with the next loop for B. The interested reader can trace the same example with FD-buffer size equal to 2 and note the differences. This algorithm only fetches and releases FDs at the level of the children of the root. For such a case, the LM for each FD is not necessary to be kept in the FD-buffer.

4.6   Many Levels FD-Buffer Join

This algorithm, follows the same basic steps, however, it releases from the FD-buffer the FDs that will no longer be needed in the current phase of the algorithm as soon as possible and fills it up. Again, in very abstract terms, this algorithm works as follows:

- Process the children of the R-tree root in FD-order.
- Read as many FDs as possible for the current child and store them in FD-buffer.
- Call recursively the Join routine for this child.
- When the Join routine returns, repeat the previous two steps until the current child has been completely checked.
- Repeat for the next child of the root.

The Join routine for a node works as follows:

- If the node is a leaf, remove from FD-buffer all FDs that will not be needed in the current phase of the algorithm, check intersections and return.
- If not (this is a non-leaf node), for each child of the node that has not been examined in relation to the FDs in FD-buffer, mark the FDs that only affect the results for this child and call the Join routine recursively.
- When all the children of the node have been examined, reorder them and repeat the previous step until all the children have been examined with the current state of the FD-buffer.
- Remove from FD-buffer all FDs that will not be needed in the current phase of the algorithm and return.

Due to the possibility to remove FDs from FD-buffer at any level, an extra variable, called NEXT, that keeps track of the pixel where fetching of FDs has stopped is needed in this algorithm. In pseudo-code form the algorithm is as follows:

```
01  insert R-tree root in path-buffer;
02  for every MBR x in R-tree root
03      START(x) := NW-corner-of(x);
04  order MBRs in R-tree root according to FD of their START;
05  for every MBR x in R-tree root, in FD-order begin
06  |   NEXT := NW-corner-of(x);
07  |   while START(x) < MAX begin
08  |   |   read-FDs-in-buffer(node of x);
09  |   |   R-Quad-Join(node of x);
10  |   end;
11  end;

12  Procedure R-Quad-Join(Z: R-tree node);
13  begin
14      if Z is not in path-buffer
15          insert Z in path-buffer;
16      if Z is internal node then begin
17      |   for every MBR x in Z
18      |       START(x) := first pixel of x ≥ START(MBR of Z),
        |       or MAX (if no such pixel exists);
19      |   order MBRs in Z according to FD of their START;
20      |   repeat-flag := True;
21      |   while repeat-flag begin;
22      |   |   repeat-flag := False;
23      |   |   for every MBR x in Z, in FD-order
24      |   |       if START(x) ≤ SE-corner-of(last FD accessed) and
        |   |       START(x) ≠ MAX begin
25      |   |       |   repeat-flag := True;
26      |   |       |   if FD-buffer not empty
27      |   |       |       for every f in FD-buffer with LM(f) = level-of(Z)
        |   |       |       such that x is intersected by f and
        |   |       |       SE-corner-of(f) < START(y), ∀ y ≠ x intersected by f
28      |   |       |           LM(f) := level-of(node of x);
29      |   |       |   if at least one FD in FD-buffer intersects x begin
30      |   |       |   |   R-Quad-Join(node of x);
31      |   |       |   |   read-FDs-in-buffer(Z);
32      |   |       |   end
33      |   |       |   else
34      |   |       |       START(x) := first pixel of x after the last FD accessed, or
        |   |       |       MAX (if no such pixel exists);
35      |   |       end;
36      |   |   order MBRs in Z according to FD of their START;
37      |   end;
38      end
```

39   **else**
40   check and report possible intersection of MBR in Z and every
          FD f in FD-buffer, such that SE-corner-of(f) $\geq$ START(MBR of Z);
41   START(MBR of Z) := first pixel of MBR of Z after the last FD accessed, or
      MAX (if no such pixel exists);
42   remove from FD-buffer every FD f with LM(f) = level-of(Z);
43 **end;**

44 **Procedure** read-FDs-in-buffer(Z: R-tree node);
45 **begin**
46   **while** NEXT < MAX **and** FD-buffer not full **begin**
47 | search in QuadTree for FD f covering NEXT
      | or for the next FD (in FD-order);
48 | **if** no FD was accessed
49 | f := MAX;
50 | **if** f intersects the active MBR in the top path-buffer node **begin**
51 | | LM(f) := level-of(top path-buffer node) - 1;
52 | | **while** LM(f) > level-of(Z) **and** f intersects only the active MBR
      | | in the next lower path-buffer node
53 | | LM(f) := LM(f) - 1;
54 | | store f in FD-buffer
55 | **end;**
56 | NEXT := first pixel of the active MBR in the top path-buffer node after f,
      | or MAX (if no such pixel exists);
57   **end;**
58 **end;**

5   Experimentation

We performed experiments with Sequoia data from the area of California. The
point data correspond to specific country sites, while regional data correspond
to three different categories: visible, emitted infrared and reflected infrared spec-
trums. We performed experiments for all the combinations of point and regional
data. The query used for all experiments is the intersection join query between
these two kinds of data: "which country sites are covered with colored regions?".

There were many parameters that varied in these experiments. Each pixel
of the regional data had a range of 256 values. Each image was converted to
black and white by choosing a threshold accordingly so as to achieve a requested
black-white analogy. This analogy ranged between 20% and 80%. The images
in our experiments were 1024×1024 and 2048×2048 pixel large. The cardinality
of the point set was 68764. The page size for both trees was 1024 bytes. Under
these conditions, both R and B$^+$ trees had 4 levels (including the leaf level). The
FD-buffer size for the two sophisticated algorithms ranged between 150 and 2500
FDs. The LRU-buffer size for each tree ranged between 0 and 40 pages.

In the following, some characteristic results of the large number of experiments performed for images of 2048×2048 pixels from the visible spectrum are

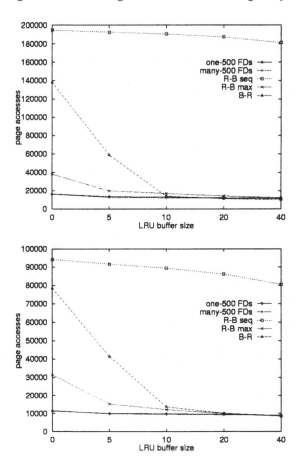

Fig. 5. The performance of the five algorithms as a function of LRU-buffer size for 50% (upper diagram) and 80% (lower diagram) black images.

depicted. Note that, since $n = 11$, an FD for such an image requires $2 \times 11 + \lceil log_2(11 + 1) \rceil = 26$ bits. In the upper (lower) part of Figure 5 the number of disk accesses during join for each of the five algorithms, when images are 50% (80%) black, as a function of LRU-buffer size is shown. More specifically, "one-500 FDs" stands for "One level FD-buffer join" with FD-buffer holding up to 500 FDs, "many-500 FDs" for "Many levels FD-buffer join" with FD-buffer holding up to 500 FDs, "R-B seq" for "R to B^+ Join with sequential FD access", "R-B max" for "R to B^+ Join with maximal block decomposition" and "B-R" for "B^+ to R Join". It is evident that R to B^+ Join with sequential FD access has the worst performance which is not improved as the LRU-buffers size increases. Nevertheless, R to B^+ Join with maximal block decomposition and B^+ to R Join gets improved and achieves comparable performance to the sophisticated algorithms as the LRU-buffers size increases. The two sophisticated algorithms perform well, even for small LRU-buffers.

To study the situation more closely, in Figure 6 we present performance results for the two simple algorithms that have better performance and two versions (for FD-buffer size equal to 500 and to 1500 FDs) of the sophisticated algorithms. The diagram on the upper (lower) part corresponds to 50% (80%) black images. We can easily see that the sophisticated algorithms perform very well for a wide range of LRU- and FD-buffers sizes, while the two simple algorithms achieve comparable, or even better, performance than the sophisticated ones, when the LRU-buffers size increases.

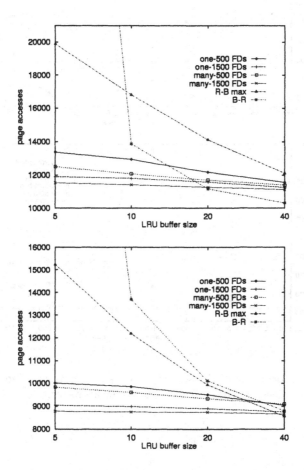

Fig. 6. The performance of two of the simple and two versions of the sophisticated algorithms as a function of LRU-buffer size for 50% (upper diagram) and 80% (lower diagram) black images.

Finally, in Figure 7 the performance of sophisticated algorithms for various kinds of images as a function of a combination of LRU-buffers size and FD-buffers size is depicted. Many levels FD-buffer join performs slightly better than

One level FD-buffer join for all cases. Note that the difference gets smaller when FD-buffer size increases.

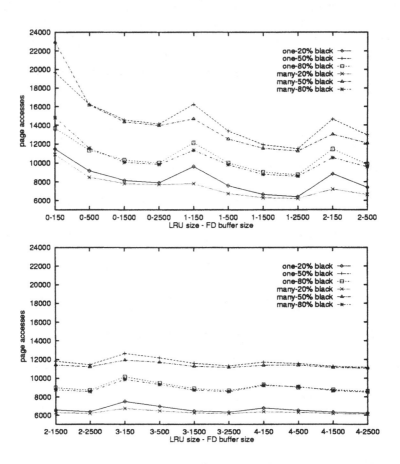

Fig. 7. The performance of the sophisticated algorithms as a function of LRU- and FD-buffers sizes for 20%, 50% and 80% black images.

6   Conclusions

In this report, we presented five algorithms for processing of joins between two popular but different structures used in spatial databases and Geographic Information Systems, R-trees and Linear Region Quadtrees. These are the first algorithms in the literature that process joins between these two structures. Three of the algorithms are simple, but suffer from unnecessary and/or repeated disk accesses, when the amount of memory supplied for buffering is limited. The other two are more sophisticated and are based on heuristics that aim at minimizing the I/O cost of join processing. That is, they try to minimize the transfer to main memory of irrelevant data or the multiple transfer of the same data. Moreover, we presented results of experiments performed with real data. These experiments investigate the I/O performance of the different join algorithms.

The presented results show that

- better performance is achieved when using the sophisticated algorithms in a system with small LRU-buffer. For example, in a system with many users, where buffer space is used by many processes, the sophisticated algorithms are the best choice. Besides, the sophisticated algorithms are quite stable. That is, their performance is not heavily dependent on the increase of available main memory.
- When there is enough main memory (when the LRU-buffer is big enough), one of the simple algorithms, the B^+ to R Join algorithm, performs very well.

Intuition leads us to believe that the sophisticated algorithms are expected to perform better for data that obey unusual distributions, since they are designed to partially adapt to the data distribution.

The presented algorithms perform only the filtering step of the join. Processing of the refinement step requires the choice and use of Computational Geometry algorithms [12,17], which is among our future plans in the area of this topic. Each choice should take into account not only worst-case complexity, but expected sizes of data sets, average case complexity and multiplicative constant of complexity as well, for each alternative. In addition, these algorithms could be tested on other kinds of data, e.g. making use of R-trees as a storage medium for region data as well. Moreover, we plan to elaborate the presented heuristics even further and/or examine different policies of page replacement (other than LRU) so as to improve performance, or master the worst case behavior of the join algorithms (when they deal with "pathological" data, deviating from the usual situations in practice). Another route of research would be to examine the parallel processing of this kind of joins, based on ideas presented in [4].

References

1. L. Arge, O. Procopiuc, S. Ramaswamy, T. Suel and J.S. Vitter, "Scalable Sweeping-Based Spatial Join", *Proceedings of the 24th VLDB conference*, New York, 1998, pp. 570-581.
2. N. Beckmann, H.P. Kriegel, R. Schneider and B. Seeger, "The R*-tree: an Efficient and Robust Access Method for Points and Rectangles", *Proceedings of the 1990 ACM SIGMOD Conference*, Atlantic City, NJ, pp. 322-331.
3. T. Brinkhoff, H.P. Kriegel and B. Seeger, "Efficient Processing of Spatial Joins Using R-trees", *Proceedings of the 1993 ACM SIGMOD Conference*, Washington, DC, pp. 237-246.
4. T. Brinkhoff, H.P. Kriegel and B. Seeger, "Parallel Processing of Spatial Joins Using R-trees", *International Conference on Data Engineering*, New Orleans, 1996, pp. 258-265.

5. E.G. Hoel and H. Samet, "Benchmarking Spatial Join Operations with Spatial Output", *Proceedings of the 21st VLDB conference*, Zurich, 1995, pp. 606-618.
6. V. Gaede and O. Guenther: "Multidimensional Access Methods", *ACM Computer Surveys*, Vol 30, No 2, 1998, pp. 170-231.
7. A. Guttman: "R-trees - a Dynamic Index Structure for Spatial Searching", *Proceedings of the 1984 ACM SIGMOD Conference*, Boston, MA, pp. 47-57.
8. N. Koudas and K.C. Sevcik, "Size Separation Spatial Join", *Proceedings of the 1997 ACM SIGMOD Conference*, Tuscon, pp. 324-335.
9. M.L. Lo and C.V. Ravishankar, "Spatial Joins Using Seeded Trees", *Proceedings of the 1994 ACM SIGMOD Conference*, Minneapolis, pp. 209-220.
10. M.L. Lo and C.V. Ravishankar, "Generating Seeded Trees From Data Sets", *Proceedings of the 4th International Symposium on Large Spatial Databases*, Portland, ME, 1995, pp. 328-347.
11. M.L. Lo and C.V. Ravishankar, "Spatial Hash-Joins", *Proceedings of the 1996 ACM SIGMOD Conference*, Montreal, Canada, pp. 247-258.
12. K. Mehlhorn, "Data Structures and Algorithms 3: Multi-dimentional Searching and Computational Geometry", *Springer Verlag*, 1984.
13. J.A. Orenstein and F.A. Manola, "PROBE Spatial Data Modeling and Query Processing in an Image Database Application", *IEEE transactions on Software Engineering*, Vol 14, No 5, 1988, pp. 611-629.
14. J.A. Orenstein, "Spatial Query Processing in an Object Oriented Database System", *Proceedings of the 1986 ACM SIGMOD Conference*, pp. 326-336.
15. J.M. Patel and D.J. DeWitt, "Partition Based Spatial Merge-Join", *Proceedings of the 1996 ACM SIGMOD Conference*, Montreal, pp. 259-270.
16. A. Papadopoulos and Y. Manolopoulos, "Similarity Query Processing using Disk Arrays", *Proceedings of the 1998 ACM SIGMOD Conference*, Seattle, pp. 225-236.
17. F.P. Preparata and M.I. Shamos, "Computational Geometry", *Springer Verlag*, 1988.
18. N. Roussopoulos and D. Leifker, "Direct Spatial Search on Pictorial Databases using Packed R-trees", *Proceedings of the 1985 ACM SIGMOD Conference*, Austin, TX, pp. 17-31.
19. H. Samet, C.A. Shaffer, R.C. Nelson, Y.G. Huang, K. Fujimura and A. Rosenfeld, "Recent Developments in Linear Quadtree-based Geographic Information Systems", *Image and Vision Computing*, Vol 5, No 3, 1987, pp. 187-197.
20. H. Samet: "The Design and Analysis of Spatial Data Structures", *Addison-Wesley*, Reading MA, 1990.
21. H. Samet: "Applications of Spatial Data Structures", *Addison-Wesley*, Reading MA, 1990.
22. T. Sellis, N. Roussopoulos and C. Faloutsos: "The R+tree - a Dynamic Index for Multi-Dimensional Objects", *Proceedings of the 13th VLDB conference*, 1987, pp. 507-518.
23. J. Sharma: "Implementation of Spatial and Multimedia Extensions in Commercial Systems", tutorial during the *6th International Symposium on Spatial Databases*, Hong Kong, July 20-23, 1999.
24. M. Vassilakopoulos and Y. Manolopoulos, "Dynamic Inverted Quadtree: a Structure for Pictorial Databases", *Information Systems (special issue on Multimedia)*, Vol 20, No 6, 1995, pp. 483-500.

Algorithms for Performing Polygonal Map Overlay and Spatial Join on Massive Data Sets

Ludger Becker, André Giesen, Klaus H. Hinrichs, and Jan Vahrenhold

FB 15, Mathematik und Informatik, Westfälische Wilhelms-Universität Münster,
Einsteinstr. 62, D–48151 Münster, Germany
{beckelu,vanessa,khh,jan}@math.uni-muenster.de

Abstract. We consider the problem of performing polygonal map overlay and the refinement step of spatial overlay joins. We show how to adapt algorithms from computational geometry to solve these problems for massive data sets. A performance study with artificial and real-world data sets helps to identify the algorithm that should be used for given input data.

1   Introduction

During the last couple of years Spatial- and Geo-Information Systems (GIS) have been used in various application areas, like environmental monitoring and planning, rural and urban planning, and ecological research. Users of such systems frequently need to combine two sets of spatial objects based on a spatial relationship.

In two-dimensional vector based systems combining two maps m_1 and m_2 consisting of polygonal objects by *map overlay* or *spatial overlay join* is an important operation. The *spatial overlay join* of m_1 and m_2 produces a set of pairs of objects (o_1, o_2) where $o_1 \in m_1, o_2 \in m_2$, and o_1 and o_2 intersect. In contrast, the *map overlay* produces a set of polygonal objects consisting of the following objects:

- All objects of m_1 intersecting no object of m_2
- All objects of m_2 intersecting no object of m_1
- All polygonal objects produced by two intersecting objects of m_1 and m_2

The map overlay operation can be considered a *special outer join*. For simplicity of presentation, we assume that all objects of map m_1 are marked red and that all objects of map m_2 are marked blue. Usually spatial join operations are performed in two steps [25]:

- In the *filter step* an—usually conservative—approximation of each spatial object, e.g., the minimum bounding rectangle, is used to eliminate objects that cannot be part of the result.
- In the *refinement step* each pair of objects passing the filter step is examined according to the spatial join condition.

R.H. Güting, D. Papadias, F. Lochovsky (Eds.): SSD'99, LNCS 1651, pp. 270–285, 1999.
© Springer-Verlag Berlin Heidelberg 1999

Although there exists a variety of algorithms for realizing the filter step of the join operator for massive data sets [3,7,19,22,28], not much research has been done in realizing the refinement step of the map overlay operation for large sets of polygonal objects, with the exception of Kriegel et al. [21]. Brinkhoff et al. [8] and Huang et al. [20] present methods to speed up the refinement step of the spatial overlay join by either introducing an additional filter step which is based on a progressive approximation or exploiting symbolic intersection detection. Both approaches allow the early identification of some pairs of intersecting objects without investigating the exact shapes. However, they do not solve the map overlay problem.

In this paper we discuss how well known main-memory algorithms from computational geometry can help to perform map overlay and spatial overlap join in a GIS. More specifically, we show how the algorithm by Nievergelt and Preparata [24] and a new modification of Chan's line segment intersection algorithm [12] can be extended to cope with massive real-world data sets. Furthermore, we present an experimental comparison of these two algorithms showing that the algorithm by Nievergelt and Preparata performs better than the modified algorithm by Chan, if the number of intersection points is sublinear, although Chan's algorithm outperforms other line segment intersection algorithms [2]. In this paper we do not consider algorithms using a network oriented representation, e.g., [11,14,16,23], since many GIS support only polygon oriented representations, e.g., ARC/INFO [15].

The remainder of this paper is structured as follows. In Section 2 we review the standard algorithm by Nievergelt and Preparata for overlaying two sets of polygonal objects and show how to modify Chan's algorithm for line intersection in order to perform map overlay. In Section 3 we discuss extensions required to use these algorithms for practical massive data sets. The results of our experimental comparison are presented in Section 4.

2   Overlaying Two Sets of Polygonal Objects

Since the K intersection points between line segments from different input maps (N line segments in total) are part of the output map, line segment intersection has been identified as one of the central tasks of the overlay process [2]. Methods for finding these intersections can be categorized into two classes: algorithms which rely on a partitioning of the underlying space, and algorithms exploiting a spatial order defined on the segments. While representatives of the first group, e.g., the method of *adaptive grids* [17], tend to perform very well in practical situations, they cannot guarantee efficient treatment of all possible configurations. The worst-case behavior of these algorithms may match the complexity of a brute-force approach, whereas the behavior for practical data sets often depends on the distribution of the input data in the underlying space rather than on the parameters N and K. Since we will examine the practical behavior of overlay algorithms under varying parameters N and K (Section 4), we restrict our study to algorithms that do not rely on a partitioning of the underlying space.

Several efficient algorithms for the line segment intersection problem have been proposed [4,6,12,13,23]; most of these are based on the *plane-sweep* paradigm [29], a framework facilitating the establishment of a spatial order.

The characterizing feature of the plane-sweep technique is an (imaginary) line that is swept over the entire data set. For sake of simplicity, this *sweep-line* is usually assumed to be perpendicular to the x-axis of the coordinate system and to move from left to right. Any object intersected by the sweep-line at $x = t$ is called *active at time t*, and only active objects are involved in geometric computations at that time. These active objects are usually stored in a dictionary called *y-table*. The status of the y-table is updated as soon as the sweep-line moves to a point where the topology of the active objects changes discontinuously: for example, an object must be inserted in the y-table as soon as the sweep-line hits its leftmost point, and it must be removed after the sweep-line has passed its rightmost point. For a finite set of objects there are only finitely many points where the topology of the active objects changes discontinuously; these points are called events and are stored in increasing order of their x-coordinates as an ordered sequence, e.g, in a priority queue. This data structure is also called *x-queue*. Depending on the problem to be solved there may exist additional event types.

In the following two sections we recall the plane-sweep algorithm by Bentley and Ottmann [6] and its extension by Nievergelt and Preparata [24] to compute the overlay of polygonal objects. These algorithms exploit an order established during the plane-sweep: for a set of line segments we can define a total order on all segments active at a given time by means of the aboveness relation. An active segment a lying above an active segment b is considered to be "greater" than b. This order allows for efficient storage of all active segments in an ordered dictionary (y-table), e.g., in a balanced binary tree which in turn provides for fast access to the segments organized by it.

2.1   The Line Segment Intersection Algorithm by Bentley and Ottmann

The key observation for finding the intersections is that two segments intersecting at $x = t_0$ must be direct neighbors at $x = t_0 - \varepsilon$ for some $\varepsilon > 0$. To find all intersections it is therefore sufficient to examine all pairs of segments that become direct neighbors during the sweep. The neighborhood relation changes only at event points and only for those segments directly above or below the segments inserted, deleted, or exchanged at that point. Thus, we have to check two pairs of segments at each *insert* event and one pair of segments at each *delete* event. If we detect an intersection between two segments we do not report it immediately but insert a new *intersection* event in the x-queue. If we encounter an intersection event we report the two intersecting segments and exchange them in the y-table to reflect the change in topology. The neighbors of these segments change, and two new neighboring pairs need to be examined. Since there is one insert and one delete event for each of the N input segments and one event for each of the K intersection points the number of events stored in the x-queue is bounded by $\mathcal{O}(N + K)$. At most N segments can be present in the y-table at any time. If we realize both the x-queue and the y-table so that they allow updates in logarithmic time, the overall running time of the algorithm can be bounded by $\mathcal{O}((N + K) \log N)$. Brown [10] showed how to reduce the space requirement to $\mathcal{O}(N)$ by storing at most one intersection event per active segment. Pach and Sharir [27] achieve the same reduction by storing only intersection points of pairs of segments that are currently adjacent on the sweep-line.

2.2   The Region-Finding Algorithm by Nievergelt and Preparata

Two neighboring segments in the y-table can also be considered as bounding edges of the region between them. This viewpoint leads to an algorithm for computing the regions formed by a self-intersecting polygon as proposed by Nievergelt and Preparata [24]. Each active segment s maintains information about the regions directly above and directly below in two lists $A(s)$ and $B(s)$. These lists store the vertices of the regions swept so far and are updated at the event points. Nievergelt and Preparata define four event types as illustrated in Figure 1.

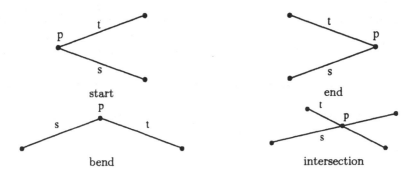

Fig. 1. Four event types in Nievergelt and Preparata's algorithm [24].

At each event point p the segments s and t are treated as in the line segment intersection algorithm described in Section 2.1. In addition, new regions are created at start and intersection events, while existing regions are closed or merged at intersection and end events. Bend events cause no topological modifications but only the replacement of s by t and the augmentation of the involved vertex lists by p. With a careful implementation the region lists can be maintained without asymptotic overhead so that the algorithm detects all regions induced by N line segments and K intersection points in $\mathcal{O}((N + K) \log N)$ time and $\mathcal{O}(N + K)$ space.

Although there are only $\mathcal{O}(N)$ active *segments* at any given time, the combinatorial complexity of all active *regions*, i.e., the total number of points on their boundaries, is slightly higher.

Lemma 1. *The combinatorial complexity of active regions at any given time during the region-finding algorithm is bounded by $\mathcal{O}(N\alpha(N))$, where $\alpha(N)$ is the inverse Ackermann function.*

Proof. [1] Consider the sweep-line at time t, cut all segments crossing the sweep-line at $x = t$, and discard all segments and fragments to the right of the sweepline. This results in an arrangement of at most N segments. All active regions now belong to one single region of this arrangement, namely to the outer region. Applying a result by Sharir and Agarwal [30, Remark 5.6] we immediately obtain an upper bound of $\mathcal{O}(N\alpha(N))$ for the combinatorial complexity of the active regions to the left of the sweep-line. The same argument obviously holds for the combinatorial complexity of the active regions to the right of the sweep-line.

Taking into account Lemma 1 we can bound the space requirement for the region-finding algorithm by $\mathcal{O}(N\alpha(N))$. To do so we modify the Bentley-Ottmann algorithm according to Brown's comments [10] and report all regions as soon as they have been swept completely.

To use this algorithm for overlaying a set of red polygonal objects and a set of blue polygonal objects we store the endpoints of the corresponding blue and red segments in the x-queue in sorted order. During the sweep we can determine for each computed region whether it is an original region, i.e., a red or blue region, or resulting from the intersection of a red and a blue region. In the latter case, we can combine the attributes of two overlapping regions by an application-specific function to obtain the attribute of each new region if we store the attribute of each region for each of its bounding segments.

2.3 The Red/Blue Line Segment Intersection Algorithm by Chan

The algorithm described in the previous section can be used to determine the intersection points induced by any set of line segments. In many GIS applications, however, we know that the line segments in question belong to two distinct sets, each of which is intersection-free. Chan [12] proposed an algorithm that computes all K intersection points between a red intersection-free set of line segments and a blue intersection-free set of line segments with a total number of N segments in optimal $\mathcal{O}(N \log N + K)$ time and $\mathcal{O}(N)$ space. Since this algorithm forms the basic building block of an overlay algorithm which we present in the following sections, we explain it in more detail than the previous algorithms.

The asymptotic improvements are achieved by computing intersection points by looking backwards at each event instead of computing intersection points in advance and storing them in the x-queue. Since intersection events are used to determine intersection points and to update the status of the sweep-line (via exchanging the intersecting segments) it is not possible to maintain both red and blue segments in a single y-table without storing intersection events. Chan solves this conflict by introducing two y-tables, one for each set of line segments. Since segments within each set do not intersect both y-tables can be maintained without exchanging segments. The events, i.e., the endpoints of the segments, are maintained in one x-queue, and depending on the kind and color of the event point the segment starting or ending at that point is inserted in or deleted from the corresponding y-table. For the analysis of the time complexity we assume that the table operations *insert*, *delete*, and *find* can be handled in $\mathcal{O}(\log N)$ time, and that the successor and predecessor of a given table item can be determined in constant time.

The algorithm can be explained best by looking at the trapezoidal decomposition of the plane induced by the blue segments and the endpoints of the blue and the red segments. Each endpoint produces vertical extensions ending at the blue segments directly above and below this point. These extensions are only imaginary and are neither stored nor involved in any geometric computation during the algorithm. Figure 2(a) shows the trapezoidal decomposition induced by a set of blue segments (solid lines) and the endpoints of both these and the red segments (dashed lines).

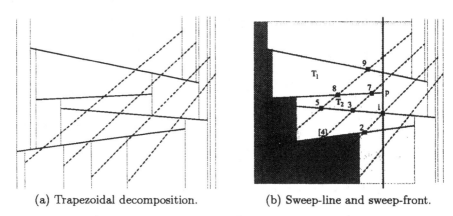

(a) Trapezoidal decomposition.   (b) Sweep-line and sweep-front.

Fig. 2. Trapezoid sweep as proposed by Chan [12].

At each event p the algorithm computes all intersections between the *active* red segments and the boundaries of the active blue trapezoids that lie to the left or on the sweep-line and have not been reported yet. The trapezoids whose right boundaries correspond to the event p are then added to the imaginary sweep-front, the collection of trapezoids that have been swept completely so far. An example is depicted in Figure 2(b): all active trapezoids are shaded light grey, and the sweep-front is shaded dark grey. After having processed the event p the trapezoids T_1 and T_2 will be added to the sweep-front.

The blue boundary segments of the trapezoid(s) ending at p are processed in ascending order: starting from the red segment directly above the current blue segment we check the active red segments in ascending order for intersections to the left of the sweep-line with the current blue segment and its predecessors in the blue y-table. The algorithm locates each red segment in the blue y-table and traces it to its left endpoint while reporting all relevant intersections with active blue segments. Note that active red segments and active blue segments might intersect outside the active trapezoids, i.e., on the boundary of trapezoids already in the sweep-front (for example, the intersection points [4] and [6] in Figure 2(b)).

To avoid multiple computation of such points, each active red segment stores its intersection point with the largest x-coordinate (if any) and stops finding intersections for a given red segment as soon as this most recent intersection point is detected again, or if no intersection between a red segment and the current blue segment has been found. In the first situation the algorithm starts examining the next active red segment, whereas in the latter situation we know that no red segment above the current segment can have an undetected intersection with this blue segment. Therefore the red segment directly below the current blue and its predecessors in the red y-table are tested for intersection with the current blue segment. If these tests are finished we process the next blue boundary segment of a trapezoid ending at p. In Figure 2(b) the intersection points are labeled according to the order in which they are detected: black squares indicate new intersection points, and white squares indicate intersection points found during the processing of earlier events.

Since each of the N segments requires one start event and one end event and is stored in exactly one y-table, the space requirement for this algorithm is $\mathcal{O}(N)$. Initializing the x-queue and maintaining both y-tables under N insertions and N deletions can be done in $\mathcal{O}(N \log N)$ time. To find the red segment directly above each event point and to locate this segment in the blue y-table we need additional $\mathcal{O}(\log N)$ time per event. Tracing a red segment through the blue table takes $\mathcal{O}(1)$ time per step. Except for the last step per segment each trace step corresponds to a reported intersection, and there are at most two segments examined per event that do not contribute to the intersections reported at that event. Thus, the number of trace steps performed at one event point is of the same order as the number of intersections reported at that event point resulting in a total number of $\mathcal{O}(K)$ trace steps during the algorithm. The time complexity of this line segment intersection algorithm is $\mathcal{O}(N \log N + K)$, which is optimal [23]. For a complete description we refer the reader to the original paper [12].

2.4   Region-Finding Based upon Red/Blue Line Segment Intersection

The key to an efficient region-finding algorithm based upon Chan's algorithm is to use the lists $A(s)$ and $B(s)$ introduced by Nievergelt and Preparata. Recall that these lists store the vertices of the regions above and below each active segment s and are updated at the event points. Updating these lists at endpoints of segments or intersection points is done analogously to the algorithm of Nievergelt and Preparata. Since there are no explicit intersection events in Chan's algorithm, we have to ensure that these updates are done in the proper order. Consider the situation in Figure 3(a): if we try to process the intersection points in the standard order we are not able to describe the region ending at intersection point 1 since we do not know the intersection points 2 and 3 also describing the region (intersection point [4] has been detected at a previous event). If we had detected the points in the order shown in Figure 3(b) we would have found the intersection points 4 and 5 prior to intersection point 6. To do so, we only need to report the intersection points in reverse order. It is not hard to show that intersection points are then reported in lexicographical order along each blue and red segment, and thus that all events describing a region are reported in the proper order. We omit this proof due to space constraints.

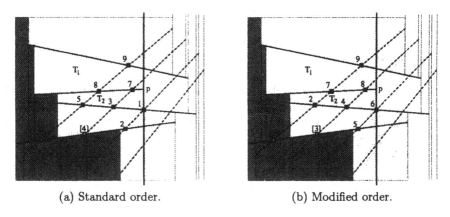

(a) Standard order.   (b) Modified order.

Fig. 3. Intersection points detected during region-finding.

As with the algorithm by Nievergelt and Preparata, maintaining the region list does not asymptotically increase the time spent per event, and thus we can compute the overlay of two polygonal layers in optimal $\mathcal{O}(N \log N + K)$ time. According to Lemma 1 the space requirement can be bounded by $\mathcal{O}(N\alpha(N))$.

3   Extensions to the Overlay Algorithms to Handle Real-World Data

In this section we show how the region-finding algorithms of Section 2 can be extended to handle real-world configurations. In general, these data sets are "degenerate" in a geometric sense, i.e., there are many segments starting or ending at a given node, endpoints may have identical x-coordinates, or segments from different layers overlap. Section 3.1 describes how to handle such situations, and in Section 3.2 we discuss how to deal with massive data sets that are too large to fit completely into main memory.

3.1   Handling Non-general Configurations

The most frequently occurring non-general configuration consists of multiple endpoints having the same x-coordinate. Such situations can be handled easily if we implicitly rotate the data set using a lexicographical order of the end-points. Other non-general configurations include multiple segments meeting at one point, overlapping segments, or endpoints lying on the interior of other segments. In the following discussion we assume that all geometric computations in the algorithms can be carried out in a numerically robust way (as for example described by Brinkmann and Hinrichs [9] or by Bartuschka, Mehlhorn, and Näher [5]) and thus, that all non-general configurations can be detected consistently. A recent approach to handle degenerate cases in map-overlay has been presented by Chan and Ng [11]. Their algorithm requires two phases: first all intersections are computed by some line segment intersection algorithm (thereby passing the main part of the care for degeneracies to that algorithm), and then all sorted event points (including intersection points) are processed in a plane-sweep. Since their algorithm does not improve the $\mathcal{O}((N+K)\log N)$ time bound of Nievergelt and Preparata's algorithm and requires a separate intersection algorithm we do not consider it in this paper.

Figure 4(a) depicts a situation where an end event and an intersection event coincide at some point p. No matter which of these events is handled first the algorithm by Nievergelt and Preparata (Section 2.2) does not construct the region bounded by s_1 and t_1 and the region bounded by t_1 and s_2. Kriegel, Brinkhoff, and Schneider [21] solved this problem by replacing the four event types by two more general situations: first, all segments ending at a given point are processed in clockwise order (end event) and then all segments starting at that point are processed in counterclockwise order (start event). Bend events and intersections events are handled as a pair of one end and one start event.

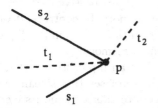
(a) Segments meeting in one point.

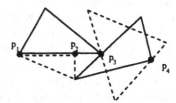
(b) Coincident points and segments.

Fig. 4. Non-general configurations.

However, for the map overlay operation there are two more details that deserve special attention: in the situation of Figure 4(b) we have two overlapping segments $\overline{p_1p_2}$ and $\overline{p_1p_3}$ of different colors and a red segment passing through a blue endpoint p_4. To avoid topological inconsistencies in the construction and attribute propagation of the empty region between $\overline{p_1p_2}$ and $\overline{p_1p_3}$ we modify these two segments such that there are two segments $\overline{p_1p_2}$ and $\overline{p_2p_3}$ by adjusting the left endpoint of the longer segment $\overline{p_1p_3}$. The resulting unique segment $\overline{p_1p_2}$ needs to store the names of both the blue region above and the red region below. To detect segments passing through an endpoint, at each event p we locate the current endpoint in the y-table of the other color and check whether it lies on an open segment. In this case we split that segment into two new segments (one ending at p and one starting at p) and proceed as usual. It is easy to see that these modifications guarantee topological consistency and require only a constant number of operations per event, thereby not affecting the asymptotic time complexity.

3.2   Handling of Massive Data Sets

The handling of massive data sets is a well-known problem in GIS and the primary focus of investigation in the field of external memory algorithms (see Vitter [31] for a recent survey). In this section we explain how the algorithms for the overlay of polygonal maps presented above can be used for the map overlay even when dealing with massive data sets.

As noticed by several authors [3,19,26] real-world data sets from several application domains seem to obey the so-called "$\sqrt{N}$-rule": for a given set of N line segments and a fixed vertical line L there are at most $\mathcal{O}(\sqrt{N})$ intersections between the line segments and the line L. As a result, data sets obeying this rule have at most $\mathcal{O}(\sqrt{N})$ active segments that need to be stored in an internal memory y-table. The TIGER/Line data set for the entire United States is estimated to contain no more than 50 million segments [3], and we have $\sqrt{50,000,000} \approx 7,000$.

Both overlay algorithms presented in Section 2 allow for space-efficient storage of the computed overlay. As soon as a region of the resulting layer has been swept completely it does not need to be present in internal memory anymore and can be transferred to disk. As long as no resulting region contains too many boundary segments, there is a fair chance that even for massive data sets both active segments and active regions fit into main memory. In contrast, if we aim at constructing a network oriented layer for the resulting partition [14,23] we are forced to keep the complete data structure in main memory.

Conceptually, the x-queue can be thought of as an input stream that feeds the algorithm, and due to the size of the input data set this stream will be resident on secondary storage. To optimize access to this x-queue its elements should be accessed sequentially, i.e., one should avoid random access as needed for update operations. The basic algorithm by Nievergelt and Preparata does not fulfil this property since intersection events are detected while sweeping and need to be inserted in the x-queue. However, by applying Brown's modification and thus storing at most one intersection event per active segment we can avoid such updates during the sweep. Instead, we store all detected intersection events in an internal memory priority queue. Whenever we find a new intersection event p, say an intersection between the segments s and t, we check whether the events possibly associated with s and t are closer to the sweep-line than p. If any of these event lies farther away from the sweep-line than p, we remove this event from the internal memory priority queue and insert p instead. At each transition the sweep algorithm has to choose the next event to be processed by comparing the first element in the x-queue and the first element in the priority queue storing the intersection events. This internal memory priority queue can be maintained in $\mathcal{O}((N+K)\log N)$ time during the complete sweep and occupies $\mathcal{O}(\sqrt{N})$ space under the assumption that the data sets obey the "$\sqrt{N}$-rule". Our modification of Chan's algorithm does not require such an additional internal memory structure since there is no need for updating the x-queue during the sweep.

If we use the extension proposed in Section 3.1 for handling non-general configurations, we are forced to modify endpoints of active segments or to split segments which triggers the need for updating the x-queue. This situation can be handled by using an internal memory priority queue similar to the one proposed above. If we encounter two overlapping segments, we adjust the left endpoint of the longer one and insert it in the internal memory event queue. Since both original segments are active we simply move this modified segment from the y-table to the priority queue without increasing internal memory space requirements. For each segment split due to handling non-general configurations, we can charge one endpoint of the active segments coincident with the split point for the new segment inserted into the priority queue. Since each active segment can contribute to at most two splits the internal memory space requirement remains asymptotically the same. Synchronizing the x-queue and the priority queue is done as described above.

4   Empirical Results

In this section we discuss our implementation of the algorithms described in Section 2.2 and Section 2.4 and the results of experiments with artificial and real-world data sets. Our focus is on the comparison of algorithms having different asymptotical complexities. The comparison is based on real-world situations and artificial situations simulating special cases. According to Section 3.2 we expect identical I/O cost for both algorithms. Hence, we ran the algorithms on data sets ranging from small to quite large comparing the absolute internal memory time complexities of both algorithms.

The algorithms have been implemented in C++, and all experiments were conducted on a Sun[TM] Ultra 2 workstation running Solaris 2.5.1. The workstation was equipped with 640 MB main memory, to ensure that even without implementing the concepts from Section 3.2 all operations are performed without causing page faults or involving I/O operations. For a more detailed description of the implementation we refer the reader to Giesen [18].

4.1   Artificial Data Sets

The first series of experiments concentrated on the behavior of both algorithms for characteristic values of K. To this end we generated three classes of input data with $K = 0$, $K \in \Theta(N)$, and $K \in \Theta(N^2)$ intersection points, respectively, as shown in Figure 5. The results of the experiments are presented in Table 1.

As a careful analysis shows the algorithm from Section 2.4 performs three times as many intersection tests as the algorithm by Nievergelt and Preparata resulting in a slow-down by a factor of three.

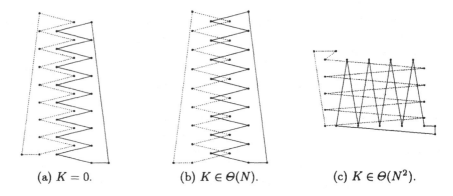

(a) $K = 0$.   (b) $K \in \Theta(N)$.   (c) $K \in \Theta(N^2)$.

Fig. 5. Artificial data sets.

For $K \in \Theta(N)$ the more efficient reporting of our algorithm cancels out that drawback, and for $K \in \Theta(N^2)$ we see a clear superiority of our algorithm reflecting the theoretical advantage of $\mathcal{O}(N \log N + N^2) = \mathcal{O}(N^2)$ versus $\mathcal{O}((N + N^2) \log N) = \mathcal{O}(N^2 \log N)$. The speed-up varies between factors of 10.4 and 17.8, and the more segments (and thus intersection points) are involved the more our algorithm improves over Nievergelt and Preparata's algorithm.

N	$K = 0$			$K \in \Theta(N)$			$K \in \Theta(N^2)$		
	[24]	Section 2.4	Ratio	[24]	Section 2.4	Ratio	[24]	Section 2.4	Ratio
256	0.3	0.7	0.4	0.7	0.7	1	51	5	10.4
512	0.6	1.6	0.4	1.6	1.7	1	228	18	12.4
1,024	1.2	3.6	0.3	3.8	3.6	1	1,033	72	14.4
2,048	2.7	8.0	0.3	8.3	8.0	1	4,401	282	15.6
4,096	5.8	17.8	0.3	18.1	17.8	1	19,939	1195	16.7
8,192	12.6	38.5	0.3	39.4	38.7	1	89,726	5,019	17.8

Table 1. Overlaying artificial maps (time in seconds).

4.2 Real-World Data Sets

After having examined the behavior of the two algorithms under varying parameters N and K, we tested the described algorithms for real-world data sets. We ran a series of experiments with polygonal maps of cities and counties from the state of North-Rhine Westphalia (NRW) and Germany (GER). Vectorized maps obtained from different distributors are likely to unveil different digitizing accuracies, and as can be seen from Figure 6 such differences bring about so-called "sliver polygons".

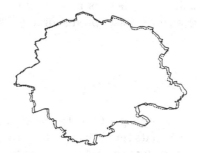

Fig. 6. Real-world data set: City of Münster, Germany.

Detecting and post-processing sliver polygons is an important issue in map overlay, and there are specialized methods for this task depending on the application at hand. At this point we only mention that the region-finding algorithms can be extended in a straightforward way to serve as a filter for sliver polygons because they produce all regions while sweeping.

To simulate additional sliver polygons we performed experiments where we overlayed a data set with a slightly translated second data set. The running times for overlaying real-world maps are summarized in Table 2.

Map 1	Map 2	N	[24]	Section 2.4	Ratio
NRW (counties)	NRW (counties) +	23,732	22	51	0.43
NRW (cities)	NRW (counties)	66,436	62	142	0.45
NRW (cities)	NRW (cities) +	109,140	113	256	0.45
GER (cities)	GER (counties) +	745,544	945	2036	0.48
GER (cities)	GER (counties)	745,544	657	1026	0.67
GER (cities)	GER (cities) +	1,238,538	2216	3559	0.63
GER (cities)	GER (cities)	1,238,538	988	1151	0.90

Table 2. Overlaying real-world maps (time in seconds, "+" indicates translated data set).

These figures show an advantage for Nievergelt and Preparata's algorithm by a factor of up to 2.3. This result could have been preempted by the results from Section 4.1, however, since there are less than linearly many intersection points in all of these experiments. In analogy to the above observations, our proposed algorithm improves with increasing number of segments involved.

Summarizing the experiments from both this section and Section 4.1 we see an advantage for the conventional method by Nievergelt and Preparata as long as there are at most linearly many intersection points. For large data sets this advantage becomes smaller, and there is a clear superiority for our proposed algorithm when dealing with data sets producing more than linearly many intersection points.

5 Conclusions

In this paper we have examined algorithms for performing map overlay operations with massive real world-data sets. We have proposed to modify plane-sweep algorithms to cope with such data sets. It is obvious that such algorithms may also be used to implement the refinement step of a spatial join efficiently. However, this approach is not useful if the filter step produces only few candidates which must be checked in the refinement step, since the plane-sweep algorithms may require to store theses candidates on disk for sorting. We have not yet addressed this problem in our research, but obviously it is not efficient to use a plane-sweep algorithm if each polygon of the two input maps only participates in at most one candidate pair.

Our experiments have shown that the modified algorithm of Chan performs better than the algorithm of Nievergelt and Preparata, if there are more than linearly many intersections points. On the other hand the algorithm of Nievergelt and Preparata performs better than the algorithm of Chan, if there are sublinearly many intersection points.

In our future work we will focus on the realization of the methods proposed in Section 3.2 and on the combination with existing algorithms for the filter step of the spatial join operation. This also includes a comparison of using plane-sweep algorithms in the refinement step with a simple checking of each pair reported by the filter step for an intersection.

References

1. P. Agarwal. Private communication. 1999.
2. D. Andrews, J. Snoeyink, J. Boritz, T. Chan, G. Denham, J. Harrison, and C. Zhu. Further comparison of algorithms for geometric intersection problems. In T. Waugh and R. Healey, editors, *Advances in GIS Research – Proceedings of the 6th International Symposium on Spatial Data Handling (SDH '94)*, volume 2, 709–724, 1994.
3. L. Arge, O. Procopiuc, S. Ramaswamy, T. Suel, and J. Vitter. Scalable sweeping-based spatial join. In A. Gupta, O. Shmueli, and J. Widom, editors, *VLDB'98: Proceedings of the 24th International Conference on Very Large Data Bases*, 570–581. Morgan Kaufmann, 1998.
4. I. Balaban. An optimal algorithm for finding segment intersections. In *Proceedings of the 11th Annual ACM Symposium on Computational Geometry*, 211–219, 1995.
5. U. Bartuschka, K. Mehlhorn, and S. Näher. A robust and efficient implementation of a sweep line algorithm for the straight line segment intersection problem. Online-Proccedings of the First Workshop on Algorithm Engineering <http://www.dsi.unive.it/~wae97/proceeedings/ONLY_PAPERS/pap13.ps.gz>, accessed 7 Jul. 1998, 1997.

6. J. Bentley and T. Ottmann. Algorithms for reporting and counting geometric intersections. *IEEE Transactions on Computers*, C-28(9):643–647, 1979.
7. T. Brinkhoff, H.-P. Kriegel, and B. Seeger. Efficient processing of spatial joins using R-trees. In P. Buneman and S. Jajoda, editors, *Proceedings of the 1993 ACM SIGMOD International Conference on Management of Data*, volume 22.2 of *SIGMOD Record*, 237–246. ACM Press, June 1993.
8. T. Brinkhoff, H.-P. Kriegel, R. Schneider, and B. Seeger. Multi-step processing of spatial joins. In R. Snodgrass and M. Winslett, editors, *Proceedings of the 1994 ACM SIGMOD International Conference on Management of Data*, volume 23.2 of *SIGMOD Record*, 197–208. ACM Press, June 1994.
9. A. Brinkmann and K. Hinrichs. Implementing exact line segment intersection in map overlay. In T. Poiker and N. Chrisman, editors, *Proceedings of the Eighth International Symposium on Spatial Data Handling*, pages 569–579. International Geographic Union, Geographic Information Science Study Group, 1998.
10. K. Brown. Comments on "Algorithms for reporting and counting geometric intersections". *IEEE Transactions on Computers*, C-30(2):147–148, 1981.
11. E. Chan and J. Ng. A general and efficient implementation of geometric operators and predicates. In M. Scholl and A. Voisard, editors, *Advances in Spatial Databases — Proceedings of the Fifth International Symposium on Spatial Databases (SSD '97)*, volume 1262 of *Lecture Notes in Computer Science*, 69–93. Springer, 1997.
12. T. Chan. A simple trapezoid sweep algorithm for reporting red/blue segment intersections. In *Proceedings of the 6th Canadian Conference on Computational Geometry*, 263–268, 1994.
13. B. Chazelle and H. Edelsbrunner. An optimal algorithm for intersecting line segments in the plane. *Journal of the ACM*, 39(1):1–54, 1992.
14. M. de Berg, M. van Kreveld, M. Overmars, and O. Schwarzkopf. *Computational Geometry: Algorithms and Applications*. Springer, Berlin, 1997.
15. Environmental Systems Research Institute, Inc. ARC/INFO, the world's GIS. ESRI White Paper Series, Redlands, CA, March 1995.
16. U. Finke and K. Hinrichs. The quad view data structure: a representation for planar subdivisions. In M. Egenhofer and J. Herring, editors, *Advances in Spatial Databases — Proceedings of the Fourth International Symposium on Spatial Databases (SSD '95)*, volume 951 of *Lecture Notes in Computer Science*, 29–46, 1995.
17. W. Franklin. Adaptive grids for geometric operations. In *Proceedings of the Sixth International Symposium on Automated Cartography (Auto-Carto Six)*, volume 2, 230–239, 1983.
18. A. Giesen. Verschneidung von Regionen in Geographischen Informationssystemen (Overlaying polygonal regions in geographic information systems). Master's thesis, University of Münster, Dept. of Computer Science, November 1998. (in German).
19. R. Güting and W. Schilling. A practical divide-and conquer algorithm for the rectangle intersection problem. *Information Sciences*, 42:95–112, 1987.
20. Y.-W. Huang, M. Jones, and E. Rundensteiner. Improving spatial intersect using symbolic intersect detection. In M. Scholl and A. Voisard, editors, *Advances in Spatial Databases — Proceedings of the Fifth International Symposium on Spatial Databases (SSD '97)*, volume 1262 of *Lecture Notes in Computer Science*, 165–177. Springer, 1997.
21. H.-P. Kriegel, T. Brinkhoff, and R. Schneider. An effecent map overlay algorithm based on spatial access methods and computational geometry. In *Proceedings of the International Workshop on DBMS's for Geographic Applications*, 194–211, Capri, May 12–17 1991.

22. M.-L. Lo and C. Ravishankar. Spatial joins using seeded trees. In R. Snodgrass and M. Winslett, editors, *Proceedings of the 1994 ACM SIGMOD International Conference on Management of Data*, volume 23.2 of *SIGMOD Record*, 209–220. ACM Press, June 1994.

23. H. Mairson and J. Stolfi. Reporting and counting intersections between two sets of line segments. In R. Earnshaw, editor, *Theoretical Foundations of Computer Graphics and CAD*, volume F40 of *NATO ASI*, 307–325. Springer-Verlag, 1988.

24. J. Nievergelt and F. Preparata. Plane-sweep algorithms for intersecting geometric figures. *Communications of the ACM*, 25(10):739–747, 1982.

25. J. Orenstein. A comparison of spatial query processing techniques for native and parameter spaces. In H. Garcia-Molina and H. Jagadish, editors. *Proceedings of the 1990 ACM SIGMOD International Conference on Management of Data*, volume 19.2 of *SIGMOD Record*, pages 343–352. ACM Press, June 1990.

26. T. Ottmann and D. Wood. Space-economical plane-sweep algorithms. *Computer Vision, Graphics and Image Processing*, 34:35–51, 1986.

27. J. Pach and M. Sharir. On vertical visibility in arrangements of segments and the queue size in the Bentley-Ottmann line sweeping algorithm. *SIAM Journal on Computing*, 20(3):460–470, 1991.

28. J. Patel and D. DeWitt. Partition based spatial-merge join. In H. Jagadish and I. Mumick, editors, *Proceedings of the 1996 ACM SIGMOD International Conference on Management of Data*, volume 25.2 of *SIGMOD Record*, 259–270. ACM Press, June 1996.

29. F. Preparata and M. Shamos. *Computational Geometry: An Introduction*. Springer, Berlin, 2nd edition, 1988.

30. M. Sharir and P. Agarwal. *Davenport-Schinzel Sequences and Their Geometric Applications*. Cambridge University Press, Cambridge, 1995.

31. J. Vitter. External memory algorithms. In *Proceedings of the 17th Annual ACM SIGMOD-SIGACT-SIGART Symposium on Principles of Database Systems (PODS '98)*, 119–128, 1998. invited tutorial.

A Performance Evaluation
of Spatial Join Processing Strategies[*]

Apostolos Papadopoulos[1], Philippe Rigaux[2], and Michel Scholl[2]

[1] Data Engineering Lab., Aristotle Univ., 54006 Thessaloniki, Greece,
[2] Cedric/CNAM, 292 rue St Martin, F-75141 Paris Cedex 03, France

Abstract. We provide an evaluation of query execution plans (QEP) in the case of queries with one or two spatial joins. The QEPs assume R*-tree indexed relations and use a common set of spatial joins algorithms, among which one is a novel extension of a strategy based on an on-the-fly index creation prior to the join with another indexed relation. A common platform is used on which a set of spatial access methods and join algorithms are available. The QEPs are implemented with a general iterator-based spatial query processor, allowing for pipelined QEP execution, thus minimizing memory space required for intermediate results.

1   Introduction

It is well known that the application of Database Management Systems (DBMS) join techniques, such as sort-merge, scan and index, hash join and join indices, to the context of spatial data is not straightforward. This is due to the fact that these techniques, as well as B-tree-based techniques, intensively rely on the domain ordering of the relational attributes, which ordering does not exist in the case of multi-dimensional data.

A large number of spatial access methods (SAM) have been proposed in the past fifteen years [VG98] as well as a number of spatial join algorithms [Ore86, GS87, Gun93, BKS93, LR96, PD96, HJR97, APR+98], some of them relying on the adaptation of well-known join strategies to the particular requirements of spatial joins.

These strategies have been validated through experiments on different platforms, with various methodologies, datasets and implementation choices. The lack of a commonly shared performance methodology and benchmarking renders difficult a fair comparison among these numerous techniques.

The methodology and evaluation are crucial not only for the choice of a few efficient spatial join algorithms but also for the optimization of complex queries involving several joins in sequence (multi-way joins). In the latter more general case, the generation and evaluation of complex query execution plans (QEP) is

---

[*] Work supported by the European Union's TMR program ("Chorochronos" project, contract number ERBFMRX-CT96-0056) and by the 1998-1999 French-Greek bilateral protocol.

R.H. Güting, D. Papadias, F. Lochovsky (Eds.): SSD'99, LNCS 1651, pp. 286–307, 1999.
© Springer-Verlag Berlin Heidelberg 1999

central to optimization. Only a few papers study the systematic optimization of spatial queries containing multi-way joins [MP99].

The objective of this paper is two fold: (i) to provide a common framework and evaluation platform for spatial query processing, and (ii) to use it to experimentally evaluate spatial join processing strategies.

A complex spatial query can be translated into a QEP with some physical operations such as data access (sequential or through an index), spatial selection, spatial join, sorting, etc. A QEP is then represented as a binary tree in which leaves are either indices or data files and internal nodes are physical operators.

We use as a model for spatial query processing the pipelined execution of such QEPs with each node (operation) being implemented as an *iterator* [Gra93]. This execution model provides a sound framework: it encompasses spatial and non-spatial queries, and allows to consider in an uniform setting simple and large complex queries involving several consecutive joins. Whenever possible, records are processed one-at-a-time and transfered from one node to the following, thereby avoiding the storage of intermediate results.

Such an execution model is not only useful to represent and evaluate complex queries, but also to specify and make a fair comparison of simple ones. Indeed, consider a query including a single spatial join between two relations. The join output is, unfortunately, algorithm dependent. Some algorithms provide as an output a set of pairs of record identifiers (one per relation), others, such as the so-called *Scan and Index* (SAI) strategy provide a set in which each element is composed of a record (of the first relation) and the identifier of a record in the second relation. Then to complete the join, the former case requires two data accesses, while only one data access is necessary in the latter case. This example illustrates the necessity for a consistent comparison framework. The above execution model provides such a framework.

Another advantage of this execution model is that it allows not only to compare two QEPs on their time performance but also on their memory space requirement. Some operations cannot be pipelined, e.g., sorting an intermediate result, and require the completion of an operation before starting the following operation. Such operators, denoted *blocking iterators* in this paper, are usually memory-demanding and raise some complex issues related to the allocation of the available memory among the nodes of a QEP. In order to make a fair comparison between several QEPs, we shall always assign the *same* amount of memory to each QEP during an experiment.

The study performed in this paper is a first contribution to the evaluation of complex spatial queries that may involve several joins in sequence (multi-way joins). Based on the above model we evaluate queries with one or two spatial joins. We make the following assumptions: (i) all relations in the database are indexed on their spatial attribute, (ii) we choose the R*-tree [BKSS90] for all indices, (iii) the index is always used for query optimization. While the first assumption is natural, the second one is restrictive. Indeed, while the R*-tree is an efficient SAM, there exists a number of other data structures that deserve some attention, among which it is worth noting the grid based structures derived from the grid file [NHS84]. The third assumption is also restrictive since it does not take into account the proposal of several techniques for joining non indexed relations.

The comparison of QEPs as defined above has been done on a common general platform developed for spatial query processing evaluation. This platform provides basic I/O and buffer management, a set of representative SAMs, a library of spatial operations, and implements a spatial query processor according to the above iterator model using as nodes the SAMs and spatial operations available.

The rest of the paper is organized as follows. Section 2 briefly surveys the various spatial join techniques proposed in the literature and summarizes related work. The detailed architecture of the query processor is presented in Section 3. Section 4 deals with our choices for spatial join processing and the generated QEPs. Section 5 reports on the experiment, the datasets chosen for the evaluation and the results of the performance evaluation. Some concluding remarks are given in Section 6.

2   Background and Related Work

We assume each relation has a spatial attribute. The spatial join between relations R_1 and R_2 constructs the pairs of tuples from $R_1 \times R_2$ whose spatial attributes satisfy a spatial predicate. We shall restrict this study to *intersect joins*, also referred to as *overlap* joins. Usually, each spatial attribute has for a value a pair (*MBR, spatial object representation*), where *MBR* is the minimum bounding rectangle of the spatial object. Intersect spatial joins are usually computed in two steps. In the *filter step* the tuples whose MBR overlap are selected. For each pair that passes the filter step, in the *refinement step* the spatial object representations are retrieved and the spatial predicate is checked on these spatial representations [BKSS94].

Many experiments only consider the filter step. This might be misleading for the following reasons: first one cannot fairly compare two algorithms which do not yield the same result (for instance if the SAI strategy is used, at the end of the filter step, part of the record value has been already accessed, which is useful for the refinement step, while it is not true with the STT strategy [Gun93, BKS93]), second by considering only the filter step, one ignores its interactions with the refinement step, for instance in terms of memory requirements. We shall include in our experiments all the operations necessary to retrieve data from disk, whether this data access is for the filter step, or the refinement step. Only the evaluation of the computational geometry algorithm on the exact spatial representation, which is equivalent whatever the join strategy, will be excluded.

SAMs can be roughly classified into two categories:

- *Space driven* structures, among which *grids* and *quadtrees* are very popular, partition the tuples according to some spatial scheme independent from the spatial data distribution of the indexed relation.

– *Data-driven* structures, on the other hand, adapt to the spatial data distribution of tuples. The most popular SAM of this category is the R-tree [Gut84]. The R^+-tree [SRF87], the R*-tree [BKSS90] and the X-tree [BKK96] are improved versions of the R-tree. These *dynamic* SAMs maintain their structure on each insertion/deletion. In the case of *static* collections which are not often updated, packing algorithms [RL85, KF93, LEL97] build optimal R-trees, called packed R-trees.

Spatial joins algorithms can be classified into three categories depending on whether each relation is indexed or not.

1. *no index*: for the case where no index exists on any relation, several *partitioning* techniques have been proposed which partition the tuples into buckets and then use either hashed based techniques or sweep-line techniques [GS87, PD96, LR96, KS97, APR^+98].
2. *two indices*: when both relations are indexed, the algorithms that have been proposed depend on the SAM used. [Ore86] is the first known work on spatial joins. It proposes a 1-dimensional ordering of spatial objects, which are then indexed on their rank in a B-tree and merge-joined. [Gun93] was the first proposal of an algorithm called Synchronized Tree Traversal (STT) which adapts to a large family of spatial predicates and tree structures. The STT algorithm of [BKS93] is the most popular one because of its efficiency. Proposed independently from [Gun93], it uses R*-trees and an efficient depth-first tree traversal of both trees for intersection joins. The algorithm is sketched below.

Algorithm STT (Node N_1, Node N_2)
begin
    for all (e_1 in N_1)
        for all (e_2 in N_2) such that $e_1.MBR \cap e_2.MBR \neq \emptyset$
            if (the leaf level is reached) **then**
                output (e_1, e_2)
            else
                $N_1' = $ readPage ($e_1.pageID$); $N_2' = $ readPage ($e_2.pageID$);
                STT(N'1, N'2)
            endif
end

Advanced variants of the algorithm apply some local optimization in order to reduce the CPU and I/O costs. In particular, when joining two nodes, the overlapping of entries is computed using a plane-sweeping technique instead of the brute-force nested loop algorithm shown above. The MBRs of each node are sorted on the x-coordinate, and a merge-like algorithm is carried out. This is shown to significantly reduce the number of intersection tests.

3. *single index* : when only one index exists, the simplest strategy is the Scan And Index (SAI) strategy, a variant of the nested loop algorithm which scans the non-indexed relation and for each tuple r delivers to the index of the other

relation a window query with $r.MBR$ as an argument. The high efficiency of the STT algorithm suggests that an "on-the-fly" construction of a second index, followed by STT, could compete with SAI. This idea has inspired the join algorithm of [LR98] which constructs a *seeded-tree* on the non indexed relation which is a R-tree whose first levels match exactly those of the existing R-tree. It is shown that the strategy outperforms SAI and the naive on-the-fly construction of an R-tree with dynamic insertion. An improvement of this idea is the SISJ algorithm of [MP99]. An alternative is to build a *packed R-tree* by using bulk-load insertions [LEL97]. Such constructions optimize the STT algorithm since they reduce the set of nodes to be compared during traversal. These algorithms are examples of strategies referred to as *Build-and-Match* strategies in the sequel.

The complexity of the spatial join operation, the variety of techniques and the numerous parameters involved in a spatial join render extremely difficult the comparison between the above proposals briefly sketched above.

Only a few attempts have been made toward a systematic comparison of spatial join strategies. [GOP+98] is a preliminary attempt to integrate in a common platform the evaluation of spatial query processing strategies. It proposes a web-based rectangle generator and gives first results on the comparison of three join strategies: nested loop, SAI and STT. The major limit of this experiment is that it is built on top of an existing DBMS. This not only limits the robustness of the results but renders impossible or inefficient the implementation of complex strategies, the tuning of numerous parameters and a precise analysis.

[MP99] is the first study on multi-way spatial joins. It proposes an iterator pipelined execution of QEPs [Gra93] for multi-way spatial joins, with three join algorithms, one per category, namely STT, SISJ, and a spatial hash-join technique [LR96]. An analytical model predicts the cost of QEPs, a dynamic programming algorithm for choosing the optimal QEP is proposed. The query optimization model is validated through experimental evaluation.

The modeling of QEPs involving one or several joins in the study reported below follows the same pipelined iterator based approach. This execution model is implemented on a platform common to all evaluations. This platform allows for fine tuning of parameters impacting the strategies performance and is general enough to implement and evaluate any complex QEP: it is not limited to spatial joins. Last but not least, such a model and its implementation allow for various implementation details generally absent from evaluations reported in the literature. The relation access after a join is an example of implementation "detail" which accounts for an extremely significant part of the query response time as shown below.

3   The Query Processor

The platform has been implemented in C++ and runs on top of UNIX or WindowsNT. Its architecture is shown in Fig. 1. It is composed of a *database* and three modules which implement some of the standard low-level services of a centralized DBMS.

The *database* is a set of binary files. Each binary file either stores a *data file* or a SAM. A data file is a sequential list of records. A SAM or index refers to records in an indexed data file through *record identifiers*. The lowest level module is the I/O module, which is in charge of reading (writing) pages from (to) the disk. The second module manages buffers of pages fetched (flushed) from (to) the disk through the I/O module. On-top of the buffer management module is the query processing module, which supports spatial queries.

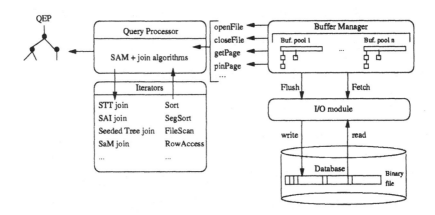

Fig. 1. Platform architecture

Database and Disk Access

The database (data files and SAM) is stored in binary files divided into *pages* whose size is chosen at database creation. A page is structured as a *header* followed by an array of fixed-size records which can be either data records or *index entries*. The header and record sizes depend on the file. By knowing the record size, one can compute the number of records per page and the data file size.

Each page is uniquely identified by its **PageID** (4 bytes). A record is identified by a **RecordID** (8 bytes) which is a pair [**PageID, offset**] where **offset** denotes the record offset in the page.

1. *Data files* are sequential collections of pages storing *data records*. In the current setting, a record basically is a binary representation of a spatial object. From the query processing point of view, the most important information stored in a record is its *geometric key*, which is, throughout this experiment, its MBR. A data file can either be accessed sequentially (*FileScan* in the sequel) , or by **RecordID** (*RowAccess* in the sequel). It is important to note that the datafiles are *not* clustered on their geometric representation (i.e., objects close in space are not necessarily close on disk).

2. *SAMs* are structured collections of `IndexEntry`. An index entry is a pair
   [`Key, RecordID`], where `Key` denotes the geometric key (here the MBR)
   and `RecordID` identifies a record in the indexed data file. The currently
   implemented SAMs are a grid file, an R-tree, an R*-tree and several packed
   R-trees. In the sequel, each datafile is indexed with an R*-tree.

Buffer Management

The buffer manager handles one or several *buffer pools*: a data file or index
(SAM) is assigned to one buffer pool, but a buffer pool can handle several indices.
This allows much flexibility when assigning memory to the different parts of a
query execution plan. The buffer pool is a constant-size cache with LRU or FIFO
replacement policy (LRU by default). Pages can be *pinned* in memory. A pinned
page is never flushed until it is unpinned.

Currently, all algorithms requiring page accesses uniformly access these pages
through the interface provided by the buffer manager. In particular, spatial join
algorithms share this module and therefore cannot rely on a tailored main memory management or a specialized I/O's policy unless it has already been implemented in this module.

Query Processing Module

One of the important design choices for the platform is to allow for any experimental evaluation of *query execution plans* (QEP) as generated by database
query optimizers with an algebraic view of query languages. During optimization,
a query is transformed into a QEP represented as a binary tree which captures
the order in which a sequence of *physical* algebraic operations are going to be
executed. The leaves represent data files or indices, internal nodes represent algebraic operations and edges represent dataflows between operations. Examples of
algebraic operations include data access (*FileScan* or *RowAccess*), spatial selections, spatial joins, etc. As mentioned above we use as a common framework for
query execution, a demand-driven process with iterator functions [Gra93]. Each
node (operation) is an iterator. This allows for a pipelined execution of multiple
operations, thereby minimizing the system resources (memory space) required
for intermediate results: data consumed by an iterator, say *I*, is generated by
its son(s) iterator(s), say *J*. Records are produced and consumed one-at-a-time.
Iterator *I* asks iterator *J* for a record. Therefore the intermediate result of an
operation is not stored in such pipelined operations except for some specific
iterators called *blocking iterators*, such as sorting.

This design allows for simple QEP creation by "assembling" iterators together. Consider the QEP for a spatial join $R \bowtie S$ implemented by the simple
scan-and-index (SAI) strategy (Fig. 2.a): scan *R* (*FileScan*); for each tuple *r* in
R, execute a window query on index I_S with key $r.MBR$. This gives a record
ID `RecordID2` [1]. Finally read the record with id `RecordID2` in *S* (*RowAccess*).

---

[1] As a matter of fact each index (leaf) access returns an `IndexEntry`, i.e., a pair [`MBR,
RecordID`]. For the sake of simplicity, we do not show this MBR on the figures.

The refinement step not represented in the figure can then be performed on the exact spatial object available in **Record1** and **Record2**.

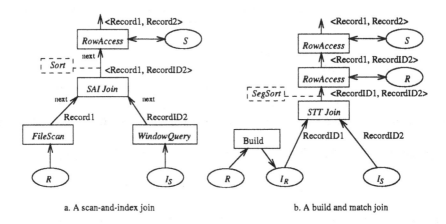

a. A scan-and-index join   b. A build and match join

Fig. 2. Query execution plans

This is a fully pipelined QEP: therefore the response time (e.g., the time to get the first record) is minimal. It is sometimes necessary to introduce *blocking iterators* in a QEP which require the consumption of all input data before any output is possible. Then significant memory is necessary for intermediate results.

As an example, one can introduce a *Sort* blocking iterator in the QEP of Fig. 2.a in order to sort the data flow output by the SAI join on the **PageID** of the **RecordID2** component. This allows to access only once the pages of S instead of issuing random **reads** to get the S records, which might lead to several accesses to the same page. However no record can be delivered to the user before the join is completely processed.

As a more complex (and realistic) example, consider the QEP of Fig. 2.b. It implements a different join strategy: an index already exists on S, another one is built on the fly on R (iterator *Build*)[2], and an STT join is executed. Such a join delivers pairs of **RecordID**, hence two further *RowAccess*, one per relation, are necessary to complete the query (refinement step). *Build* is blocking: the join cannot be started before index I_R has been completely built. In addition, the maximal amount of available memory should be assigned to this iterator to avoid as much as possible to flush pages on disk during index construction.

It may happen that a QEP relies on several *blocking iterators*. In that case the management of memory is an important issue. Consider the QEP of Fig. 2.b. The STT node delivers pairs of **RecordID**, $[r_i, s_i]$, which resembles the join index upon non clustered data, as described in [Val87]. The naive strategy depicted in Fig. 2.b alternates random accesses on the datafiles R and S; then the same page (either in R or S) will be accessed several times, which leads to a large

---

[2] R can be an intermediate result delivered by a sub-QEP.

number of page faults. The following preprocessing algorithm is proposed in [Val87] and denoted *segmented sort (SegSort)* in the sequel: (1) allocate a buffer of size B; (2) compute the number n of pairs (Record1,RecordID2) which can fit in B; (3) load the buffer with n pairs (RecordID1, RecordID2); (4) sort on RecordID1; (5) access relation R and load records from R (now the buffer is full); (6) sort on RecordID2; load records from S, one at a time, and perform the refinement step. Repeat from step (3) until the source (STT join in that case) is exhausted. Hence, this strategy, by ordering the pairs of records to be accessed, saves numerous page faults.

The resulting QEP includes two blocking iterators (*Build* and *SegSort*) between which the available buffer memory must be split. Basically there are two strategies for memory allocation for such QEPs:

1. *Flush intermediate results.* This is the simplest solution: the total buffer space M allocated to the QEP is assigned to the *Build* iterator, the result of the join (STT) is flushed onto disk and the total space M is then reused for *SegSort*. The price to be paid is a possibly very large amount of write and read operations onto (from) disk for intermediate results.
2. *Split memory* among iterators and avoid intermediate materialization. Each of the iterators of the QEP is assigned part of the global memory space M. Then intermediate results are kept in memory as much as possible but less memory is available for each iterator [BKV98, ND98].

4 Spatial Join Query Processing

Using the above platform, our objective is to experimentally evaluate strategies for queries involving one or several spatial joins in sequence.

Fig. 3 illustrates two possible QEPS for processing query $R_1 \bowtie R_2 \ldots \bowtie R_n$, using index $I_1, I_2, \ldots I_n$., which both assume (i) the optimizer tries to use as much as possible existing spatial indices when generating QEPs and (ii) that the n-way join is first evaluated on the MBRs (filter step) and then on the exact geometry: an n-way join is performed on a limited number of tuples of the cartesian product $R_1 \times R_2 \times \ldots R_n$ (refinement step, requiring n row accesses). Both QEPS are left-deep trees [Gra93]. In such trees the right operand of a join is always an index, as well as the left operand for the left-most node. Another approach, not investigated here, would consists in an n-way STT, i.e., a synchronized traversal of n R-trees down to the leaves. See [MP99] for a comprehensive study.

The first strategy (Fig. 3.a) is fully pipelined: a STT join is performed as the left-most node, and a SAI join is executed for the following joins: at each step a new index entry [MBR, RecordID] is produced. The MBR is the argument of a window query for the following join. The result is a tuple $i_1, i_2, \ldots i_n$ of record id: the records are then retrieved with *RowAccess* iterators, one for each relation, in order to perform the refinement step on the n-way join but on a limited number of records. The second strategy (Fig. 3.b) uses instead of SAI the Build-and-Match strategy.

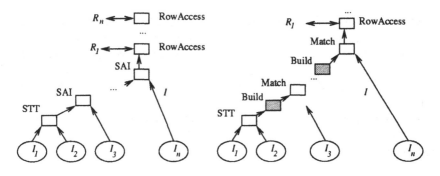

a. A left-deep tree with pipelined iterators   b. A left-deep tree with blocking iterators

Fig. 3. Two basic strategies for left-deep QEPs

Evidently, the QEPs shown in Fig. 3 are extreme cases. Depending on the estimated size of output(s), a merge of both strategies can be used for a large number of joins. More importantly, the refinement step can be done prior to the completion of the query if it is expected that the candidate set contains a large number of false hits. By computing the refinement step in a lazy mode, as suggested in Fig. 3, the cardinality of intermediate results is larger (because of false hits) but the size of records is smaller.

We do not consider the case of bushy trees since they involve joins algorithms upon non-indexed relations. As an example of bushy-tree QEP, consider the following QEP for the join of 4 relations R_1, R_2, R_3, R_4: R_1 and R_2 are joined using STT as well as R_3 and R_4. The two (non-indexed) intermediate results must then be joined. In the case of $n \leqslant 3$ (only one or two joins) which will be considered here, only left-deep trees can be generated.

Join Strategies

We describe in this section the three variants of the same strategy called *Build-and-Match* (Fig. 3.b) which consists in building on the fly an index on a non indexed intermediate relation and to join the result with an indexed relation. When the structure built is an R-tree, then the construction is followed by a regular STT join. The rationale of such an approach is that even though building the structure is time consuming, the join behind is so efficient that the overall time performance is better than applying SAI. Of course the building phase is implemented by a blocking iterator and requires memory space.

STJ

The first one is the Seeded Tree Join (STJ) [LR98]. This technique consists in building from an existing R-tree, used as a *seed*, a second R-tree called *seeded R-tree*. The motivation behind this approach is that tree matching during the join phase should be more efficient than if a regular R-tree were constructed. During the *seeding phase*, the top k levels of the seed are copied to become the top k levels of the seeded tree. The entries of the lowest level are called *slots*.

During the *growing phase*, the objects of the non indexed source are inserted in one of the slots: a rectangle is inserted in the slot that contains it or needs the least enlargement. Whenever the buffer is full, all the slots which contain at least one full page are written in temporary files.

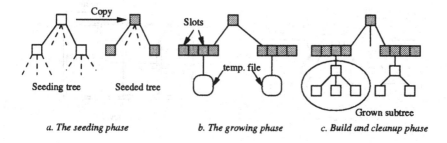

a. The seeding phase   b. The growing phase   c. Build and cleanup phase

Fig. 4. Seeded tree construction

When the source has been exhausted, the construction of the tree begins: for each slot, the objects inserted in the associated temporary files (as well as the objects remaining in the buffer) are loaded to build an R-tree (called a *grown subtree*): the slot entry is then modified to point to the root of this grown subtree. Finally a *cleanup phase* adjusts the bounding boxes of the nodes (Fig. 4), as in classical R-trees.

The grown subtrees may have different heights: hence the seeded tree is not balanced. It can be seen as a forest of relatively small R-trees: one of the expected advantages of the method is that the construction of each grown subtree is done in memory.

There is however an important condition to fulfill: the buffer must be large enough to provide at least one page to each slot. If this is not the case, the pages associated to a slot will be read and written during the growing phase, thus rendering the method ineffective.

STR

The second Build-And-Match variant implemented, called Sort-Tile-Recursive (STR), constructs on the fly a STR packed R-tree [LEL97]. We also experimented the Hilbert packed R-tree [KF93], but found that the comparison function (based on the Hilbert values) was more expensive than the centroid comparison of STR.

The algorithm is as follows. First the rectangles from the source are sorted[3] by x-coordinate of their centroid. At the end of this step, the size N of the dataset is known: this allows to estimate the number of leaf pages as $P = \lceil N/c \rceil$ where c is the page capacity. The dataset is then partitioned into $\lceil \sqrt{P} \rceil$ vertical slices. The $\lceil \sqrt{P} \rceil.c$ rectangles of each slice are loaded, sorted by the y-coordinate

---

[3] The sort is implemented as an iterator which carries out a sort-merge algorithm according to the design presented in [Gra93].

of their center, grouped into runs of length c and packed into the R-tree leaves. The upper levels are then constructed according to the same algorithm. At each level, the nodes are roughly organized in horizontal or vertical slices (Fig .3).

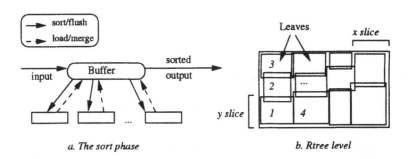

a. The sort phase   b. Rtree level

Fig. 5. STR tree construction

SaM

The third Build-And-Match variant called Sort-and-Match (SaM) is novel. It uses the STR algorithm but the construction is stopped at the leaf level, and the pages are not written onto disk. As soon as a leaf l has been produced, it is joined to the existing R-tree I_R: a window query with the bounding box of l is generated which retrieves all I_R leaves l' such that $l.MBR$ intersects $l'.MBR$. l and l' are then joined with the plane-sweep algorithm already used in the STT algorithm.

An interesting feature of this algorithm is that, unlike the previous ones, it does not require the entire structure to be built before the matching phase thus saving the flushing of this structure onto disk, resulting in much faster response time.

5   Performance Evaluation

The machine used throughout the experiments is a SUN SparcStation 5 with 32 MB of memory, running SunOS 4.1. We use in our experiments synthetic datasets, created with the ENST rectangle generator [GOP$^+$98]. This tool[4] generates a set of rectangles according to a statistical model whose parameters (size, coverage, distribution) can be specified. The 3 following statistical models were used sharing the same 2D universe (map):

1. *Counties* (called *Biotopes* in [GOP$^+$98]) simulates a map of counties; rectangles have a shape and location uniformly distributed, and the overlap (ratio between sum of the areas of the rectangles and map area) is 100%.

---

[4] Available at **http://www-inf.enst.fr/ bdtest/sigbench/**.

Nb records	Pages	Size (MB)
20K	769	3.1
40K	1539	6.3
60K	2308	9.4
80K	3078	12.6
100K	3846	15.7

Dataset	Pages	Levels	Root entries
COUN20	169	2	166
COUN40	330	3	3
COUN60	483	3	4
COUN80	650	3	6
COUN100	802	3	8

Datafiles   R*trees

Fig. 6. Database sample

2. *Cities* [GOP+98] simulates a map of cities: the map contains small rectangles whose shape is normally distributed (around the square shape) and whose location is uniformly distributed. The overlap is equal to 5%.
3. *Roads* simulates a map of roads: rectangles location and shape is uniform as in *Counties* but overlap is 200%.

For each of the statistical models, 5 datasets have been generated, with a size ranging from 20 000 to 100 000 objects referred to as DATxx, where DAT is in COUN, CIT, ROA and xx ranges from 20 to 100. For example, COUN20 stands for *Counties* with 20 000 rectangles.

Join strategies are evaluated on the query *Cities* ⋈ *Counties* in the case of single joins and the query *Cities* ⋈ *Counties* ⋈ *Roads* for two-way joins.

We assume a page size of 4K and a buffer size ranging from 400K (100 pages) to 2.8MB (700 pages). The record size is 158 bytes and the buffer policy is LRU. Fig. 6 gives some statistics on the generated database (data file and index). Only the information on *Counties* is reported. Indeed the sizes do not depend on the statistical model, so *Cities* and *Roads* have almost identical characteristics. The fanout (maximum number of entries/page) of an R*tree node is 169. We give the number of entries in the root since it is an important parameter for the seeded tree construction.

The main performance criteria are (i) the number of I/O, i.e., the number of calls (page faults) to the I/O module and (ii) the CPU consumption.

The latter criteria depends on the algorithm. It is either measured as the number of comparisons (when sorting occurs), or the number of rectangle intersections (for join) or the number of unions (for R-tree construction): see Appendix A. The parameters chosen are the buffer size, the data set size and of course, the variants in the query execution plan and the join algorithms.

Single Join

When there is a single join in the query and both relations are indexed, a good candidate strategy is STT. Part of our work below is related to a closer assessment of this choice. To this end, we investigate the behavior of the candidate algorithms for single joins, namely SAI and STT. Fig. 7 gives for each algorithm, the number of I/Os as well as the number of rectangle intersection tests (NBI), for a buffer set to 250 pages (1 MB). STT_{RA} stands for a QEP where the join is followed by a *RowAccess* operator, while STT is a stand alone join. Indeed, SAI and STT_{RA} deliver exactly the same result, namely pairs of [Record, RecordID], while STT only yields pairs of RecordID.

As expected, the larger the dataset, the worse is SAI performance, both in I/Os and NBIs. There is a significant overhead as the R*tree size is larger than the available buffer. This is due to the repeated execution of window queries with randomly distributed window arguments.

STT outperforms SAI with respect to both I/Os and NBI. But as explained above, the comparison to be done is not between SAI and STT but between SAI and STT_{RA}. Then, looking at Fig. 7, the number of I/Os is of the same order for the two algorithms. Furthermore, it is striking that the *RowAccess* cost is more than one order of magnitude larger than the join itself for STT (e.g., for a dataset size of 100K, there are 104 011 I/Os while the join phase costs only 1 896 I/Os)!

The *RowAccess* iterator in the QEP implementing STT_{RA}, reads the pages at random. Then a large number of pages are read more than once. The number of I/Os depends both on the buffer size and on the record size (here 158, which is rather low) and can be estimated according to the model in [Yao77].

Since STT's performance (without *RowAccess*) is not very sensitive to an increase in the index size, it should not be very sensitive to a decrease in memory space. This justifies that most of the buffer space available should be dedicated to the *RowAccess* iterator in order to reduce its extremely large cost.

	Dataset size							
	40 000		60 000		80 000		100 000	
	I/Os	NBI	I/Os	NBI	I/Os	NBI	I/Os	NBI
SAI	19 761	11 741	49 885	20 399	81 206	27 032	114 805	33 855
STT_{RA}	37 557	2 576	59 826	4 000	81 165	5 431	104 011	6 898
STT	755	2 576	1 084	4 000	1 570	5 431	1 896	6 898
Result size:	116 267		171 343		228 332		288 846	

Fig. 7. Left-most node: join Cities-Counties, buffer size = 250 pages

To reduce the number of datafile accesses, we insert in the QEP a *SegSort* iterator before the *RowAccess*. Pages whose ids are loaded in the *SegSort* buffer can then be read in order rather than randomly. The efficiency depends on the size SGB of this buffer.

Fig. 8 displays the number of I/Os versus the data size, for SAI and STT, for several values of the parameter SGB. The total buffer size is 250 pages, and is split into a buffer dedicated to *SegSort* and a 'global' buffer whose size is 250 - SGB. STT-xx stands for the STT join where SGB=xx. In order to compare with the results of Fig. 7, we only access one relation. The larger SGB, the larger the gain. This is due to the robustness of STT performance with respect to buffer size: its performance is not significantly reduced with a small dedicated buffer size.

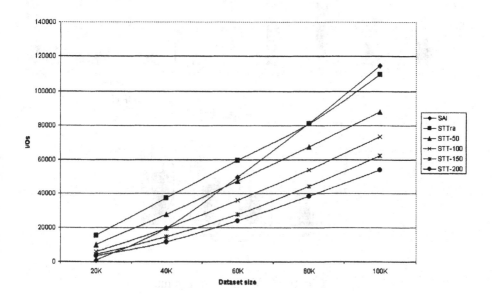

Fig. 8. SegSort experiment

Figure 8 illustrates the gain from sorting the pages to be accessed: for a large data set size, the gain with STT-200 is almost 3, compared to STT$_{RA}$.

In conclusion, the combination of STT with a *SegSort* operator (or any other mean to reduce the cost of random I/Os, for instance spatial data clustering) outperforms SAI.

We now compare the performance of the 3 Build-And-Match candidate algorithms (STJ, STR and SaM). Both the *Build* and the *Match* phases are considered, but we do not account for any *FileScan*. In other words, as stressed above, we restrict to the case where the join is executed on an intermediate result in which each tuple is produced one at a time.

Figure 9.a displays the cost of the 3 algorithms for 4 data set sizes. The case of STJ deserves some discussion. Note first that it is very unlikely that we can copy more than the root of the seeding tree because of the large fanout (169) of the R*tree. Indeed, in copying the first level also, the number of slots would largely exceed the buffer size.

In copying only the root, the number of slots may vary between 2 and 169. Actually, in our database the root is either almost full (dataset size 20K) or almost empty (dataset size ≥ 40K). See Figure 6.

When the number of slots is large, one obtains a large number of grown R-trees (one per slot) whose size is *small*. Then the memory utilization is very low:

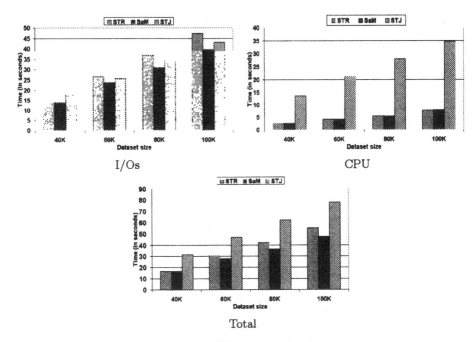

I/Os

CPU

Total

Fig. 9. Build and match joins

an almost empty root with 2 or 3 leaves[5]. If the number of slots is small, then there is a small number of large R-trees, each of them requiring a significant construction time. In all cases, the CPU construction cost is high, although the I/Os cost is low because each grown subtree can be built in memory.

STR and SaM are extremely efficient with small dataset sizes (40K). Indeed the construction of the index is entirely done in main memory. Even for large data sets, SaM is very efficient. Compared to STR, the number of rectangle intersections is the same, but since the tree is not constructed the number of I/Os is smaller, the more the data set size increases (it is 20 % smaller than for STR with a dataset size greater than 80K).

During the match phase, SaM is also efficient: in fact it can be seen as a sequence of window queries, with two major improvements: (i) leaves are joined, and not entries, hence one level is saved during tree traversal, and (ii) more importantly, two successive leaves are located in the same part of the search space. Therefore the path in the R-tree is likely to be already loaded in the buffer.

We now test the robustness of the algorithms performance with respect to the buffer size. In Figure 10, we measure the performance of the algorithms by

---

[5] We do not pack the roots of grown subtrees, as proposed in [LR98]. This renders the data structure and implementation complex, and has some further impact on the design of the STT.

joining two 100K datasets and letting the buffer size vary from 100 pages (400K) to 700 pages (2.8 MB). *RowAccess* is not taken into account. We do not include the cost of STJ for the smallest buffer size since buffer thrashing cannot be avoided in that case.

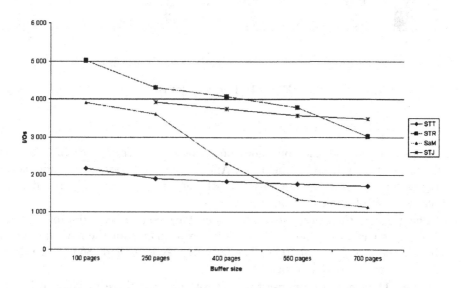

Fig. 10. JOIN Cities 100K - Counties 100K, varying buffer

Looking at Figure 10, the following remarks are noteworthy: (i) the sort-based algorithms benefit from large buffers; this is less clear for STJ; (ii) as expected, STT performance is robust with respect to buffer size; this is important since algorithms whose memory requirement is known and reasonable in size allow for more flexibility when assigning memory among several operators, as shown in the next section. Observe also that when the *Build* phase can be performed in memory, the *Join* phase of SaM outperforms STT; (iii) the larger the buffer size, the more SaM outperforms the two other Build-And-Match strategies: while its gain over STR is only 20% for small buffer size, it reaches three for a buffer capacity of 700 pages.

Two Way Joins

This section relies on the above results for the evaluation of QEPs involving two joins. In the sequel, the left-most node of the QEP is always an STT algorithm performed on the two existing R*trees (on *Cities* and *Counties*) which delivers pairs of index entries $[i_1, i_2]$. The name of a join algorithm denotes the second join algorithm, which takes the result of STT, builds a structure and performs the join with the index on *Roads*.

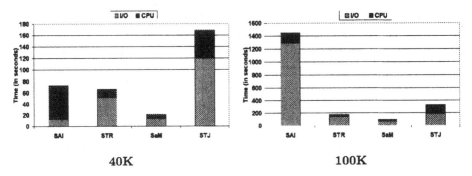

Fig. 11. Two way joins

Note that in that case one does not save a *RowAccess* with SAI for the refinement step. Indeed as the Build-And-Match strategies, SAI reads as an entry only an index entry [RecordID,MBR] from the STT join. The result is, in all cases, a set of triplets of index entries.

The datasets *Counties*, *Cities* and *Roads*, have equal size, and a fixed buffer of 500 pages has been chosen. We make the experiments for the medium size of 40K and the larger size of 100K. The latter 2 way-join yields 865 473 records, while the former 314 617 records.

Figure 11 gives the response time for SAI and the three variants of Build-And-Match algorithms. Let us look first at SAI performance. For a small dataset size (40K), the index fits in memory, and only few I/Os are generated by the algorithm. However the CPU cost is high because of the large number of inter-section tests. For large dataset sizes, the number of I/Os is huge, rendering this algorithm definitely not the right candidate.

STJ outperforms SAI for large datasets. But its performance is always much below that of SaM and STR. The explanation of this discrepancy is the following. For a 40K size, the first level of the seeding tree could be copied, resulting into 370 slots. The intermediate result consists of 116 267 entries. So, there is an average of 314 entries per slot: each subtree includes a root with an average of two leaves, leading to a very bad space utilization. A large number of window queries are generated due to the unbalance of the matched R-tree. In the case of 100K datasets, only 8 slots can be used, and the intermediate result consists of 288 846 records. Hence we must construct a few, large R-trees, which is very time consuming.

SaM significantly outperforms STR, mostly because it saves the construction of the R-tree structure, and also because the join phase is very efficient. It is worth noting, finally, that SAI is a good candidate for small datasets sizes, although its CPU cost is still larger. One should not forget that SAI is, in that case, the only fully pipelined QEP. Therefore the response time is very short, a parameter which can be essential when the regularity of the data output is more important than the overall resource consumption.

Discussion

By considering complete QEPs, including the I/O operations for the refinement step, we were able to identify the bottlenecks and the interactions between the successive parts of a QEP.

The efficiency of the commonly accepted STT algorithm is natural: an index is a small, structured collection of data, so joining two indices is more efficient than other strategies involving the data files. The counterpart, however, is the cost of accessing the records after the join for the refinement step, whose cost is often ignored in evaluations, although several papers report the problem (see for instance [PD96] and the recent work of [AGPZ99]). It should be noted that in pure relational optimization, the manipulation of RecordID lists has been considered for a long time to be less efficient than the (indexed) nested loop join [BE77]. Even nowadays, the ORACLE DBMS does use a SAI strategy in the presence of two indices [Ora]. In the context of spatial databases, though, SAI provides a prohibitive cost as soon as the index size is larger than the buffer and the number of window queries is high. Whenever STT is chosen, we face the cost of accessing the two relations for the refinement step. When data is not spatially clustered, the present experiment suggests to introduce a scheduling of row accesses through a specific iterator. We used the algorithm of [Val87], but other techniques are available [Gra93]. The combination of STT and *SegSort* outperforms SAI for large datasets, in part because of the robustness of STT with respect to the buffer size.

For two-way joins, the same guidelines should apply. Whenever we intend to build an index for subsequent matching with an existing R-tree, the build algorithm performance should not degrade when there is a shortage of buffer space, since most of the available space should be dedicated to the costly access to records after the join. We experimented three such Build-And-Match strategies: a top-down index construction (STJ), a bottom-up index construction (STR) and an intermediate strategy which avoids the full index construction (SaM). Several problems were encountered with STJ, while the classical solutions based on sorting appear quite effective. They provide a simple, robust and efficient solution to the problem of organizing an intermediate result prior to its matching with an existing index. The SaM algorithm was shown to be a very good candidate: it can be carried out with reasonably low memory space and provides the best response time since its *Build* phase is not completely blocking: records are produced before the build phase is completed.

6   Conclusion and Future Work

The contribution of this paper is three fold: (i) provide an evaluation platform general enough to experimentally evaluate complex plans for processing spatial queries and to study the impact on performance of design parameters such as buffer size, (ii) show that in build-and-match strategies for spatial joins it was not necessary to completely build the index before the join: this resulted into a join strategy called SaM that was shown in our experiment to outperform the other known build-and-match strategies, (iii) show that physical operations that occur in the query execution plan associated with a join strategy have a large impact on performance. For example, we studied the impact of record access after the join, which is a very costly operation.

The performance evaluation stressed the importance of memory allocation in the optimization of complex QEPs. The allocation of available buffer space among the (blocking) operators of a QEP, although it has been addressed at length in a pure relational setting, it is still an open problem. We intend to refine our evaluation by studying the impact of selectivity and relation size on the memory allocation. Some other parameters such as the data set distribution or the placement of the record access in the QEP may also have some impact. The aim is to exhibit a cost model simple enough to be used in an optimization phase to decide for memory allocation.

References

[AGPZ99]   D. Abel, V. Gaede, R. Power, and X. Zhou. Caching Strategies for Spatial Joins. *GeoInformatica*, 1999. To appear.

[APR+98]   L. Arge, O. Procopiuc, S. Ramaswami, T. Suel, and J. Vitter. Scalable Sweeping Based Spatial Join. In *Proc. Intl. Conf. on Very Large Data Bases*, 1998.

[BE77]   M. Blasgen and K. Eswaran. Storage and access in relational databases. *IBM System Journal*, 1977.

[BKK96]   S. Berchtold, D. Keim, and H.-P. Kriegel. The X-tree: An Index Structure for High-Dimensional Data. In *Proc. Intl. Conf. on Very Large Data Bases*, 1996.

[BKS93]   T. Brinkhoff, H.-P. Kriegel, and B. Seeger. Efficient Processing of Spatial Joins Using R.-Trees. In *Proc. ACM SIGMOD Symp. on the Management of Data*, 1993.

[BKSS90]   N. Beckmann, H.P. Kriegel, R. Schneider, and B. Seeger. The R.*tree : An Efficient and Robust Access Method for Points and Rectangles. In *Proc. ACM SIGMOD Intl. Symp. on the Management of Data*, pages 322–331, 1990.

[BKSS94]   T. Brinkhoff, H.P. Kriegel, R. Schneider, and B. Seeger. Multi-Step Processing of Spatial Joins. In *Proc. ACM SIGMOD Symp. on the Management of Data*, pages 197–208, 1994.

[BKV98]   L. Bouganim, O. Kapitskaia, and P. Valduriez. Memory Adaptative Scheduling for Large Query Execution. In *Proc. Intl. Conf. on Information and Knowledge Management*, 1998.

[GOP+98]   O. Gunther, V. Oria, P. Picouet, J.-M. Saglio, and M. Scholl. Benchmarking Spatial Joins À La Carte. In *Proc. Intl. Conf. on Scientific and Statistical Databases*, 1998.

[Gra93]   G. Graefe. Query evaluation techniques for large databases. *ACM Computing Surveys*, 25(2):73–170, 1993.

[GS87]   R.H. Güting and W. Schilling. A Practical Divide-and-Conquer Algorithm for the Rectangle Intersection Problem. *Information Sciences*, 42:95–112, 1987.

[Gun93]   O. Gunther. Efficient Computation of Spatial Joins. In *Proc. IEEE Intl. Conf. on Data Engineering*, pages 50–59, 1993.

[Gut84]   A. Guttman. R-trees : A Dynamic Index Structure for Spatial Searching. In *Proc. ACM SIGMOD Intl. Symp. on the Management of Data*, pages 45–57, 1984.

[HJR97]   Y.-W. Huang, N. Jing, and E.A. Rudensteiner. Spatial Joins Using R-trees: Breadth-first Traversal with Global Optimizations. In *Proc. Intl. Conf. on Very Large Data Bases*, 1997.

[KF93]   I. Kamel and C. Faloutsos. On Packing Rtrees. In *Proc. Intl. Conf. on Information and Knowledge Management (CIKM)*, 1993.

[KS97]   N. Koudas and K. C. Sevcik. Size separation spatial join. In *Proc. ACM SIGMOD Symp. on the Management of Data*, 1997.

[LEL97]   S. Leutenegger, J. Edgington, and M. Lopez. STR: a Simple and Efficient Algorithm for Rtree Packing. In *Proc. IEEE Intl. Conf. on Data Engineering (ICDE)*, 1997.

[LR96]   M.-L. Lo and C.V. Ravishankar. Spatial Hash-Joins. In *Proc. ACM SIGMOD Symp. on the Management of Data*, pages 247–258, 1996.

[LR98]   M.-L. Lo and C.V. Ravishankar. The Design and Implementation of Seeded Trees: An Efficient Method for Spatial Joins. *IEEE Transactions on Knowledge and Data Engineering*, 10(1), 1998. First published in SIGMOD'94.

[MP99]   N. Mamoulis and D. Papadias. Integration of spatial join algorithms for joining multiple inputs. In *Proc. ACM SIGMOD Symp. on the Management of Data*, 1999.

[ND98]   B. Nag and D. J. DeWitt. Memory Allocation Strategies for Complex Decision Support Queries. In *Proc. Intl. Conf. on Information and Knowledge Management*, 1998.

[NHS84]   J. Nievergelt, H. Hinterger, and K.C. Sevcik. The Grid File: An Adaptable Symmetric Multikey File Structure. *ACM Transactions on Database Systems*, 9(1):38–71, 1984.

[Ora]   Oracle 8 Server Concepts, Chap. 19 (The Optimizer). Oracle Technical Documentation.

[Ore86]   J. A. Orenstein. Spatial Query Processing in an Object-Oriented Database System. In *Proc. ACM SIGMOD Symp. on the Management of Data*, pages 326–336, 1986.

[PD96]   J.M. Patel and D. J. DeWitt. Partition Based Spatial-Merge Join. In *Proc. ACM SIGMOD Symp. on the Management of Data*, pages 259–270, 1996.

[RL85]   N. Roussopoulos and D. Leifker. Direct Spatial Search on Pictorial Databases Using Packed R-Trees. In *Proc. ACM SIGMOD Symp. on the Management of Data*, pages 17–26, 1985.

[SRF87]   T. Sellis, N. Roussopoulos, and C. Faloutsos. The R+Tree: A Dynamic Index for Multi-Dimensional Objects. In *Proc. Intl. Conf. on Very Large Data Bases (VLDB)*, pages 507–518, 1987.

[Val87]   P. Valduriez. Join Indices. *ACM Trans. on Database Systems*, 12(2):218–246, 1987.

[VG98]   V.Gaede and O. Guenther. Multidimensional Access Methods. *ACM Computing Surveys*, 1998. available at http://www.icsi.berkeley.edu/ oliverg/survey.ps.Z.

[Yao77]   S. B. Yao. Approximating Block Accesses in Data Base Organizations. *Communication of the ACM*, 20(4), 1977.

Appendix A

We give below a simple cost model for estimating the response time of an algorithm (query), which includes both I/Os and CPU time. For the I/O time calculation, we just assume that each I/O, i.e., that each disk access has a fixed cost of 10msec. Therefore, if nb_io denotes the number of I/Os, the time cost (in seconds) due to the disk is:

$$T_{disk} = nb\_io \cdot 0.01 \qquad (1)$$

In order to estimate CPU time, we restricted to the following operations: rectangle intersections, rectangle unions and sort comparisons. The parameters are then: (a) the number of rectangle intersections nb_inter, (b) the number of number comparisons nb_comp and (c) the number of rectangle unions nb_union. Since we consider a two-dimensional address space (generalizations are straightforward), each test for rectangle intersection costs four CPU instructions (two comparisons per dimension). Also, each rectangle union costs four CPU instructions. Finally, each comparison between two numbers costs one CPU instruction. If MIPS denotes the number of instructions executed in the CPU per second, then the time for each operation is calculated as:

$$T_{inter} = \frac{nb\_inter \cdot 4}{MIPS} \cdot 10^{-6} \qquad (2)$$

$$T_{union} = \frac{nb\_union \cdot 4}{MIPS} \cdot 10^{-6} \qquad (3)$$

$$T_{comp} = \frac{nb\_comp}{MIPS} \cdot 10^{-6} \qquad (4)$$

The CPU cost is thus estimated as

$$T_{proc} = T_{inter} + T_{union} + T_{comp} \qquad (5)$$

In addition to the above CPU costs, we assume that each read or write operation contributes to a CPU overhead of 5000 CPU instructions for pre and post processing of the page:

$$T_{prep} = \frac{nb\_io \cdot 5000}{MIPS} \cdot 10^{-6} \qquad (6)$$

The total CPU cost is then

$$T_{CPU} = T_{prep} + T_{proc} \qquad (7)$$

The response time of a query is then estimated as:

$$T_{response} = T_{CPU} + T_{disk} \qquad (8)$$

Uncertainty and Geologic Hypermaps

Abduction and Deduction
in Geologic Hypermaps

Agnès Voisard

Computer Science Institute, Freie Universität Berlin
D-14195 Berlin, Germany
voisard@inf.fu-berlin.de
http://www.inf.fu-berlin.de/~voisard

Abstract. A geologic map is a 2-dimensional representation of an *interpretation* of 3-D phenomena. The work of a geologist consists mainly in (i) inferring subsurface structures from observed surface phenomena and (ii) building abductive models of events and processes that shaped them during the geologic past. In order to do this, chains of explanations are used to reconstruct the Earth history step-by-step. In this context, many interpretations may be associated with a given output. In this paper, we first present the general contexts of geologic map manipulation and design. We then propose a framework for geologic map designers which supports multiple interpretations.

1   Introduction

A geologic map of a given area is a 2-dimensional representation of accepted models of its 3-dimensional subsurface structures. It contains geologic data which allow an understanding of the distribution of rocks that make up the crust of the Earth as well as the orientation of the structures they contain. It is based on a geologist's model explaining the observed phenomena and the processes that shaped them in the geologic past. In the current analog approach, this model is recorded in a geologic map with an explanatory booklet that describes the author's conclusions as well as relevant field and other observations (e.g., tectonic measurements, drill-hole logs, fossil records). Today, this variety of information is handled in a digital and even hypermedial form. This necessitates to conceive, develop and implement a suitable geologic hypermap model beforehand. The main objective of our project is the design of models and tools well-suited for the interaction between users and geologic hypermaps. Geologic hypermaps are a family of hyperdocuments [CACM95] with peculiar requirements due to the richness (in terms of semantics and structure) of the information to be stored. In these applications, users in general are both endusers (e.g., engineers or geology researchers) and designers (map makers).

This contribution focuses on the handling and representation of geologic knowledge within hypermap applications and addresses the needs for a data model that supports multiple interpretations from one or many geologists. When

R.H. Güting, D. Papadias, F. Lochovsky (Eds.): SSD'99, LNCS 1651, pp. 311–329, 1999.

designing geologic maps, the objectives are twofold: (i) inferring subsurface structures from observed surface phenomena and (ii) building abductive models of events and processes that shaped them in the geologic past. For this, chains of explanation are used to reconstruct the Earth history step by step.

Basic requirements for handling geologic applications in DBMS environments are described in [Voi98]. Those have been studied within a joint project with geologists at the Free University of Berlin. We focus here on the task of the map maker. To our knowledge, this is one of the first attempts to define tools for geologists. The US Geological Survey [USGS97] is currently defining a database to store geologic information in a relational context. A few authors (e.g., [Hou94, BN96, BBC97]) have also studied the 3-dimensional geometric aspects of geologic applications. Note however that, even though our goal is to build tools for the next generation of geologic maps (i.e., stored in a database), the map creation cannot be fully automated as some knowledge is difficult to express in terms of rules: There are no real laws that work at all time in all possible situations. Hence some steps still need to be performed manually.

Building the next generation of tools for geologists is a challenging task. The underlying geologic models and all the possible ways of manipulating the information make the supporting systems extremely complex. In addition, these systems borrow from many disciplines such as geospatial databases of course but also sophisticated visualization, simulation models, or artificial intelligence.

Here we restrict our attention to the reasoning process of geologists. As a first step, we present an explanation model for map designers, which is meant to be used during the abduction and the deduction processes. The model is based on complex explanation patterns (e.g., simulation models, similarities), and confidence coefficients. Even though geologic map interpretation has been studied thoroughly in the field of pure geology (e.g., [Bly76, Pow92]) for many years, to the best our knowledge, the mechanism behind geologic map making was not studied by computer scientists. Beside its complexity, a reason for that is the current lack of data. At present, not many complete geologic maps are stored in a digital form. So far, publishing a geologic map was a time-consuming task. These maps are now in the process of being digitized, but it will take many years before several documented versions of a map are available. The ultimate goal we are aiming at is the participation in the definition of a digital library of geologic maps whose elements may serve as starting points for further geologic map definitions.

This paper is organized as follows. Section 2 gives examples of geologic map manipulation and shows three main categories of users. Objects of interest and representative queries are given for each category. In Section 3, we present a reasoning model based on explanations and coefficients of confidence. Finally, Section 4 draws our conclusions, relates our work to other disciplines, and presents the future steps of our project.

2   Geologic Hypermaps Manipulation

According to the degree of expertise of the enduser, manipulating geologic hypermaps can be understood at many levels of abstraction. Three main categories of users can be identified, from the naive user to the application designer. They all communicate with the geologic maps database in a different way. While a naive user would like to know, for instance, the nature of soils in a given area, a more sophisticated user would like to find out why a particular geologic object was defined a certain way. An even more sophisticated user would like to access knowledge on other areas, for instance for comparison. Note that the most sophisticated users also have the requirements of the less qualified ones. They all access the basic data of their level together with metadata, which is understood differently for each category.

In this section, each category of users is studied separately. For each one we give the main objects of interests. To illustrate our discourse, examples of queries and a description of the tools needed are also given. This leads to a hierarchy of tools for manipulating such maps.

2.1   Traditional Enduser (User Type 1)

These users want to get straightforward information stored in a geologic database. A typical user of this category is an engineer who would like to find out the type of the soil in a given area in order to install a water pipe. These endusers need to access such basic information as well as metainformation. Metainformation has two aspects here. In a geospatial sense, it denotes information regarding the origin of data. It is for instance the date and the conditions under which some measurements were performed. This metadata is invariant in the three categories. The other kind of metadata is a high level description of data in the database, as for instance the possible values of a given attribute at a naive user level.

Objects of Interest. The major objects of interest at this level are geospatial, textual and multimedia objects (e.g., pictures and videos). These objects denoted *HO* are linked together via hyperlinks. A geologic object is a complex entity of the real world. It has a description as well as spatial attributes. The description may be elaborate, as we will see later. In particular, structured information, possibly of a multimedia type, can be attached to a geologic object. This information can be easily accessed in a hypermap system. What the users see and interact with are hypermaps in a graphical window. These are manipulated in a straightforward manner, through both mouse clicks on cartographic objects and a basic user interface that allows to access data values. More information on the *HO*'s manipulation and structure can be found in [Voi98].

Definition 2.1 A geologic map GM is a directed weakly connected graph (V_{HO}, E_{HO}, F), where

1. V_{HO} is a set of hyperobjects HO
2. E_{HO} is a set of edges.
3. $F : E_{HO} \rightarrow \mathcal{P}(V_{HO})$ is a labeling function.

A prototype for this category of users was coded using ArcViews [ESRI96a] and its programming language Avenue [ESRI96b], as reported in [KSV98]. In addition to basic visualization and querying, features such as flexible attribute display and specialized tools for specialists (e.g., a soil scientist) were also implemented. Figure 1 gives an example of a screenshot of the prototype.

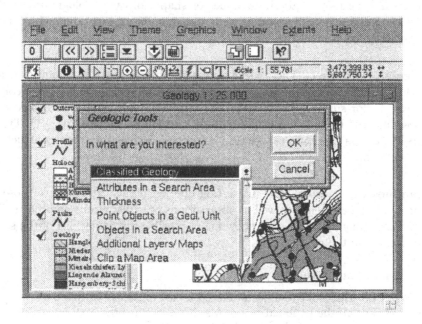

Fig. 1. Screenshot from the prototype

Examples of Queries. Queries to be posed to the system may include:

At a basic data level (instance level)

⋄ What are the characteristics (alphanumeric description) of this object (*point query*)?
⋄ What are all geo-objects in this area? (*region query*)
⋄ What is the nature of the soil here?

⋄ Where on this map do I find type of fossils FS22?
⋄ How much of geologic layer l is represented at depth d?

At a metadata level

⋄ What are the existing soil classifications in this map?
⋄ When was this map created?

2.2 Geology Researcher (User Type 2)

A geologic map is the basic tool of geologists. Hence these sophisticated users need to access basic information as well but also to understand why objects relate to each other and thus access the theories that justify the existence or the particular aspect (e.g., attribute values or shape) of each geologic object that constitutes the map. For instance, they may require explanations behind given phenomena.

Objects of Interest. These users access all objects described above but also another type of hyperobjects. As we will see in the next section, some geologic objects are for sure part of a map (with probability 1), and other objects are "guesstimated" by the map maker (with $0 \leq$ existence-probability ≤ 1). Type 2 users can access these two categories of objects as well as the complex structures behind them.

Examples of Queries. The interesting queries in this category are those related to object existence as well as assumptions or explanations on objects. The list is quite long as we want to show many possible manipulations of the structure defined in the next section.

At a basic data level

⋄ What are the geologic objects related to the existence of this one?
⋄ What are all the explanations of Ms. XYZ in this area?
⋄ What assumptions led to the definition of this object (if any)?
⋄ Which objects are not based on assumptions?
⋄ Which objects are defined exclusively under assumptions?
⋄ Which objects are defined using *this* explanation?
⋄ What is the explanation of the genesis of this layer?

At a metadata level

⋄ What are the maps designed after October 1990?
⋄ What are all the geologic maps using Explanation E?
⋄ What are the names of the geologists who studied this area?
⋄ How many versions do I have for geologic map GM231?
⋄ Where should we install drill-holes in this map?

2.3 Designer (User Type 3)

One peculiarity of this level is that, for a given map (i.e., covering a given area), many versions based on different interpretations can be defined, as described below.

The map making process can be described by two basic mechanisms, *abduction* and *deduction*. The geologist starts with a map of a given area. Currently, this map is often a topographic map. From such a map, s/he first looks at broad topographic features such as valleys and summits. Relationships between geologic borders and topography is then inferred. With knowledge of geologic structures at the surface (from drill-holes, for instance), the geologist also infers other geologic structures and groups them together. Groups are obtained by interpolation at the surface but also at the subsurface. Besides, s/he associates explanations or justification of the presence of certain geologic features together with a coefficient of confidence (abduction mechanism).

Geologic map making is an iterative process. The map maker draws a first version of a map based of course on observed features but also on his/her interpretation of the whole scene. Then s/he verifies the hypothesis by running simulation models and going to the field. A new version is then inferred, and so on and so forth. Eventually, s/he will obtain a map that corresponds to his/her interpretation but which is not "frozen" as things may still evolve in the future (e.g., with new explanation or new observed facts). However, the delta between his/her successive versions usually tends to become smaller and smaller. What we want to provide the designer with, beside assistance in this process, is a possibility of storing many underlying theories. Thus many explanations can be associated with one object, with different coefficients of confidence.

Explanations can be combined to form chains of explanations. Different combinations of explanations lead to many *interpretations* for the same output. A given interpretation of a geologic region together with all its interpreted geologic features is a *version* of a geologic map (in the database sense) for that region. Hence there are two ways to create new versions of a geologic map *GM* in the database:

1. Various *explanations* can be associated with *one* geologic object *o* ($o \in GM$) by one or many geologists.
2. Various *definitions* (identification and attribute values assignments) of a *collection* of geologic objects within the same region can be given by one or many geologists.

In addition, the novelty of the approach that underpins our new generation of tools for map makers is the ability for them to start with *existing* geologic maps. Then those will be transformed and customized according to different interpretations. Hence there is no need anymore to create a geologic map from scratch.

Objects of Interest. The objects manipulated by these users are geologic maps composed of documented geologic features (in our model, to geologic object, we prefer the term geologic feature in order to stay closer to the geologist's reasoning, although both terms are used interchangeably in the paper). Geologic features are simple or complex as defined below:

```
geologic-feature = (AttributeList, SpatialExtent, UndergroundExtent)
                   /* atomic geologic feature */
                 | (AttributeList, SpatialExtent, UndergroundExtent,
                   {geologic-feature}) /* complex geologic feature */
```

Complex geologic features are for instance stratigraphic or tectonic structures. Note that some values of a complex geologic feature (e.g., `SpatialExtent` and `UndergroundExtent`) can be inferred from those of subobjects [EF89].

There is a clear distinction between *observed* objects, hence existing for sure in a map (probability 1) and objects that are "guesstimated". In the sequel we refer to observed objects as "hard objects" as opposed to "soft objects". A soft object can become a hard object when an hypothesis is verified, while a hard object cannot turn into a soft object. The probability attached to the existence of a soft object can change according to a different interpretation.

Note that within a given map, some external events can change the nature of geologic objects, for instance the introduction of a drill-hole introduces a new geologic object in the map. Associated with these objects are explanations together with coefficients of confidence. An explanation can justify the presence of an object or document the value of an attribute (e.g., the soil concentration of a component).

Objects Manipulation and Querying. Obviously, querying does not play a crucial rule for this category of users. What is important is to provide designers with tools for defining and manipulating the underlying structure (see Section 3). However below are a few examples of queries and structure manipulation. As we can see from the selected collection of queries, many concerns are of interest for the designer, from the representation of a given attribute to the number of maps defined in a given area.

At a basic data level

⬦ What are the interpretations using Geologic Model M?
⬦ What if assumptions of Ms. XYZ are not justified anymore?
⬦ What if model M turns out to be unapplicable in this area?
⬦ I know there should be cobalt here. Where should it be (shape/location)?

At a metadata level

⋄ What geologic model could I use to justify this tectonic structure?
⋄ What maps were defined with geologic map GM324 as a starting point?
⋄ How should I represent a layer with iron?
⋄ What are the maps published in area *a*?
⋄ What is the difference between interpretation I1 and interpretation I2?

3 Supporting Geologic Map Making

This section presents the kernel of a tool that assists map designers in the geologic map making process. We place ourselves in the context of a *geologic map factory* that communicates with a geologic map library. Eventually, objects defined in the map factory will be validated and stored in the geologic map library. Such a library is extremely useful in this context as it allows in particular geologic maps to be built with other maps as starting points. Browsing the library is hence a key functionality of the environment offered to geologic map makers. The generic geologic map library is not presented here. It is a special kind of geospatial library [SF95, FE99] that supports multiversioning based on various interpretations.

A geologic map factory contains many modules, among them a reasoning module. Components such as cartographic module, help module, or validation module are not our focus here. Rather we study the major task of the reasoning module, namely the support for abduction and deduction in the map making process.

3.1 Reasoning Model

In order to assist the designer in the map making process, the main object to consider is a reasoning structure, i.e., a *documented geologic map* (*DGM*). All the elements of a DGM, i.e., geological features (atomic or complex) and explanations are described thereafter, first in general terms and then using an O_2-like [BDK92] specification language. For the sake of simplicity and legibility, the specifications we give are kept as simple as possible.

Geologic Features. The basic elements to consider are geologic features. These can be atomic (e.g., a fault) or complex (e.g., a tectonic structure). In any case they have a *description* (alphanumeric attributes), a *spatial extension* (spatial part that gives the shape and the location) and an *underground extent*. A geologic feature is of one of the two following types:

⋄ *Hard type.* Such objects are part of the map with probability one. The feature was for instance seen on the fields and the numeric values of its attribute could be computed.

◇ *Soft type.* These objects belong to the map with a probability (confidence) between 0 and 1. We will see further how they are created.

Many kinds of relationships exist among geologic features. Beside composition, topological relationships such as adjacency are of course of prime importance. These relationships are beyond the scope of our study. In addition, fuzzyness plays a crucial role in these applications as geological features can have (i) a fuzzy description (e.g., concentration of cobalt between 70 and 100 %), (ii) a fuzzy spatial part (location not precise, shape with fuzzy borders), and (iii) a fuzzy underground extent.

Schema definition:

```
Class GeologicFeature type tuple
                (description: Attributelist,
                 spatialextent: Spatial,
                 undergroundextent: Solid)

Class SimpleGeologicFeature inherits GeologicFeature
Class ComplexGeologicFeature inherits GeologicFeature
        type tuple (geolfeatures: { GeologicFeature})


Class HardGF inherits GeologicFeature
Class SoftGF inherits GeologicFeature
        type tuple (origin: Text)
```

Classes `AttributeList`, `Spatial`, `Solid` are not detailed here. `Spatial` and `Solid` embody both geometry (coordinate location and shape) and topology. For a possible definition of `Spatial` (basically points, lines, curves, polygons and regions) in a database context, see [EF89, SV92, Wor94, GS95, Sch97]. Solid (volume) modeling (3D-geometric modeling) appears typically in CAD/CAM applications (see for instance [BN96, DH97]).

Explanations. We are interested here in the process of documenting such maps, i.e., in the possible collections of explanations to justify:

1. The existence of a geologic feature.
2. The particular values of some attributes of a geologic feature.
3. The presence of other explanations (when for instance an explanation contains references. Typically when a bibliographic reference is given in an explanatory text).

In this context, an explanation can be of three different types:

1. *Provable* (reliable). It can be justified by a hard fact, such as a drill-hole, or a geologic simulation model. Note the hierarchy in the reliability.

2. *Similarity-based* (training areas). This occurs when some part of a map seems to be similar to a part of either the same map or of another map.
3. *Experience-based* (or feeling). Such explanations can also mutate to become provable if an underlying assumption is verified.

An explanation is a complex object composed of structured text, and possibly (bibliographic) references, references to simulation models, geologic features in the map, and other areas (coordinates) from either the same or a different map. Moreover, it contains the geologic features that serve as justification for the argumentation as well as the geologic features that could be further consequences of this explanation through the deduction mechanism. Such arguments could be defined as *query expressions* over a geologic map (set of geologic features). The simple specification of an explanation is given below. Note the basic superclass Explanation which is further specialized into various explanation classes such as ProvableExplanation, SimilarityExplanation, and ExpertExplanation. In addition, an explanation is either basic or complex, which leads to classes BasicExplanation and ComplexExplanation.

Schema definition:

```
Class  Explanation type tuple
                     (author: string,
                      argument: {HardGF},
                      consequence: {SoftGF})


Class BasicExplanation inherits Explanation ()
Class ComplexExplanation inherits Explanation
                 type tuple (all-explanations: {Explanation})


Class ProvableExplanation inherits Explanation
                          /* e.g., models or drill-holes */
Class SimilarityExplanation inherits Explanation
                          /* similarity-based explanation */
Class ExpertExplanation inherits Explanation
                          /* experience-based explanation */


Class BasicExplanation type tuple (text: string)

Class ModelExplanation inherits BasicExplanation
                     type tuple (argument: string)
                         /* argument = modelref */


Class BiblioExplanation inherits BasicExplanation
                     type tuple (argument: bibitem)
                   /* argument = bibliographic reference */
```

```
Class HardObjectExplanation inherits BasicExplanation
                type tuple (argument: {GeologicFeature})
                /* argument =  geologic features */

Class SoftObjectExplanation inherits BasicExplanation
                type tuple (argument: {GeologicFeature})
                /* argument based on geologic features */

Class AreaExplanation inherits BasicExplanation
                type tuple (area: zone)
                /* argument: Area, the coordinate of a region */
```

Environmental processes that lead to a given geologic feature are found out by looking recursively at all the explanations of type ModelExplanation.

Complete and Documented Geologic Map (DGM). A complete geologic map is a 5-tuple (r, d, c, l, dgm), where r is the reference of the map in the map library, d the date of creation, c the coordinates of the covered area $(dom(c) = (\Re^2 \times \Re^2))$, l the legend used and dgm a documented geologic map defined as follows.

A *documented geologic map DGM* is a set of directed acyclic weakly connected labeled graphs with the following characteristics: The source (entry) of each graph is a geologic feature (hard or soft). The (explanation) structures, EG attached to it with a certain coefficient of confidence is defined below.

Definition 3.2. An *explanation graph EG* is a pair (E, P), where

- E is a set of explanations (inner nodes and drains of the graph). An explanation is atomic or complex, as described above.
- P is a set of weighted edges. There is a weighted edge we with weight $p_{n_i,n_{i+1}}$ from a node (including the source) n_i to a node n_{i+1} if n_{i+1} is an explanation of n_i with confidence $p_{n_i,n_{i+1}}$ $((0 \le p_{n_i,n_{i+1}} \le 1)$. The semantics of an edge $we(p_{n_i,n_{i+1}})$ between n_i and n_{i+1} corresponds to "n_{i+1} *is an explanation of* n_i *with confidence* $p_{n_i,n_{i+1}}$".

Figure 2 shows a documented geologic map composed of geologic features {GF1, ..., GFn}. In this figure, source objects of the form GF_i correspond to hard geologic features whereas objects of the form gf_i correspond to soft geologic features. As we can see, GF1 can be explained by E1 with confidence $p_{GF1,E1}$ or by E2 with confidence $p_{GF1,E2}$. In turn, E2 is explained by E3 with confidence $p_{E2,E3}$, which may itself be justified by a complex explanation containing both E4 and E5 (with confidence coefficient $p_{E3,(E4,E5)}$) or by E6 (confidence coefficient $p_{E3,E6}$). The symbol | | denotes drains of the graphs ("terminal nodes").

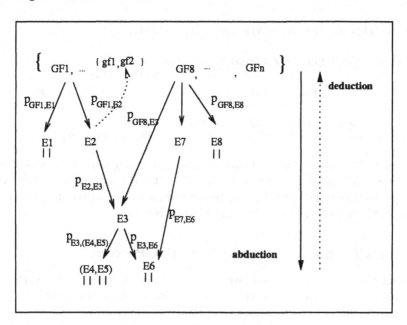

Fig. 2. A documented geologic map

Schema definition:

```
Class CompleteGeologicMap  type tuple
                   (reference: string,
                    date: date,
                    legend: Legend,
                    explgeomap:  DocumentedGeologicMap)


Class DocumentedGeologicMap type tuple
                   (explhardGF: {ExplHardGF},
                    explsoftGF: {ExplHSoftGF})

Method create-DGM (author: string) on class DocumentedGeologicMap
                   /* Method create-DGM creates a map dynamically */

Class ExplHardGeologicFeatures type tuple
                   (hardGF: HardGF,
                      explanation-graph: ExplanationGraph)

Class ExplSoftGeologicFeatures type tuple
                   (softgeolfeature: SoftGF,
                      explanation-graph: ExplanationGraph)

Class ExplanationGraph type set (graph-element: WeightedExplanation}
```

```
Class WeightedExplanation type tuple (pi: integer)

Class SimpleWE inherits WeightedExplanation
                         type tuple (expl: Explanation)
    /* the explanation itself can be complex cf. (E4,E5) in Figure 3 */

Class ComplexWE inherits WeightedExplanation
                         type tuple (expl:  ExplanationGraph)
```

The useful substructures of this documented geologic map are sets of paths associated with geologic features. Eventually, after the map making step, with a geologic feature is associated many possible chains of explanations (union of all the possible paths starting with this object as a source).

3.2   Specifying Abduction and Deduction Processes

Assuming the map maker starts with a collection of hard geologic features, s/he first associates explanations with the geologic features and then builds the structure recursively. The map defined by method
create-DGM below is then linked to a complete geologic map (with area coordinates, reference, date, etc.)

The abduction step is described by Procedure **create-DGM**. Once this is finished (i.e., explanations defined with coefficient of confidence), method **infer-all** is invoked on the resulting structure (deductive mechanism). Method **create-DGM** can be called many times in order to assign many explanations to an object.

Building a DGM. The following method generalizes the abduction process. It creates an explanation chain associated with each geologic feature of a map.

```
Method create-DGM (author: string)  on class DocumentedGeologicMap
for each GF in +(self.explgeomap.explhardGF,self.explgeomap.expsoftGF)
              /* union set */
        add-explanation-chain(0)
                  @GF.explanation-graph.graph-elelement.explanation
    /* invokes method add-explanation-chain on all geologic features */


Method add-explanation-chain (done:Boolean)  on class Explanation
      {if (not done)
       e = new(Explanation)
       for each ge in self.explanation
        e.explanation = add-explanation@ge.explanation
             Print('Done?')
       add-explanation-chain(done)@e.explanation
       }
```

```
Method add-explanation on class Explanation
  /* This method creates one weighted explanation.
     It can be called separately in other contexts (e.g., change
     of explanation).
     The coefficient of confidence and the contents of the explanation
     are  entered dynamically. ''*'' denotes a call for a function
     interacting with the user (input parameter) */
     {e = new(Explanation)
      write ('Value for p?') read(p)
      e.pi = p
      e.text = *textinput
      e.argument = *variableinputs
     }
```

The deduction step that follows is realized by:

```
Method infer-all on Class DGM
  /* This method goes recursively through an interpretation.
     It runs on all explanations and infer all possible consequent
     objects (Explanation.consequence) by propagating all chosen
     explanations*/
```

3.3 Operations on DGM

In addition to classical graph manipulation, operations on this structure include:

- ◇ Change a coefficient of confidence.
  `method change-confidence (pi: integer)`
  `on class WeightedExplanation`
- ◇ Change an explanation.
  `method change-explanation (Ei: Explanation) on class Explanation.`
  In this method, the explanation receiving the method has to be looked for
  everywhere in the DGM and replaced by Ei.
- ◇ Change the type of an explanation from *similarity-* or *experience-based* to
  provable. This is done by removing an explanation and inserting a new one
  of type *provable*.
- ◇ Add a soft object (source) with explanations.

Queries on this structure are for instance:

- ◇ Retrieve all interpretations of documented map DM that use explanation
  E1.
- ◇ Retrieve all interpretations of documented map DM with a threshold of 0.5
  in the definition of all its geologic features.

Other operations concern a change of definition of soft geologic objects, and
more precisely of the structuring of complex objects. Suppose that two layers
l_i and l_{i+1} were assigned to the same stratigraphic unit (for instance because

they contain the same kind of rocks). Suppose also that this unit was inferred to be an anticline, which means in particular that $age(l_i) < age(l_{i+1})$. Assume that a new layer l' in between is discovered with $age(l') < age(l_i)$. This shows a contradiction and the two original layers should be part of different stratigraphic units.

3.4   Support for Multiple Interpretations

Multiple interpretations can be defined by extracting paths in the structure defined above. Thus a given set of objects can be associated with various interpretations. A more general concept is the one of *version* which is associated with a complete geologic map (see Section 3.2). It hence allows to customize the definition and structuring of the original set of geologic objects. The definitions are given below.

Definition 3.5.1 An *explanation chain EC* is a 3-tuple (GF, P, p) where GF is a geologic object, P a path $(P \subset EG$, where EG is an explanation graph) and p a coefficient of confidence. p is defined as the multiplication of the coefficients of confidence of all edges of the chain. In Figure 3, explanation E1 is associated to GF1 with confidence $p_{GF1,E1}$. This path is likely to be part of a relevant version for the user compared to the second explanation for GF1 (Explanation E2) which has for confidence $(p_{GF1,E2} \times p_{E2,E3} \times p_{E3,(E4,E5)})$ or $(p_{GF1,E2} \times p_{E2,E3} \times p_{E6})$.

Definition 3.5.2 An *interpretation I_{GM}* of a geologic map GM is a set of explanation chains.

Definition 3.5.3 A *weighted interpretation WI_{GM}* of a geologic map GM is a set of pairs (EC, p) where EC is an explanation chain and p a coefficient of confidence.

Definition 3.5.4 A *version V_{CGM,I_G}* of a complete geologic map CGM is a pair (CGM, I_G) where CGM is a complete geologic map composed in particular of the set of geologic features G, and I_G a corresponding interpretation over this set.

Figure 3 shows a documented geologic map M1 with two different interpretations, I1 and I2. The first interpretation at the top uses explanation chain (E2, E3, E6) for GF1, (E7,E6) for GF2 and (E9,E10) for GF3. The interpretation $I_{\{GF1,GF2,GF3\}}$ is then
$I1_{GF1,GF2,GF3} = \{(E2, E3, E6), (E7, E6), (E9, E10)\}$.

At the bottom of Figure 3, the second interpretation also uses explanation chain (E2,E3,E6) attached to GF1, but different explanation chains are associated with objects GF2 (Chain (E3,(E4,E5))) and GF3 (Chain (E14,E15)). Hence $I2_{\{GF1,GF2,GF3\}} = \{(E2, E3, E6), (E3, (E4, E5)), (E14.E15)\}$.

Finally, following Definition 3.5.3, we can define the weighted interpretations $WI1_{\{GF1,GF2,GF3\}}$ and $WI2_{\{GF1,GF2,GF3\}}$.

$WI1_{\{GF1,GF2,GF3\}} = \{((E2, E3, E6), p_{GF1,E2} \times p_{E2,E3} \times p_{E3,E6}),$
$((E7, E6), p_{GF2,E7} \times p_{E7,E6}), ((E9, E10), p_{GF3,E9} \times p_{E9,E10})\}.$
$WI2_{\{GF1,GF2,GF3\}}$ is computed as
$WI2_{\{GF1,GF2,GF3\}} = \{((E2, E3, E6), p_{GF1,E2} \times p_{E2,E3} \times p_{E3,E6})$
$((E3, (E4, E5)), p_{GF2,E3} \times p_{E3,(E4,E5)}), ((E14, E15), p_{GF3,E14} \times p_{E14,E15})\}.$

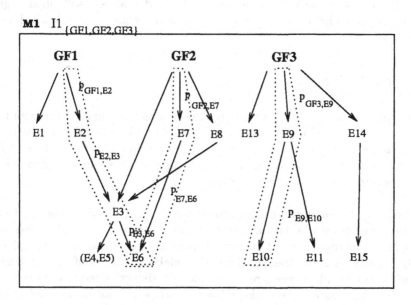

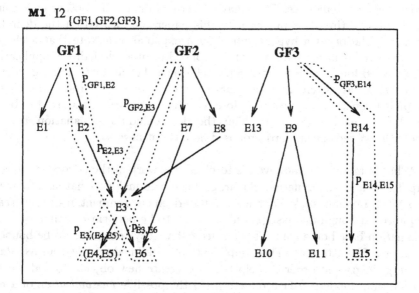

Fig. 3. Support for multiple interpretations

Check for Consistency within an Interpretation. Consistency can be checked within an interpretation once all objects are frozen in the world of geologic features and explanations for a given version. Consistency will ensure that (i) relationships among geologic structures are not violated, (ii) explanations do not contradict each other. Defining consistency among explanations of one documented map is a complex problem. However, it is handled easily when justifications (edges in the graph) are based on the negation of explanations. For instance, suppose that E1 explains GF1 and that one soft object sgf2 is also supposed to belong to the map (e.g., as a consequent object) based on the negation of E1.

4 Conclusion

In this paper, we set a framework to manipulate geologic maps at different levels of abstraction. We identified three categories of users, from naive to expert, and we started with a description of their needs when communicating with a geologic hypermap system. The most sophisticated class of users, the designers, clearly need tools to define new geologic maps.

We then explained the mechanism behind the map making process. To sum up, the geologist starts with a map (e.g., a topographic map) of a given area. This map contains observed surface phenomena. From this map, complex geologic structures are extracted. This is an elaborate mechanism that leads to (i) inferences of subsurface structures, and (ii) abductive models of events and processes that shaped the geologic features. Explanations or a justification of the presence of certain geologic features are associated with them together with a coefficient of confidence. They may be verified or not, and other conclusions may be drawn (iterative process). In this paper, we focused on the definition and manipulation of a hypergraph of such explanations. Note that such general structures and mechanisms can be applied to other domains of application, and are used for instance in artificial intelligence. The abduction and deduction mechanisms have been used for years in medical diagnosis, route planning or equipment repair, among others. More recently, these technics started to be applied by the genome community. Thus the idea behind the explanation structure (documented geological map) proposed here is not new.

What is innovative, however, is to place this work in the context of geologic map design and manipulation. So far, geologic map making (that lead to analog maps) was a tedious task. Within a computerized environment, it can be greatly simplified, although many steps will still have to be performed manually, as full automation based on a complete knowledge-based system seems to be bound to fail after a few attempts. This simplification is due mainly to (i) an *assistance* to the geologists in order to help them to create new objects (based on map consultation, model verification, etc.), (ii) the possible *customization* of a map through the reuse of another version based on different interpretations. Both aspects were shown is the paper. One of the future directions of our work is to specify the versioning process (i.e., definition and storage of complete geologic map versions) in a database environment.

A geologic hypermap prototype for user types 1 and 2 (as defined in the first section) was coded using ArcView. We plan to implement a prototype based on the explanation model described in Section 3, and to link it eventually to our hypermap prototype. This prototype will eventually be linked to a geologic map library that is currently under specification.

This work was done within the framework of a joint-project with geologists. Understanding the complex requirements of a foreign area of applications is a challenging task. Defining the appropriate tools for the next generation of geologists who will use only elaborate digitized map versions is certainly an ambitious project that will take a long time. We believe, however, that several problems can be isolated and studied step-by-step. The purpose of the work presented here is to define a brick necessary to the realization of such a large and complex system.

Acknowledgments. I wish to thank Claudia Bauzer Medeiros, Michel Scholl and Marwan Abu-Khalil for insightful discussions, as well as Stefanie Kübler and Wolfdietrich Skala for their cooperation on the project. The development of the GeoHyp prototype by Stefanie Kübler and Heike Friedrichs is gratefully acknowledged. Part of this work was funded by the German Research Foundation (DFG). Thanks also to the anonymous referees for their useful comments.

References

[BBC97]   O. Balovnev, M. Breunig, and A. B. Cremers. From GeoStore to GeoToolKit: The second step. In *Advances in Spatial Databases, Lecture Notes in Computer Science*, volume 1262, pages 223–238, Heidelberg/Berlin/New York, 1997. Springer Verlag.

[BDK92]   F. Bancilhon, C. Delobel, and P. C. Kanellakis, editors. *Building an Object-Oriented Database Systems: The Story of O₂*. Morgan Kaufmann, San Francisco, California, 1992.

[Bly76]   F. Blyth. *Geological Maps and Their Interpretation*. Edward Arnold, London, 1976.

[BN96]   J.-D. Boissonnat and S. Nullans. Reconstruction of Geological Structures from Heterogeneous and Sparse Data. In *Proc. 4th ACM GIS Workshop*, 1996.

[CACM95]   Association for Computing Machinery (ACM), editor. *Communications of the ACM, Special issue on Designing Hypermedia Applications*, New York, August 1995.

[DH97]   J. Döllner and K. Hinrichs. Object-Oriented 3D Modeling, Animation and Interaction. *Journal of Visualization and Computer Animation*, 8(1):33–64, 1997.

[EF89]   M. Egenhofer and A. Frank. Object-Oriented Modeling in GIS: Inheritance and Propagation. In *Proc. Auto-Carto 9*, pages 588–598, 1989.

[ESRI96a]   Inc. Environmental Systems Research Institute. Arcview GIS Version 2 , 1996.

[ESRI96b]   Inc. Environmental Systems Research Institute. Avenue: Customization and Application Development for Arcview, 1996.

[FE99]   D. Flewelling and M. Egenhofer. Using Spatial Archives Effectively. *Int. Journal on Geographical Information Science (IJGIS)*, 13(1):1–8, 1999.

[GS95]   R. H. Güting and M. Schneider. Realm-Based Spatial Data Types: The ROSE Algebra. *The VLDB Journal*, 4(2):243–286, April 1995.

[Hou94]   S. Houlding. *3D Geoscience Modeling: Computer Techniques for Geological Characterization*. Springer-Verlag Inc., Berlin Heidelberg New-York, 1994.

[KSV98]   S. Kübler, W. Skala, and A. Voisard. The Design and Development of a Geologic Hypermaps Prototype. In *Proceedings of ISPRS Conference*, 1998.

[Pow92]   D. Powell. *Interpretation of Geological Structures Through Maps*. Langman Scientific and Technical, 1992.

[Sch97]   M. Schneider. *Spatial Data Types for Database Systems: Finite Resolution Geometry for Geographic Information systems*, volume 1288 of *Lecture Notes in Computer Science*. Springer-Verlag, Berlin/Heidelberg/New York, 1997.

[SF95]   T. R. Smith and J. Frew. Alexandria Digital Library. *Communications of the ACM*, 38(4):61–62, April 1995.

[USGS97]   US Geological Survey. Digital Geologic Maps Data Model, 1997. URL = http://www.usgs.gov/.

[SV92]   M. Scholl and A. Voisard. Object-Oriented Database Systems for Geographic Applications: An Experiment With O_2, 1992. Chapter 28 of [BDK92].

[Voi98]   A. Voisard. Geologic Hypermaps are more than Clickable Maps! In *Proceedings of the International ACM GIS Symposium*, pages 14–19, 1998.

[Wor94]   M. Worboys. Object-Oriented Approaches to Geo-Referenced Information. *Int. Journal on Geographical Information Systems (IJGIS)*, 8(4), 1994.

Uncertainty Management for Spatial Data in Databases: Fuzzy Spatial Data Types

Markus Schneider

FernUniversität Hagen, Praktische Informatik IV
D-58084 Hagen, Germany
Markus.Schneider@fernuni-hagen.de

Abstract. In many geographical applications there is a need to model spatial phenomena not simply by sharply bounded objects but rather through vague concepts due to indeterminate boundaries. Spatial database systems and geographical information systems are currently not able to deal with this kind of data. In order to support these applications, for an important kind of vagueness called *fuzziness*, we propose an abstract, conceptual model of so-called *fuzzy spatial data types* (i.e., a *fuzzy spatial algebra*) introducing *fuzzy points*, *fuzzy lines*, and *fuzzy regions*. This paper* focuses on defining their structure and semantics. The formal framework is based on fuzzy set theory and fuzzy topology.

1 Introduction

Representing, storing, quering, and manipulating spatial information is important for many non-standard database applications. Specialized systems like geographical information systems (GIS) and spatial database systems to a certain extent provide the needed technology to support these applications. So far, spatial data modeling has implicitly assumed that the extent and hence the borders of spatial phenomena are precisely determined, homogeneous, and universally recognized. From this perspective, spatial phenomena are typically represented by sharply described *points* (with exactly known coordinates), *lines* (linking a series of exactly known points), and *regions* (bounded by exactly defined lines which are called *boundaries*). Special data types called *spatial data types* (see [Sch97] for a survey) have been designed for modeling these spatial data. We speak of *spatial objects* as instances of these data types. The properties of the space at the points, along the lines, or within the regions are given by attributes whose values are assumed to be constant over the total extent of the objects. Well known examples are especially man-made spatial objects representing engineered artifacts like highways, houses, or bridges and some predominantly immaterial spatial objects exerting social control like countries, districts, and land parcels

---

* This research was partially supported by the CHOROCHRONOS project, funded by the EU under the Training and Mobility of Researchers Programme, contract no. ERB FMRX-CT96-0056.

R.H. Güting, D. Papadias, F. Lochovsky (Eds.): SSD'99, LNCS 1651, pp. 330–351, 1999.
© Springer-Verlag Berlin Heidelberg 1999

with their political, administrative, and cadastral boundaries. We will denote this kind of entities as *crisp* or *determinate* spatial objects.

Increasingly, researchers are beginning to realize that the current mapping of spatial phenomena of the real world to exclusively crisp spatial objects is an insufficient abstraction process for many spatial applications and that the feature of *spatial vagueness* or *spatial indeterminacy* is inherent to many geographic data [BF96]. Moreover, there is a general consensus that applications based on this kind of indeterminate spatial data are not covered by current GIS and spatial database systems. In this paper we focus on a special kind of spatial vagueness called *fuzziness*. Fuzziness captures the property of many spatial objects in reality which do not have sharp boundaries or whose boundaries cannot be precisely determined. Examples are natural, social, or cultural phenomena like land features with continuously changing properties (such as population density, soil quality, vegetation, pollution, temperature, air pressure), oceans, deserts, English speaking areas, or mountains and valleys. The transition between a valley and a mountain usually cannot be exactly ascertained so that the two spatial objects "valley" and "mountain" cannot be precisely separated and defined in a crisp way. We will designate this kind of entities as *fuzzy* spatial objects.

The goal of this paper is to present a formal object model for *fuzzy points*, *fuzzy lines*, and *fuzzy regions* in two-dimensional Euclidean space, an effort which is to lead to a *fuzzy spatial algebra*. We propose *fuzzy set theory* and *fuzzy topology* as appropriate conceptual tools for modeling indeterminate spatial data. Fuzzy set theory is an extension and generalization of classical set theory; the approach of fuzzy sets replaces the crisp boundary of a classical set by a gradual transition zone and permits partial and multiple set membership. For fuzzy regions, different views give a better understanding of their nature and also demonstrate how these objects can be represented as (collections of) crisp regions. Consequently, the current exact object models for crisp spatial objects can be considered as simplified special cases of a richer class of models for general spatial objects. It turns out that this is exactly the case for the model to be presented.

Section 2 explains different aspects of spatial vagueness and presents related work. Section 3 introduces some basic definitions of fuzzy set theory and fuzzy topology as far as they are needed in this paper. Sections 4, 5 and 6 formally define fuzzy points, fuzzy lines, and fuzzy regions, respectively. Since the definition for fuzzy regions does not expose their geometric structure, Section 7 provides several structured views of fuzzy regions based on collections of crisp regions. Section 8 draws some conclusions and gives a prospect of future research.

2   Aspects of Spatial Vagueness and Related Work

In current spatial data modeling, the entity-oriented view of spatial phenomena, which we will take in this paper, considers determinate spatial objects as conceptual and mathematical abstractions of real-world entities which can be identified and distinguished from the rest of space. For example, a crisp region partitions space into an interior, a boundary, and an exterior part which are mutually exclusive and cover the whole space. Hence, the notion of a crisp region is intrinsically related to the notion of a boundary. This view fits very well with the mathematical concepts given by the Jordan Curve Theorem and ordinary point set topology.

Boundaries are considered as sharp lines that represent abrupt changes of spatial phenomena and that describe and thereby distinguish regions with different characteristic features. The assumption of crisp boundaries harmonizes very well with the internal representation and processing of spatial objects in a computer which requires precise and unique internal structures. Hence, in the past, there has been a strong tendency to force reality into crisp objects. In practice, however, there is no apparent reason for the whole boundary of a region to be determined. There are a lot of geographical application examples illustrating that the boundaries of spatial objects can be partially or totally indeterminate or blurred. For instance, boundaries of geological, soil, and vegetation units [Alt94, Bur96, KV91, LAB96] are often sharp in some places and vague in others; many human concepts like "the Indian Ocean" are implicitly vague.

In the real world, there are essentially two categories of indeterminate boundaries: sharp boundaries whose position and shape are unknown or cannot be measured precisely, and boundaries which are not well-defined or which are useless (e.g., between a mountain and a valley) and where essentially the topological relationship between spatial objects is of interest. According to these two categories, mainly two kinds of spatial vagueness can be identified: uncertainty and fuzziness. *Uncertainty* is traditionally equated with randomness and chance occurrence and relates either to a lack of knowledge about the position and shape of an object with an existing, real boundary (*positional* uncertainty) or to the inability of measuring such an object precisely (*measurement* uncertainty). *Fuzziness* is an intrinsic feature of an object itself and describes the vagueness of an object which certainly has an extent but which inherently cannot or does not have a precisely definable boundary.

The subject of modeling spatial vagueness has so far been predominantly treated by geographers but rather neglected by computer scientists. At least three alternatives are proposed as general design methods: (1) *exact models* [CF96, CG96, ES97b, Sch96] which transfer type systems and concepts for spatial objects with sharp boundaries to objects with unclear boundaries and which model both uncertainty and fuzziness but in a restricted way, (2) *probabilistic models* [Bla84, Bur96, Fin93, Shi93] which are based on probability theory and predominantly model positional and measurement uncertainty, and (3) *fuzzy models* [Alt94, Bur96, Dut89, Dut91, KV91, LAB96, Use96, Wan94, WHS90] which are all based on fuzzy set theory and predominantly model fuzziness.

The exact object model approach profits from existing definitions, techniques, data structures, algorithms, etc. which need not be redeveloped but only modified and extended. Except for [ES97b], the approaches are based on some kind of *zone* concept. Vague boundaries of a region are modeled as zones expressing the minimal and maximal possible extent of a region. *Vague regions* [ES97b] are a generalization of these models. A vague region is defined as a pair of disjoint, crisp regions. The first region called the *kernel* describes the area which definitely and always belongs to the vague region. The second region called the *boundary* describes the *area* for which we cannot say with any certainty whether it or parts of it belong to the vague region or not. *Maybe* it is the case, *maybe* it is not. Or we could say that this is *unknown*. Vague regions are based on a *three-valued logic*, and boundaries need not necessarily be one-dimensional structures but can be regions.

Probability theory is able to represent uncertainty and defines the membership grade of an entity in a set by a statistically defined probability function. It deals with the *expectation* of a future event, based on something known now. Examples are the uncertainty about the spatial extent of regions defined by some property such as temperature, or the water level of a lake.

Fuzzy set theory deals only with fuzziness. It describes the *admission of the possibility* (given by a so-called *membership function*) that an individual is a member of a set or that a given statement is true. Hence, the vagueness represented by fuzziness is not the uncertainty of expectation. It is the vagueness resulting from the imprecision of meaning of a concept. Examples of fuzzy spatial objects include mountains, valleys, biotopes, oceans, and many other geographic features which cannot be rigorously bounded by a sharp line.

Another difference between fuzzy set theory and probability theory is that in the first case the possibility that an individual belongs to a set depends on *subjective* factors (e.g., expert knowledge) whereas in the second case probability can be computed formally or determined empirically and is thus more *objective*. Moreover, fuzzy set theory enables vague statements about one concrete object whereas probability theory makes statements about a collection of objects from which one is selected. Hence, fuzzy set theory models *local vagueness* while probability theory models *global* vagueness.

The only proposal of a fuzzy data type relates to fuzzy regions [Alt94] defined as a fuzzy set over $\mathbb{N}^2$. Each coordinate $(x, y) \in \mathbb{N}^2$ is associated with a value between 0 and 1 and describes the concentration of some feature attribute at that point. Unfortunately, the simple set property is insufficient since geometric anomalies can arise, as we will see later. The possible importance of fuzzy sets for geographical applications is demonstrated in [Bur96, LAB96, Use96] where also examples of application-specific membership functions are given. The benefits of fuzzy set theory for approximate spatial reasoning and fuzzy query languages is shown in [Dut89, Dut91, KV91, Wan94]. [WHS90] models fuzzy objects by means of the relational data model.

3 Fuzzy Sets and Fuzzy Topology

Crisp regions have been formally defined on the basis of point sets and point set topology (e.g., [ES97b, Gaa64, Sch97]) which mainly rest on the set operations of union, intersection, and difference. In a straightforward way we will now describe extensions of these two concepts to fuzzy set theory and fuzzy topology.

Fuzzy set theory [Zad65] is an extension and generalization of Boolean set theory. Let X be a classical (crisp) set of objects, called the *universe* (*of discourse*). Membership in a classical subset A of X can then be described by the *characteristic function* $\chi_A : X \to \{0, 1\}$ such that for all $x \in X$ holds:

$$\chi_A(x) = \begin{cases} 1 & \text{if and only if } x \in A \\ 0 & \text{if and only if } x \notin A \end{cases}$$

This function, which discriminates sharply between members and non-members of a set, can be generalized such that all elements of X are mapped to the real interval $[0,1]$ indicating the *degree of membership* of these elements in the set in question. Hence, fuzzy set theory permits an element to have partial and multiple membership. Larger values designate higher grades of set membership. Let X again be the universe. Then

$$\mu_{\tilde{A}} : X \to [0, 1]$$

is called the *membership function* of $\tilde{A}$, and the set

$$\tilde{A} = \{(x, \mu_{\tilde{A}}(x)) \mid x \in X\}$$

is called a *fuzzy set* in X. All elements of X receive a valuation with respect to their membership in $\tilde{A}$. Those elements $x \in X$ that in the classical sense do not belong to $\tilde{A}$ get the membership value $\mu_{\tilde{A}}(x) = 0$; elements $x \in X$ that completely belong to $\tilde{A}$ get the membership value $\mu_{\tilde{A}}(x) = 1$.

There are many ways of extending the set inclusion as well as the basic crisp set operations to fuzzy sets. We will comply with the definitions in [Zad65]. Let $\tilde{A}$ and $\tilde{B}$ be fuzzy sets in X. Then

(i) $\neg \tilde{A} = \{(x, \mu_{\neg \tilde{A}}(x)) \mid x \in X, \mu_{\neg \tilde{A}}(x) = 1 - \mu_{\tilde{A}}(x)\}$

(ii) $\tilde{A} \subseteq \tilde{B} \Leftrightarrow \forall\, x \in X : \mu_{\tilde{A}}(x) \leq \mu_{\tilde{B}}(x)$

(iii) $\tilde{A} \cap \tilde{B} = \{(x, \mu_{\tilde{A} \cap \tilde{B}}(x)) \mid x \in X \wedge \mu_{\tilde{A} \cap \tilde{B}}(x) = \min(\mu_{\tilde{A}}(x), \mu_{\tilde{B}}(x))\}$

(iv) $\tilde{A} \cup \tilde{B} = \{(x, \mu_{\tilde{A} \cup \tilde{B}}(x)) \mid x \in X \wedge \mu_{\tilde{A} \cup \tilde{B}}(x) = \max(\mu_{\tilde{A}}(x), \mu_{\tilde{B}}(x))\}$

(v) $\tilde{A} - \tilde{B} = \tilde{A} \cap \neg \tilde{B}$

A [*strict*] α-*cut* or [*strict*] α-*level set* of a fuzzy set $\tilde{A}$ for a specified value α is the crisp set

$$A_\alpha\,[A_\alpha^*] = \{x \in X \mid \mu_{\tilde{A}}(x) \geq [>] \alpha \wedge 0 \leq \alpha \leq [<] 1\}$$

The strict α-cut for $\alpha = 0$ is called *support* of $\tilde{A}$, i.e., $supp(\tilde{A}) = A_0^*$. For a fuzzy set $\tilde{A}$ and $\alpha, \beta \in [0, 1]$ holds

(i) $X = A_0$

(ii) $\alpha < \beta \Rightarrow A_\alpha \supseteq A_\beta$

The set of all levels $\alpha \in [0,1]$ that represent distinct α-cuts of a given fuzzy set $\tilde{A}$ is called the *level set* $\Lambda_{\tilde{A}}$ of $\tilde{A}$:

$$\Lambda_{\tilde{A}} = \{\alpha \in [0,1] \mid \exists\, x \in X : \mu_{\tilde{A}}(x) = \alpha\}$$

Fuzzy (point set) topology [Cha68] is a straightforward extension and generalization of ordinary point set topology and allows one to distinguish specific topological structures of a fuzzy set like its closure or interior.

A *fuzzy topology* on a universe X is a family $\tilde{T}$ of fuzzy sets in X satisfying the following conditions:

(i) $X \in \tilde{T}, \emptyset \in \tilde{T}$
(ii) $\tilde{A} \in \tilde{T}, \tilde{B} \in \tilde{T} \Rightarrow \tilde{A} \cap \tilde{B} \in \tilde{T}$
(iii) $\tilde{S} \subseteq \tilde{T} \Rightarrow \bigcup_{\tilde{A} \in \tilde{S}} \tilde{A} \in \tilde{T}$

The pair $(X, \tilde{T})$ is said to be a *fuzzy topological space*. The elements of $\tilde{T}$ are called *open fuzzy sets*. Note that X and $\emptyset$ are crisp sets and simultaneously special fuzzy sets. X corresponds to the fuzzy set $\tilde{X} = \{(x, \mu_X(x)) \mid x \in X \wedge \mu_X(x) = 1\}$. The empty set $\emptyset$ corresponds to the empty fuzzy set $\tilde{\emptyset} = \{(x, \mu_X(x)) \mid x \in X \wedge \mu_X(x) = 0\}$. We will identify X and $\tilde{X}$ as well as $\emptyset$ and $\tilde{\emptyset}$ and use the crisp notations for these two sets.

The family $\tilde{T}'$ of all *closed fuzzy sets* in a fuzzy topological space $(X, \tilde{T})$ is given by

$$\tilde{T}' = \{\neg \tilde{A} \mid \tilde{A} \in \tilde{T}\}$$

The *closure [interior]* of a fuzzy set $\tilde{A}$ in a fuzzy topological space $(X, \tilde{T})$ is the smallest closed [largest open] fuzzy set containing $\tilde{A}$ [contained in $\tilde{A}$], i.e.,

$$cl_{\tilde{T}}(\tilde{A}) = \bigcap \{\tilde{S} \mid \tilde{S} \in \tilde{T}' \wedge \tilde{A} \subseteq \tilde{S}\}$$
$$[int_{\tilde{T}}(\tilde{A}) = \bigcup \{\tilde{S} \mid \tilde{S} \in \tilde{T} \wedge \tilde{S} \subseteq \tilde{A}\}]$$

4   Fuzzy Points

Due to our assumption of the point set paradigm, an understanding of the nature of a point, or more precisely a *fuzzy point*, is necessary. There are at least two meaningful definitions for a fuzzy point.

The first definition views a fuzzy point as a point in two-dimensional Euclidean space with a membership value greater than 0, since 0 documents the non-existence of a point. A *fuzzy point* $\tilde{p}$ at (a, b) in $\mathbb{R}^2$, written $\tilde{p}(a, b)$, is a fuzzy singleton in $\mathbb{R}^2$ defined by

$$\mu_{\tilde{p}(a,b)}(x, y) = \begin{cases} m & \text{if } (x, y) = (a, b) \\ 0 & \text{otherwise} \end{cases}$$

with $0 < m \leq 1$. Point $\tilde{p}$ is said to have support (a, b) and value m. Let P_f be the set of all fuzzy points. P_f is, of course, a proper superset of P_c, the set of all crisp points in $\mathbb{R}^2$. For $\tilde{p} = p = (a, b) \in P_c$, we obtain $\mu_{\tilde{p}(a,b)}(x, y) = \chi_p(x, y) = 1$, if $(x, y) = (a, b)$, and 0 otherwise.

The second definition uses a membership function that returns the *degree of proximity* of a point to a reference point $\tilde{p}$. That is, we consider that the point (x, y) is "approximately (a, b)" or "about (a, b)" to the degree $\mu_{\tilde{p}(a,b)}(x, y)$. A fuzzy point $\tilde{p}(a, b)$ is then generally defined by

(i) $\mu_{\tilde{p}(a,b)}$ is upper semicontinuous[1]
(ii) $\mu_{\tilde{p}(a,b)}(x, y) = 1$ if and only if $(x, y) = (a, b)$
(iii) $\forall\, 0 \leq \alpha \leq 1 : \tilde{p}_\alpha$ is a convex[2] subset of $\mathbb{R}^2$

The concrete "distance-based" membership function

$$\mu_{\tilde{p}(a,b)}(x, y) = c^{-\lambda\,((x-a)^2 + (y-b)^2)}$$

with $c \in \mathbb{R}^+, c > 1$, and $\lambda > 0$ illustrates this definition. The degree of proximity decreases as (x, y) moves further away from (a, b). It reaches 1 if $(x, y) = (a, b)$.

Unfortunately, this membership function with unbounded support is difficult to represent. Alternatively, we can employ the following, restricted but more practical function which defines a circle around (a, b) with radius $r \in \mathbb{R}^+$:

$$\mu_{\tilde{p}(a,b)}(x, y) = \begin{cases} 1 - \dfrac{\sqrt{(x-a)^2 + (y-b)^2}}{r} & \text{if } (x - a)^2 + (y - b)^2 \leq r^2 \\ 0 & \text{otherwise} \end{cases}$$

Next, we define three *geometric primitives* on fuzzy points which are valid for both definitions of fuzzy points. Let $\tilde{p}(a, b), \tilde{q}(c, d) \in P_f$ with $a, b, c, d \in \mathbb{R}$. Then

(i) $\tilde{p}(a, b) = \tilde{q}(c, d) :\Leftrightarrow a = c \wedge b = d \wedge \mu_{\tilde{p}(a,b)} = \mu_{\tilde{q}(c,d)}$
(ii) $\tilde{p}(a, b) \neq \tilde{q}(c, d) :\Leftrightarrow \neg(\tilde{p}(a, b) = \tilde{q}(c, d))$
(iii) $\tilde{p}(a, b)$ and $\tilde{q}(c, d)$ are *disjoint* $:\Leftrightarrow supp(\tilde{p}(a, b)) \cap supp(\tilde{q}(c, d)) = \emptyset$

In contrast to crisp points, for fuzzy points we also have a predicate for disjointedness. We are now able to define an object of the fuzzy spatial data type *fpoint* as a set of disjoint fuzzy points:

$$fpoint = \{Q \subseteq P_f | \forall\, \tilde{p}, \tilde{q} \in Q : \tilde{p}(a, b) \text{ and } \tilde{q}(c, d) \text{ are } disjoint \wedge Q \text{ is finite}\}$$

5   Fuzzy Lines

In this section we define a data type for *fuzzy lines*. For that, we first introduce a *simple* fuzzy line as a continuous curve with smooth transitions of membership grades between neighboring points of the line. We assume a total order on $\mathbb{R}^2$ which is given by the lexicographic order "$<$" on the coordinates (first x, then y). The membership function of a *simple fuzzy line* $\tilde{l}$ is then defined by

$$\mu_{\tilde{l}} : f_{\tilde{l}} \to [0, 1] \text{ with } f_{\tilde{l}} : [0, 1] \to \mathbb{R}^2 \text{ such that}$$

---

[1] A function $f : X \to \mathbb{R}$ is *upper semicontinuous* $:\Leftrightarrow \forall\, r \in \mathbb{R} : \{x \mid f(x) < r\}$ is open.
[2] A set $X \subseteq \mathbb{R}^2$ is called *convex* $:\Leftrightarrow \forall\, p, q \in \mathbb{R}^2\ \forall\, \lambda \in \mathbb{R}^+$ with $0 < \lambda < 1 : r = \lambda p + (1 - \lambda) q \in X$ (p, q, and r are here regarded as vectors)

(i) $\mu_{\tilde{l}}$ is continuous
(ii) $f_{\tilde{l}}$ is continuous
(iii) $\forall\, a, b \in (0, 1) : a \neq b \Rightarrow f_{\tilde{l}}(a) \neq f_{\tilde{l}}(b)$
(iv) $\forall\, a \in \{0, 1\} \;\forall\, b \in (0, 1) : f_{\tilde{l}}(a) \neq f_{\tilde{l}}(b)$
(v) $f_{\tilde{l}}(0) < f_{\tilde{l}}(1) \vee (f_{\tilde{l}}(0) = f_{\tilde{l}}(1) \wedge \forall a \in (0, 1) : f_{\tilde{l}}(0) < f_{\tilde{l}}(a))$

Function $f_{\tilde{l}}$ on its own models a continuous, simple *crisp* line (a *simple curve*). The points $f_{\tilde{l}}(0)$ and $f_{\tilde{l}}(1)$ are called the *end points* of f. The definition allows loops ($f_{\tilde{l}}(0) = f_{\tilde{l}}(1)$) but prohibits equality of interior points and thus self-intersections (condition (iii)) and equality of an interior with an end point (condition (iv)). The last condition ensures uniqueness of representation, i.e., in a *closed simple line* $f_{\tilde{l}}(0)$ must be the leftmost point.

Let S be the set of fuzzy simple lines, and let $T \subseteq S$. An *S-complex* C over T is a finite set $C = \{\tilde{l}_1, \dots, \tilde{l}_n\} \subseteq T$ such that[3]

(i) $\forall\, 1 \leq i < j \leq n : f_{\tilde{l}_i}((0,1)) \cap f_{\tilde{l}_j}((0,1)) = \emptyset$
(ii) $\forall\, 1 \leq i < j \leq n : \{f_{\tilde{l}_i}(0), f_{\tilde{l}_i}(1)\} \cap f_{\tilde{l}_j}((0,1)) = \emptyset$
(iii) $\forall\, 1 \leq i \leq n \;\exists\, 1 \leq j \leq n, j \neq i : \{f_{\tilde{l}_i}(0), f_{\tilde{l}_i}(1)\} \cap \{f_{\tilde{l}_j}(0), f_{\tilde{l}_j}(1)\} \neq \emptyset$
(iv) For all $1 \leq i, j \leq n$ and for all $a, k \in \{0, 1\}$ let $V_{\tilde{l}_i}^a = \{(j, k) \mid f_{\tilde{l}_i}(a) = f_{\tilde{l}_j}(k)\}$. Then we require: $\forall\, 1 \leq i \leq n \;\forall\, a \in \{0, 1\} : (|V_{\tilde{l}_i}^a| = 1) \vee (|V_{\tilde{l}_i}^a| > 2)$
(v) $\forall\, 1 \leq i \leq n \;\forall\, a \in \{0, 1\} \;\forall\, (j, k) \in V_{\tilde{l}_i}^a : \mu_{\tilde{l}_i}(f_{\tilde{l}_i}(a)) = \mu_{\tilde{l}_j}(f_{\tilde{l}_j}(k))$

Condition (i) requires that the elements of an S-complex do not intersect or overlap within their interior. Moreover, they may not be touched within their interior by an endpoint of another element (condition (ii)). Condition (iii) ensures the property of connectivity of an S-complex; isolated fuzzy simple lines are disallowed. Condition (iv) expresses that each endpoint of an element of C must belong to exactly one or more than two incident elements of C (note that always $(i, a) \in V_{\tilde{l}_i}^a$). This condition supports the requirement of maximal elements and hence achieves minimality of representation. Condition (v) requires that the membership values of more than two elements of C with a common end point must have the same membership value; otherwise we get a contradiction saying that a point of an S-complex has more than one different membership value.

All conditions together define an S-complex over T as a *connected planar fuzzy graph* with a unique representation. The corresponding point set of C is $points(C) = \bigcup_{\tilde{l} \in C} f_{\tilde{l}}([0, 1])$. The set of all S-complexes over T is denoted by $SC(T)$. The disjointedness of any two S-complexes $C_1, C_2 \in SC(T)$ is defined as follows:

$$C_1 \text{ and } C_2 \text{ are } disjoint :\Leftrightarrow points(C_1) \cap points(C_2) = \emptyset$$

A fuzzy spatial data type for fuzzy lines called *fline* can now be defined in two equivalent ways. The "structured view" is based on S-complexes:

$$fline = \{D \subseteq SC(S) \mid \forall\, C_1, C_2 \in D : C_1 \text{ and } C_2 \text{ are } disjoint \wedge D \text{ is finite}\}$$

---

[3] The application of a function f to a set X of values is defined as $f(X) = \{f(x) \mid x \in X\}$.

Let $1_r = \mathbb{R}^2 \times \{1\}$. The "flat view" emphasizing the point set paradigm is:

$$fline = \{\tilde{Q} \subseteq 1_r \mid \exists \, D \subseteq SC(S) : \bigcup_{C \in D} points(C) = supp(\tilde{Q})\}$$

6   Fuzzy Regions

The aim of this section is to develop and formalize the concept of a *fuzzy region*. Section 6.1 informally discusses the intrinsic features of fuzzy regions, classifies them, gives application examples for them, and compares them to classical crisp regions. After this motivation, Section 6.2 provides their formal definition. Finally, Section 6.3 gives examples of possible membership functions for them.

6.1   What Are Fuzzy Regions?

The question what a *crisp region* is has been treated in many publications. A very general definition defines a crisp region as a set of disjoint, connected areal components, called *faces*, possibly with disjoint *holes* [ES97b, GS95, Sch97] in the Euclidean space $\mathbb{R}^2$. This model has the nice property that it is closed under (appropriately defined) geometric union, intersection, and difference operations. It allows crisp regions to contain holes and islands within holes to any finite level.

By analogy with the generalization of crisp sets to fuzzy sets, we strive for a generalization of crisp regions to fuzzy regions on the basis of the point set paradigm and fuzzy concepts. At the same time we would like to transfer the structural definition of crisp regions (i.e., the component view) to fuzzy regions. Thus, the structure of a fuzzy region is supposed to be the same as for a crisp region but with the exception and generalization which amounts to a relaxation and hence greater flexibility of the strict belonging or non-belonging principle of a point in space to a specific region and which enables a partial membership of a point in a region. This is just what the term "fuzzy" means here.

There are at least three possible, related interpretations for a point in a fuzzy region. First, this situation may be interpreted as the *degree of belonging* to which that point is *inside* or *part of* some areal feature. Consider the transition between a mountain and a valley and the problem to decide which points have to be assigned to the valley and which points to the mountain. Obviously, there is no strict boundary between them, and it seems to be more appropriate to model the transition by partial and multiple membership. Second, this situation may indicate the *degree of compatibility* of the individual point with the attribute or concept represented by the fuzzy region. An example are "warm areas" where we must decide for each point whether and to which grade it corresponds to the concept "warm". Third, this situation may be viewed as the *degree of concentration* of some attribute associated with the fuzzy region at the particular point. An example is air pollution where we can assume the highest concentration at power stations, for instance, and lower concentrations with increasing distance from them. All these related interpretations give evidence of *fuzziness*.

When dealing with crisp regions, the user usually does not employ point sets as a method to conceptualize space. The user rather thinks in terms of sharply determined *boundaries* enclosing and grouping areas with *equal* properties or attributes and separating different regions with different properties from each other; he or she has purely *qualitative* concepts in mind. This view changes when fuzzy regions come into play. Besides the qualitative aspect, in particular the *quantitative* aspect becomes important, and boundaries in most cases disappear (between a valley and a mountain there is no strict boundary!). The distribution of attribute values within a region and transitions between different regions may be *smooth* or *continuous*. This feature just characterizes fuzzy regions.

We now give a classification of fuzzy regions from an application point of view. The classification extends from fuzzy regions with highest vagueness and lowest gradation of attribute values to fuzzy regions with lowest vagueness and highest gradation of attribute values. The given application examples are basically valid for each class. How to model areal features as fuzzy regions depends on the application and on the "preciseness" and quality of information.

Core-Boundary Fuzzy Regions. If there is only insufficient knowledge about the grade of indeterminacy of the vague parts of a region, a first approach is to differentiate between its *core*, its *boundary*, and its *exterior* which relate to those parts that definitely belong, *perhaps* belong, and definitely do not belong, respectively, to the region. This extension just corresponds to the approach of *vague regions* where core and boundary are modeled by crisp regions. It can also be simply modeled by a fuzzy region by assigning the membership function value 1 to each point of the core, value 0 to each point of the exterior, and value $\frac{1}{2}$ (halfway between completely true and completely false) to each point of the boundary. It is important to note that a boundary in this sense can be a region and has thus a different and generalized meaning compared to traditional, crisp boundaries[4]. We will denote fuzzy regions based on a three-valued logic as *core-boundary (fuzzy) regions*.

An application example is a lake which has a minimal water level in dry periods (core) and a maximal water level in rainy periods (boundary given as the difference between maximal and minimal water level). Dry periods can entail puddles. Small islands in the lake which are less flooded by water in dry and more (but never completely) flooded in rainy periods can be modeled through holes surrounded by a boundary. If an island like a sandbank can be flooded completely, it belongs to the boundary part.

Finite-Valued Fuzzy Regions. The next step lifts the restriction of having only one degree of fuzziness. The introduction of different degrees leads from fuzzy regions based on a three-valued logic to fuzzy regions based on a *finite-valued* and thus *multivalued logic*. This enables us to describe more precisely the

---

[4] Nevertheless, core, boundary, and exterior are separated from each other by ordinary, strict "boundaries" as we know them from ordinary point set topology.

degree of membership of a point in a fuzzy region. The membership function value $\frac{3}{4}$ ($\frac{1}{4}$) could express that it is mostly true (false) and only a little false (true) that a point is an element of a specific fuzzy region. We will call this kind of fuzzy regions *finite-valued (fuzzy) regions*. If $n \in \mathbb{N}$ is the number of possible "truth values", an n-valued membership function turns out to be quite useful for representing a wide range of belonging of a point to a fuzzy region.

An application example are regions of different possibilities for virus infections. Regions could be categorized by n different risk levels extending from areas with extreme risk of infection over areas with average risk of infection to safe areas.

The two classes of fuzzy regions described so far have predominantly a *qualitative* character. This means, that the numbers involved in membership functions of a fuzzy region only play a symbolic role and that their size is of lower importance. Essentially, a total and bijective mapping is defined between n possible categories expressing different degrees of fuzziness and n discrete values out of the range $[0, 1]$. Although the selection of the n discrete values is arbitrary (they only must be disjoint from each other, and there is no order needed between them), they are usually chosen in a way that agrees with our intuition.

Interval-Based Fuzzy Regions. The following two classes emphasize a more *quantitative* character of fuzzy regions. Consider an ordered set of n arbitrary but disjoint values of the interval $[0, 1]$ and the assignment of exactly one of these values, let us say, v, to all points of a specific connected component c (a face) of a fuzzy region. We can then interpret such a value v for all points of c as their guaranteed minimal degree of belonging to c. Hence, v represents a lower bound. Since the set of values is ordered, each value v (except for the highest value) has a successor w with respect to the defined order, i.e., $v < w$. This implies that no point of c can have a value greater than w, since otherwise these points would have to be labeled with the value w. This justifies to implicitly map all points of c to the label $[v, w]$, i.e., to a closed interval. The meaning is that the degree of membership of each point of c is somewhere between v and w (we do not have more information). We denote this kind of fuzzy regions as *interval-based (fuzzy) regions*. Each pair of the $n - 1$ possible intervals is either disjoint or adjacent with common bounds. All intervals together form a finite covering of the unit interval $[0, 1]$.

An application example is a map about the population density of a country. According to a predefined interval classification, the country is subdivided into regions showing the minimal guaranteed population density per km^2 for each region. The density values of different regions can be rather different. Another example are weather maps on television which usually show single reference temperatures as sample data spread over the map and representing temperature zones. Here we assume that a direct path from a lower to a higher reference temperature is accompanied by smoothly increasing temperatures. Transitions between different regions are here smooth.

Smooth Fuzzy Regions. A last and very important class of fuzzy regions, which has so far not been treated in the literature, takes advantage of available knowledge about the distribution of attribute values within a fuzzy region. This knowledge can be gained by an expert through appropriate membership functions. We require that the distribution of attribute values within a fuzzy region is *smooth* (with a finite number of exceptions). This can be achieved by so-called *predominantly continuous* membership functions. We call this kind of fuzzy regions *predominantly smooth (fuzzy) regions*. As a special case we obtain *(totally) smooth (fuzzy) regions* with no continuity gaps.

There are a lot of spatial phenomena showing a smooth behavior. Application examples are air pollution (Figure 1), temperature zones, magnetic fields, storm intensity, and sun insolation. Predominantly smooth regions are the most general class of fuzzy regions and comprise all other aforementioned classes which are obviously (predominantly) continuous. This especially means that combinations of different classes are possible without any problems.

Fig. 1. This figure demonstrates a possible visualization of a fuzzy region which could model the expansion of air pollution caused by a power station. The left image shows a radial expansion where the degree of pollution concentrates in the center (darker locations) and decreases with increasing distance from the power station (brighter locations). The right image has the same theme but this time we imagine that the power station is surrounded by high mountains to the north, the south, and the west. Hence, the pollution cannot escape in these directions and finds its way out of the valley in eastern direction. In both cases we can recognize the smooth transitions to the exterior.

6.2   Formal Definition of Fuzzy Regions

Since our objective is to model two-dimensional fuzzy areal objects for spatial applications, we consider a fuzzy topology $\tilde{T}$ on the Euclidean space (plane) $\mathbb{R}^2$. In this spatial context we denote the elements of $\tilde{T}$ as *fuzzy point sets*. The membership function for a fuzzy point set $\tilde{A}$ in the plane is then described by $\mu_{\tilde{A}} : \mathbb{R}^2 \to [0, 1]$.

From an application point of view, there are two observations that prevent a definition of a fuzzy region simply as a fuzzy point set. We will discuss them now in more detail and at the same time elaborate properties of fuzzy regions.

Avoiding Geometric Anomalies: Regularization. The first observation refers to a necessary *regularization* of fuzzy point sets. The first reason for this measure is that fuzzy (as well as crisp) regions that actually appear in spatial applications in most cases cannot be just modeled as arbitrary point sets but have to be represented as point sets that do not have "geometric anomalies" and that are in a certain sense *regular*. Geometric anomalies relate to isolated or dangling line or point features and missing lines and points in the form of cuts and punctures. Spatial phenomena with such degeneracies never appear as entities in reality. The second reason is that, from a data type point of view, we are interested in fuzzy spatial data types that satisfy closure properties for (appropriately defined) geometric union, intersection, and difference.

We are, of course, confronted with the same problem in the crisp case where the problem can be avoided by the concept of *regularity* [ES97b, Sch97, Til80]. It turns out to be useful to appropriately transfer this concept to the fuzzy case. Let $\tilde{A}$ be a fuzzy set of a fuzzy topological space $(\mathbb{R}^2, \tilde{T})$. Then

$\tilde{A}$ is called a *regular open fuzzy set* if $\tilde{A} = int_{\tilde{T}}(cl_{\tilde{T}}(\tilde{A}))$

Whereas crisp regions are usually modeled as *regular closed crisp sets*, we will use *regular open fuzzy sets* due to their vagueness and their usual lack of boundaries. Regular open fuzzy sets avoid the aforementioned geometric anomalies, too. Since application examples show that fuzzy regions can also be partially bounded, we admit *partial boundaries* with a crisp or fuzzy character. For that purpose we define the following fuzzy set:

$$frontier_{\tilde{T}}(\tilde{A}) := \{((x,y), \mu_{\tilde{A}}(x,y)) \mid (x,y) \in supp(\tilde{A}) - supp(int_{\tilde{T}}(\tilde{A}))\}$$

A fuzzy set $\tilde{A}$ is now called a *spatially regular fuzzy set* iff

(i) $int_{\tilde{T}}(\tilde{A})$ is a regular open fuzzy set
(ii) $frontier_{\tilde{T}}(\tilde{A}) \subseteq frontier_{\tilde{T}}(cl_{\tilde{T}}(int_{\tilde{T}}(\tilde{A})))$
(iii) $frontier_{\tilde{T}}(\tilde{A})$ is a partition of n connected boundary parts (fuzzy sets)

We can conclude that $frontier_{\tilde{T}}(\tilde{A}) = \emptyset$ if $\tilde{A}$ is regular open. We will base our definition of fuzzy regions on spatially regular fuzzy sets and define a *regularization* function reg_f which associates the interior of a fuzzy set $\tilde{A}$ with its corresponding regular open fuzzy set and which restricts the partial boundary of $\tilde{A}$ (if it exists at all) to a part of the boundary of the corresponding regular closed fuzzy set of $\tilde{A}$:

$$reg_f(\tilde{A}) := int_{\tilde{T}}(cl_{\tilde{T}}(\tilde{A})) \cup (frontier_{\tilde{T}}(\tilde{A}) \cap frontier_{\tilde{T}}(cl_{\tilde{T}}(int_{\tilde{T}}(\tilde{A}))))$$

The different components of the regularization process work as follows: the *interior* operator $int_{\tilde{T}}$ eliminates dangling point and line features since their interior is empty. The *closure* operator $cl_{\tilde{T}}$ removes cuts and punctures by appropriately adding points. Furthermore, the *closure* operator introduces a fuzzy boundary (similar to a crisp boundary in the ordinary point-set topological sense) separating the points of a closed set from its exterior. The operator *frontier*$_{\tilde{T}}$ supports the restriction of the boundary.

The following statements about set operations on regular open fuzzy sets are given informally and without proof. The intersection of two regular open fuzzy sets is regular open. The union, difference, and complement of two regular open fuzzy sets are not necessarily regular open since they can produce anomalies. Correspondingly, this also holds for spatially regular fuzzy sets. Hence, we introduce *regularized set operations* on spatially regular fuzzy sets that preserve regularity. Let $\tilde{A}$, $\tilde{B}$ be spatially regular fuzzy sets of a fuzzy topological space $(\mathbb{R}^2, \tilde{T})$, and let $a \dot{-} b = a - b$ for $a \geq b$ and $a \dot{-} b = 0$ otherwise $(a, b \in \mathbb{R}_0^+)$. Then

(i) $\tilde{A} \cup_r \tilde{B} := reg_f(\tilde{A} \cup \tilde{B})$

(ii) $\tilde{A} \cap_r \tilde{B} := reg_f(\tilde{A} \cap \tilde{B})$

(iii) $\tilde{A} -_r \tilde{B} := reg_f(\{((x,y), \mu_{\tilde{A}-_r\tilde{B}}(x,y) \mid (x,y) \in \tilde{A} \ \wedge$
$$\mu_{\tilde{A}-_r\tilde{B}}(x,y) = \mu_{\tilde{A}}(x,y) \dot{-} \mu_{\tilde{B}}(x,y)\})$$

(iv) $\neg_r \tilde{A} := reg_f(\neg \tilde{A})$

Note that we have changed the meaning of difference (i.e., $\tilde{A} -_r \tilde{B} \neq \tilde{A} \cap_r \neg\tilde{B}$) since the right side of the inequality does not seem to make great sense in the spatial context. Regular open fuzzy sets, spatially regular fuzzy sets, and regularized set operations express a natural formalization of the desired dimension-preserving property of set operations. In the crisp case this is taken for granted but mostly never fulfilled by spatial type systems, geometric algorithms, spatial database systems, and GIS.

Whereas the subspace *RCCS* of regular *closed* crisp sets together with the crisp regular set operations "$\cup$" and "$\cap$" and the set-theoretic order relation "$\subseteq$" forms a Boolean lattice [ES97b], this is not the case for *SRFS* denoting the subspace of spatially regular *fuzzy* sets. Here we obtain the (unproven but obvious) statement that *SRFS* together with the regularized set operations "$\cup_r$" and "$\cap_r$" and the fuzzy set-theoretic order relation "$\subseteq$" is a pseudo-complemented distributive lattice.

This implies that (i) (*SRFS*, $\subseteq$) is a partially ordered set (reflexivity, antisymmetry, transitivity), (ii) every pair $\tilde{A}$, $\tilde{B}$ of elements of *SRFS* has a least upper bound $\tilde{A} \cup_r \tilde{B}$ and a greatest lower bound $\tilde{A} \cap_r \tilde{B}$, (iii) (*SRFS*, $\subseteq$) has a maximal element $1_r := \{((x,y), \mu(x,y)) \mid (x,y) \in \mathbb{R}^2 \wedge \mu(x,y) = 1\}$ (identity of "$\cap_r$") and a minimal element $0_r := \{((x,y), \mu(x,y)) \mid (x,y) \in \mathbb{R}^2 \wedge \mu(x,y) = 0\}$ (identity of "$\cup_r$"), and (iv) algebraic laws like idempotence, commutativity, associativity, absorption, and distributivity hold for "$\cup_r$" and "$\cap_r$".

(*SRFS*, $\subseteq$) is not a complementary lattice. Although the algebraic laws of involution and dualization hold, this is not true for the laws of complementarity. If we take the standard fuzzy set operations presented in Section 3 as a basis, the law of excluded middle $\tilde{A} \cup_r \neg\tilde{A} = 1_r$ and the law of contradiction $\tilde{A} \cap_r \neg\tilde{A} = 0_r$

do *not* hold in general. This fact explains the term "pseudo-complemented" from above and is no weakness of the model but only an indication of fuzziness.

Modeling Smooth Attribute Changes: Predominantly Continuous Membership Functions. The second observation is that according to the application cases shown in Section 6.1 the mapping $\mu_{\tilde{A}}$ itself may not be arbitrary but must take into account the intrinsic smoothness of fuzzy regions. This property can be modeled by the well known mathematical concept of *continuity* and results in special continuous membership functions for fuzzy regions. We say that a function f contains a *continuity gap* at a point x_0 of its domain if f is semicontinuous but not continuous at x_0. Function f is called *predominantly continuous* if f is continuous and has at most a *finite* number of continuity gaps.

Defining Fuzzy Regions. The type *fregion* for fuzzy regions can now be defined in the following way:

$$fregion = \{\tilde{R} \in SRFS \mid \mu_{\tilde{R}} \text{ is predominantly continuous}\}$$

6.3   Examples of Membership Functions for Fuzzy Regions

In this section we give some simple examples of membership functions which fulfil the properties required in Section 6.2. The determination of suitable membership functions is the difficulty in using the fuzzy set approach. Frequently, expert and empirical knowledge is necessary and used to design appropriate functions. We start with an example for a smooth fuzzy region. By taking a crisp region A with boundary B_A as a reference object, we can construct a fuzzy region on the basis of the following distance-based membership function:

$$\mu_{\tilde{A}} = \begin{cases} 1 & \text{if } (x,y) \in A \\ a^{-\lambda\,d((x,y),B_A)} & \text{if } (x,y) \notin A \end{cases}$$

where $a \in \mathbb{R}^+$ and $a > 1$, $\lambda \in \mathbb{R}^+$ is a constant, and $d((x,y), B_A)$ computes the distance between point (x,y) and boundary B_A in the following way:

$$d((x,y), B_A) = \min\{dist((x,y), (x',y')) \mid (x',y') \in B_A\}$$

where $dist(p,q)$ is the usual Euclidean distance between two points $p, q \in \mathbb{R}^2$. Unfortunately, this membership function leads to an unbounded spatially regular fuzzy set (regular open fuzzy set) which is impractical for implementation. We can also give a similar definition of a membership function with bounded support:

$$\mu_{\tilde{A}} = \begin{cases} 1 & \text{if } (x,y) \in A \\ a^{1-\frac{1}{\lambda}d((x,y),B_A)} & \text{if } (x,y) \notin A,\ d((x,y),B_A) \leq \lambda \\ 0 & \text{otherwise} \end{cases}$$

In the same way as the distance from a point outside of A to B_A increases to λ, the degree of membership of this point to $\tilde{A}$ decreases to zero.

[Use96] also presents membership functions for smooth fuzzy regions. The applications considered are air pollution defined as a fuzzy region with membership values based on the distance from a city center and a hill with elevation as the controlling value for the membership function. [LAB96] models the transition of two smooth regions for soil units with symmetric membership functions.

A method to design a membership function for a finite-valued region with n possible membership values (truth values) is to code the n values by rational numbers in the unit interval $[0, 1]$. For that purpose, the unit interval is evenly divided into $n - 1$ subintervals and takes their endpoints as membership values. We obtain the set $T_n = \{\frac{i}{n-1} \mid n \in \mathbb{N}, 0 \leq i \leq n - 1\}$ of truth values. Assuming that we intend to model air pollution caused by a power station located at point $p \in \mathbb{R}^2$, we can define the following (simplified) membership function for $n = 5$ degrees of truth representing, for instance, areas of extreme, high, average, low, and no pollution $(a, b, c, d \in \mathbb{R}^+)$:

$$\mu_{\tilde{A}}(x, y) = \begin{cases} 1 & \text{if } dist(p, (x, y)) \leq a \\ \frac{3}{4} & \text{if } a < dist(p, (x, y)) \leq b \\ \frac{1}{2} & \text{if } b < dist(p, (x, y)) \leq c \\ \frac{1}{4} & \text{if } c < dist(p, (x, y)) \leq d \\ 0 & \text{if } d < dist(p, (x, y)) \end{cases}$$

7 Structured Views of Fuzzy Regions

The formal definition of a fuzzy region given in Section 6.2 is conceptually somehow "structureless" in the sense that only "flat" point sets are considered and no structural information is revealed. In the following four subsections some "semantically richer" characterizations of fuzzy regions are presented which enable a better understanding of fuzzy regions. On the one hand they subdivide fuzzy regions into fuzzy components and on the other hand they describe them as collections of crisp regions. Moreover, they give hints for a possible implementation.

7.1 Fuzzy Regions as Multi-component Objects

The first structured view considers a fuzzy region as a set of *fuzzy components*. For a definition we need a notion of *connectedness* for fuzzy regions. A *separation* of a fuzzy region $\tilde{Y}$ is a pair $\tilde{A}, \tilde{B}$ of fuzzy subregions satisfying the following four conditions:

(i) $\tilde{A} \neq \emptyset, \tilde{B} \neq \emptyset$
(ii) $\tilde{Y} = \tilde{A} \cup_r \tilde{B}$
(iii) $\tilde{A} \cap int_{\tilde{T}}(\tilde{B}) = \emptyset \wedge int_{\tilde{T}}(\tilde{A}) \cap \tilde{B} = \emptyset$
(iv) $|\tilde{A} \cap_r \tilde{B}|$ is finite

If a separation of $\tilde{Y}$ into $\tilde{A}$ and $\tilde{B}$ exists, then $\tilde{Y}$ is said to be *separated*, and we call $\tilde{A}$ and $\tilde{B}$ to be *disjoint*. Otherwise $\tilde{Y}$ is said to be *connected*. Note that condition (iii) of the definition uses the usual fuzzy intersection operation and not the one on spatially regular fuzzy sets since the latter requires two fuzzy regions as operands. The property of disjointedness (condition (iv)) requires that the two fuzzy subregions $\tilde{A}$ and $\tilde{B}$ may at most share a finite number of boundary points; this makes sense since otherwise they could be simply merged into one fuzzy subregion. We now continue this separation process and decompose a fuzzy region $\tilde{Y}$ into its maximal set of pairwise disjoint fuzzy components $\tilde{Y} = \{\tilde{A}_1, \ldots, \tilde{A}_n\}$ (in the spatial context this decomposition is always finite) so that we obtain with $I = \{1, \ldots, n\}$:

(i) $\forall\, i \in I : \tilde{A}_i \neq \emptyset$

(ii) $\tilde{Y} = \bigcup_{r\, i \in I} \tilde{A}_i$

(iii) $\forall\, i, j \in I, i \neq j : \tilde{A}_i \cap int_{\tilde{T}}(\tilde{A}_j) = \emptyset \wedge int_{\tilde{T}}(\tilde{A}_i) \cap \tilde{A}_j = \emptyset$

(iv) $\forall\, i, j \in I, i \neq j : |\tilde{A}_i \cap_r \tilde{A}_j|$ is finite

(v) $\forall\, i \in I : (\tilde{A}_i$ is connected $\wedge\, \not\exists\, \tilde{B} \supset \tilde{A}_i : \tilde{B}$ is connected$)$

We call each fuzzy component $\tilde{A}_i$ a *fuzzy face*. Hence, we obtain:

A *fuzzy region* is a set of pairwise disjoint *fuzzy faces*.

A question arises whether also *fuzzy holes* can be identified from the point set view of a fuzzy region. This question has to be negated. Let us briefly consider the crisp case. If A is a crisp region, its faces can have holes which belong to the complement (exterior) of A, i.e., to $\mathbb{R}^2 - A$, and are "enclosed" by A. Unfortunately, ordinary point set topology offers no method to extract holes from a (regular closed) point set as separate components; they are simply part of the complement. Note that this does not mean that regions with holes cannot be modeled. Some research work in [ECF94, Sch97, WB93], for example, shows that this is possible by selecting a constructive approach. Roughly speaking, the idea is to assume that the holes of A are already given as regions and to subtract these holes from a "generalized region A*" being isomorphic to a closed disc and being the union of A and the holes. But since this a pure set operation, afterwards A "forgets" how it was produced and cannot reconstruct its past. Similarly to the crisp case, holes cannot be identified from a (spatially regular) fuzzy point set, since fuzzy topology also offers no concept of holes.

Moreover, we are here faced with the problem of the nature of a fuzzy hole. By analogy with the crisp case, we could say that the fuzzy holes of a fuzzy region $\tilde{A}$ exclusively contain all points that are enclosed by any fuzzy face of $\tilde{A}$ and that have membership grade 0 in $\tilde{A}$. But then, a fuzzy hole is crisp and a subset of the set

$$H = \{((x, y), 1) \mid (x, y) \in supp(\neg\tilde{A})\}$$

This model of a fuzzy hole is unsatisfactory in the sense that it only deals with those points enclosed by $\tilde{A}$ that definitely do not belong to $\tilde{A}$. It does not take into account the complement of those points of $\tilde{A}$ belonging only partially to $\tilde{A}$, i.e., the model does not consider the set

$$\underline{\tilde{A}} = \{((x,y),m) \mid (x,y) \in supp(\tilde{A}) \wedge m = 1 - \mu_{\tilde{A}}(x,y)\}$$

called the *anti-fuzzy region* of $\tilde{A}$.

One could argue that the points of $\underline{\tilde{A}}$ also belong to the fuzzy holes. And indeed, we will take this view. The consequence is that for a fuzzy face there exists exactly one fuzzy hole.

7.2 Fuzzy Regions as Three-Part Crisp Regions

The second structured view leads to a simplification of an originally smooth fuzzy region to a core-boundary region and thus to a change from a quantitative to a qualitative perspective. It distinguishes between the kernel, the boundary, and the exterior as the three parts of a fuzzy region. For a fuzzy region $\tilde{A}$, these parts are defined as crisp regions (regular closed sets)[5]:

$$kernel(\tilde{A}) = reg_c(\{(x,y) \in \mathbb{R}^2 \mid \mu_{\tilde{A}}(x,y) = 1\})$$
$$boundary(\tilde{A}) = reg_c(\{(x,y) \in \mathbb{R}^2 \mid 0 < \mu_{\tilde{A}}(x,y) < 1\})$$
$$exterior(\tilde{A}) = reg_c(\{(x,y) \in \mathbb{R}^2 \mid \mu_{\tilde{A}}(x,y) = 0\})$$

The *kernel* identifies the part that definitely belongs to $\tilde{A}$. The *exterior* determines the part that definitely does not belong to $\tilde{A}$. The indeterminate character of $\tilde{A}$ is summarized in the *boundary* of $\tilde{A}$ in a unified and simplified manner. Kernel and boundary can be adjacent with a common border, and kernel and/or boundary can be empty. This view corresponds exactly to the already described concept of *vague regions* with its three-valued logic [ES97b].

All in all, this view presents only a very coarse and restricted description of fuzzy regions since it differentiates only between three parts. The original gradation in the membership values of the points of the boundary gets lost. The benefit of this view lies in the implementation since efficient representation methods and algorithms for crisp regions can be used.

7.3 Fuzzy Regions as Collections of Crisp α-Level Regions

The third structured view attempts to diminish the drawbacks of the three-part view of fuzzy regions and to avoid the great information loss in this representation. It describes a fuzzy region in terms of nested α-level sets. Let $\tilde{A}$ be a fuzzy region. Then we represent a region A_α for an $\alpha \in [0,1]$ as

$$A_\alpha = reg_c(\{(x,y) \in \mathbb{R}^2 \mid \mu_{\tilde{A}}(x,y) \geq \alpha\})$$

---

[5] Correspondingly, for a crisp set A, the regularization function reg_c is defined as $reg_c(A) = cl_T(int_T(A))$ where T is a topology for a universe X and cl_T and int_T are the *closure* and *interior* operators on a topological space (X, T).

We call A_α an α-*level region*. Clearly, A_α is a crisp region whose boundary is defined by all points with membership value α. Note that A_α can have holes. The kernel of $\tilde{A}$, as it has been defined in Section 7.2, is then equal to $A_{1.0}$. A property of the α-level regions of a fuzzy region is that they are nested, i.e., if we select membership values $1 = \alpha_1 > \alpha_2 > \cdots > \alpha_n > \alpha_{n+1} = 0$ for some $n \in \mathbb{N}$, then

$$A_{\alpha_1} \subseteq A_{\alpha_2} \subseteq \cdots \subseteq A_{\alpha_n} \subseteq A_{\alpha_{n+1}}$$

We here describe the finite, discrete case that enables us to model and implement finite-valued and interval-based regions. If $\Lambda_{\tilde{A}}$ is infinite, then there are obviously infinitely many α-level regions which can only be finitely represented within this view if we make a finite selection of values. In the discrete case, if $|\Lambda_{\tilde{A}}| = n + 1$ and we take all these occurring membership values of a fuzzy region, we can replace "$\subseteq$" by "$\subset$" in the inclusion relationships above. This follows from the fact that for any $p \in A_{\alpha_i} - A_{\alpha_{i-1}}$ with $i \in \{2, \ldots, n + 1\}$, $\mu_{\tilde{A}}(p) = \alpha_i$. For the continuous case, we get $\mu_{\tilde{A}}(p) \in [\alpha_i, \alpha_{i-1})$ which leads to interval-based regions. As a result, we obtain:

A *fuzzy region* is a (possibly infinite) set of α-level regions, i.e., $\tilde{A} = \{A_{\alpha_i} \mid 1 \leq i \leq |\Lambda_{\tilde{A}}|\}$ with $\alpha_i > \alpha_{i+1} \Rightarrow A_{\alpha_i} \subseteq A_{\alpha_{i+1}}$ for $1 \leq i \leq |\Lambda_{\tilde{A}}| - 1$.

From the implementation perspective, one of the advantages of using (a finite collection of) α-level sets to describe fuzzy regions is that existing geometric data structures and geometric algorithms known from Computational Geometry [PS85] can be applied.

7.4   Fuzzy Regions as α-Partitions

The fourth structured view is partially motivated by the previous one and describes a fuzzy region as a partition. A partition in the spatial context, called a *spatial partition* [ES97a], is a subdivision of the plane into pairwise disjoint (crisp) regions (called *blocks*) where each block is associated with an attribute and where adjacent blocks are not allowed to be labeled with the same attribute. It differs from the set-theoretic notion of a partition in the sense that it, of course, relates to space and that it incorporates a treatment of common boundary points which at the same time may belong to two adjacent blocks.

From an application point of view, different blocks of a spatial partition are often marked differently, i.e., different *labels* of some set L are assigned to different blocks. Thus, in a certain way, L determines the type of a partition. This leads to spatial partitions of type L that are functions $\pi : \mathbb{R}^2 \to L$. In most cases, partitions are defined only partially, i.e., there are blocks (frequently called the *exterior* of a partition) which have no explicitly assigned labels. To complete π to a total function, we assume a label $\perp_L$ (called *undefined* or *unknown*) for each label type L and require that the exterior of a partition is labeled by $\perp_L$.

Like for crisp regions, we also desire regularity for the blocks of a spatial partition. We require the interiors of blocks to be regular open sets. Since points on the boundary cannot be uniquely assigned to either adjacent block, we cannot simply map them to single L-values. Instead, boundary points are mapped to the set of values given by the labels of all adjacent blocks. This leads to the definition of a *spatial mapping* of type L as a total mapping $\pi : \mathbb{R}^2 \to L \cup 2^L$. The *range* of a spatial mapping π yields the set of labels actually used in π and is denoted by $range(\pi)$. The blocks of a spatial mapping π are point sets that are mapped to the same labels. The block for a single label l (or a set S of labels) is given[6] by $f^{-1}(l)$ ($f^{-1}(S)$). The common label of a block b of π is denoted by $\pi[b]$, i.e., $\pi(b) = \{l\} \Rightarrow \pi[b] = l$. Obviously, the cardinality of block labels identifies different parts of a partition. A *region* of π is any block of π that is mapped to a single element of L, and a *border* of π is given by a block that is mapped to a set of L-values, or formally for a spatial mapping π of type L:

(i) $\rho(\pi) = \pi^{-1}(range(\pi) \cap L))$ (*regions*)
(ii) $\beta(\pi) = \pi^{-1}(range(\pi) \cap 2^L))$ (*borders*)

Now we can finally define a spatial partition by topologically constraining regions to regular open sets and by semantically constraining boundary labels to those of adjacent regions.

A *spatial partition* of type L is a spatial mapping π of type L with

(i) $\forall\, r \in \rho(\pi) : r$ is a regular open set (i.e., $r = int_T(cl_T(r))$)
(ii) $\forall\, b \in \beta(\pi) : \pi[b] = \{\pi[r] \mid r \in \rho(\pi) \wedge b \subseteq cl_T(r)\}$

The set of all spatial partitions of type L is denoted by $[L]$, i.e., $[L] \subseteq \mathbb{R}^2 \to L \cup 2^L$.

Using the representation based on α-level regions defined in the preceding subsection, we are now able to define a fuzzy region as a spatial partition. In our case $L = \Lambda_{\tilde{A}}$, i.e., the labels are formed by all possible membership values α. We have now to determine the different blocks for regions and borders. The regions of $\tilde{A}$ are given[7] by the set $\{int_T(A_{\alpha_i} -_c A_{\alpha_{i-1}}) \mid i \in \{2, \ldots, n+1\}\}$, and the borders of $\tilde{A}$ are represented by the set $\{bound_T(A_{\alpha_i} -_c A_{\alpha_{i-1}}) \mid i \in \{2, \ldots, n+1\}\}$. The object $A_{\alpha_i} -_c A_{\alpha_{i-1}}$ is a region possibly with holes. Each region is uniquely associated with an $\alpha \in \Lambda_{\tilde{A}}$, and each border has all α-labels of adjacent regions.

A *fuzzy region* $\tilde{A}$ is a spatial partition of type $\Lambda_{\tilde{A}}$ (i.e., $\tilde{A} \in [\Lambda_{\tilde{A}}]$), called an α-*partition*.

If $\Lambda_{\tilde{A}}$ is infinite, we get an infinite spatial partition.

---

[6] We use the following definition of function inverse: for $f : X \to Y$ and $\forall\, y \in Y$: $f^{-1}(y) := \{x \in X \mid f(x) = y\}$. Note that f^{-1} applied to a set yields a set of sets.
[7] In the following, the operation "$-_c$" denotes the regular difference operation on regular closed sets. The operation $bound_T$ applied to a regular closed set yields its point-set topological boundary.

8   Conclusions and Future Work

This paper lays the conceptual and formal foundation for the treatment of spatial data blurred by the feature of fuzziness. It is also a contribution to bridge the gap between the entity-oriented and field-oriented view of spatial phenomena since the transitions between both views now become more and more flowing. The paper focuses on the design of a type system for fuzzy spatial data and leads to three fuzzy spatial data types for fuzzy points, fuzzy lines, and fuzzy regions whose structure and semantics is formally defined. The characteristic feature of the design is the modeling of smoothness and continuity which is inherent to the objects themselves and to the transitions between different fuzzy objects. This is achieved by the framework of fuzzy set theory and fuzzy topology which allow partial and multiple membership and hence different membership degrees of an element in sets. Different structured views of fuzzy regions as special collections of crisp regions enable us to obtain a better understanding of their nature and to decrease their complexity.

Future work will have to deal with the formal definition of fuzzy spatial operations and predicates, with the integration of fuzzy spatial data types into query languages, and with implementation aspects leading to sophisticated data structures for the types and efficient algorithms for the operations.

References

[Alt94]   D. Altman. Fuzzy Set Theoretic Approaches for Handling Imprecision in Spatial Analysis. *Int. Journal of Geographical Information Systems*, 8(3):271–289, 1994.

[BF96]   P.A. Burrough and A.U. Frank, editors. *Geographic Objects with Indeterminate Boundaries*, volume 2 of *GISDATA Series*. Taylor & Francis, 1996.

[Bla84]   M. Blakemore. Generalization and Error in Spatial Databases. *Cartographica*, 21, 1984.

[Bur96]   P.A. Burrough. Natural Objects with Indeterminate Boundaries. *In [BF96]*, pages 3–28, 1996.

[CF96]   E. Clementini and P. di Felice. An Algebraic Model for Spatial Objects with Indeterminate Boundaries. *In [BF96]*, pages 153–169, 1996.

[CG96]   A.G. Cohn and N.M. Gotts. The 'Egg-Yolk' Representation of Regions with Indeterminate Boundaries. *In [BF96]*, pages 171–187, 1996.

[Cha68]   C.L. Chang. Fuzzy Topological Spaces. *Journal of Mathematical Analysis and Applications*, 24:182–190, 1968.

[Dut89]   S. Dutta. Qualitative Spatial Reasoning: A Semi-Quantitative Approach Using Fuzzy Logic. *1st Int. Symp. on the Design and Implementation of Large Spatial Databases (SSD'89)*, Springer-Verlag, LNCS 409:345–364, 1989.

[Dut91]   S. Dutta. Topological Constraints: A Representational Framework for Approximate Spatial and Temporal Reasoning. *2nd Int. Symp. on Advances in Spatial Databases (SSD'91)*, Springer-Verlag, LNCS 525:161–180, 1991.

[ECF94]   M.J. Egenhofer, E. Clementini, and P. di Felice. Topological Relations between Regions with Holes. *Int. Journal of Geographical Information Systems*, 8(2):128–142, 1994.

[ES97a]   M. Erwig and M. Schneider. Partition and Conquer. *3rd Int. Conf. on Spatial Information Theory (COSIT'97)*, Springer-Verlag, LNCS 1329:389–408, 1997.

[ES97b]   M. Erwig and M. Schneider. Vague Regions. *5th Int. Symp. on Advances in Spatial Databases (SSD'97)*, Springer-Verlag, LNCS 1262:298–320, 1997.

[Fin93]   J.T. Finn. Use of the Average Mutual Information Index in Evaluating Classification Error and Consistency. *Int. Journal of Geographical Information Systems*, 7(4):349–366, 1993.

[Gaa64]   S. Gaal. *Point Set Topology.* Academic Press, 1964.

[GS95]   R.H. Güting and M. Schneider. Realm-Based Spatial Data Types: The ROSE Algebra. *VLDB Journal*, 4:100–143, 1995.

[KV91]   V.J. Kollias and A. Voliotis. Fuzzy Reasoning in the Development of Geographical Information Systems. *Int. Journal of Geographical Information Systems*, 5(2):209–223, 1991.

[LAB96]   P. Lagacherie, P. Andrieux, and R. Bouzigues. Fuzziness and Uncertainty of Soil Boundaries: From Reality to Coding in GIS. *In [BF96]*, pages 275–286, 1996.

[PS85]   F.P. Preparata and M.I. Shamos. *Computational Geometry.* Springer Verlag, 1985.

[Sch96]   M. Schneider. Modelling Spatial Objects with Undetermined Boundaries Using the Realm/ROSE Approach. *In [BF96]*, pages 141–152, 1996.

[Sch97]   M. Schneider. *Spatial Data Types for Database Systems - Finite Resolution Geometry for Geographic Information Systems*, volume LNCS 1288. Springer-Verlag, Berlin Heidelberg, 1997.

[Shi93]   R. Shibasaki. A Framework for Handling Geometric Data with Positional Uncertainty in a GIS Environment. *GIS: Technology and Applications*, pages 21–35, World Scientific, 1993.

[Til80]   R.B. Tilove. Set Membership Classification: A Unified Approach to Geometric Intersection Problems. *IEEE Transactions on Computers*, C-29:874–883, 1980.

[Use96]   E. L. Usery. A Conceptual Framework and Fuzzy Set Implementation for Geographic Features. *In [BF96]*, pages 71–85, 1996.

[Wan94]   F. Wang. Towards a Natural Language User Interface: An Approach of Fuzzy Query. *Int. Journal of Geographical Information Systems*, 8(2):143–162, 1994.

[WB93]   M.F. Worboys and P. Bofakos. A Canonical Model for a Class of Areal Spatial Objects. *3rd Int. Symp. on Advances in Spatial Databases (SSD'93)*, Springer-Verlag, LNCS 692:36–52, 1993.

[WHS90]   F. Wang, G.B. Hall, and Subaryono. Fuzzy Information Representation and Processing in Conventional GIS Software: Database Design and Application. *Int. Journal of Geographical Information Systems*, 4(3):261–283, 1990.

[Zad65]   L.A. Zadeh. Fuzzy Sets. *Information and Control*, 8:338–353, 1965.

Industrial and Visionary Applications Track

Oracle8i Spatial: Experiences with Extensible Databases

Siva Ravada and Jayant Sharma

Spatial Products Division
Oracle Corporation
One Oracle Drive
Nashua NH-03062
{sravada,jsharma}@us.oracle.com

1 Introduction

Conventional relational databases often do not have the technology required to handle complex data like spatial data. Unlike the traditional applications of databases, spatial applications require that databases understand complex data types like points, lines, and polygons. Typically, operations on these types are complex when compared to the operations on simple types. Hence relational database systems need to be extended in several areas to facilitate the storage and retrieval of spatial data. Several research reports have described the requirements from a database system and prioritized the research needs in this area.

A broad survey of spatial database requirements and an overview of research results is provided in [3,4,6,10]. Research needed to improve the performance of spatial databases in the context of object relational databases was listed in [4]. The primary research needs identified were extensible indexing and optimizer, concurrency control techniques for spatial indexing methods, development of cost models for query processing, and the development of new spatial join algorithms. Many of the system requirements identified in [4] have since been addressed in some commercial systems [1,8,9]. In this context, we describe our experiences in implementing a spatial database on top of Oracle's extensible architecture.

1.1 Requirements of a Spatial Database System

Any database system that attempts to deal with spatial applications has to provide the following features:

- A set of spatial data types to represent the primitive spatial data types (point, line, area), complex spatial data types (polygons with holes, collections) and operations on these data types like intersection, distance, etc.
- The spatial types and operations on top of them should be part of the standard query language that is used to access and manipulate non spatial data in the system. For example, SQL in case of relational database systems should be extended to be able to support spatial types and operations.

R.H. Güting, D. Papadias, F. Lochovsky (Eds.): SSD'99, LNCS 1651, pp. 355–359, 1999.

– The systems should also provide performance enhancements like indexes to process spatial queries (range and join queries), parallel processing, etc. which are available for non spatial data.'

2   Oracle's Spatial

Oracle8i Spatial [7] provides a completely open, standards based architecture for the management of spatial data within a database management system. Users can use the same query language (industry standard SQL) to access the spatial data and all other data in the database. The functionality provided by Oracle8i Spatial is completely integrated within the Oracle database server. Users of spatial data gain access to standard Oracle8i features, such as a flexible client/server architecture, object capabilities, and robust data management utilities, ensuring data integrity, recovery, and security features that are virtually impossible to obtain with other architectures. Oracle8i Spatial enables merging GIS (Geographic Information System) and MIS (Management Information System) data stores and implementing a unified data management architecture for all data across the enterprise. The Oracle8i Spatial provides a scalable, integrated solution for managing structured and spatial data inside the Oracle server.

2.1   Spatial Data Modeling

Oracle Spatial supports three primitive geometric types and geometries composed of collections of these types. The three primitive types are: (i) Point, (ii) Line String, (iii) and N-point polygon where all these primitive types are in 2-Dimensions. A 2-D point is an element composed of two ordinates, X and Y. Line strings are composed of one or more pairs of points that define line segments. Any two points in the line segment can be connected either by a straight line or a circular arc. That means line strings can be composed of straight line segments, arc segments or a mixture of both. Polygons are composed of connected line strings that form a closed ring and the interior of the polygon is implied. A geometry is the representation of a spatial feature, modeled as a set of primitive elements. A geometry can consist of a single element or a homogeneous or heterogeneous collection of primitive types. A layer is a collection of geometries which share the same attribute set. For example, one layer in a GIS might include topographical features, while another describes population density, and a third describes network of roads and bridges in an area.

2.2   Operations on the Spatial Data Types

The binary topological relationships between two spatial objects A and B in the euclidean space is based on how the two objects A and B interact with respect to their interior, boundary and exterior. This is called the 9-intersection model [2] for the topological relationships between two objects. In this model, one can theoretically distinguish between $2^9 = 512$ binary relationships between

A and B. In case of 2-dimensional objects, only eight relations can be realized which provide mutually exclusive and complete coverage for A and B. These relationships are contains, coveredby, covers, disjoint, equal, inside, overlap, touch. Oracle Spatial supports this 9-intersection [2] model for determining the topological relationships between two objects. In addition the system can also support other relationships derived as a combination of the above 8 relations. For example, OVERLAPBDYDISJOINT can be defined as the relation where the objects overlap but the boundaries are disjoint. Oracle Spatial also provides a within distance function where the distances are calculated in the Euclidean space. In addition, this system also provides set theoretical operations like UNION, INTERSECTION, DIFFERENCE and SYMMETRIC-DIFFERENCE. For example, given two spatial objects A and B, one can compute and return a new object C which is the UNION of A and B.

2.3   SQL Support for Spatial Data

Query language is the principal interface to the data stored in a relational database system. A popular commercial language used for accessing data in a RDBMS is SQL. Traditional SQL has been extended recently to be able to support access for new data types. In case of Oracle8i Spatial, SQL is extended in two ways: SQL can be used to define and create objects of spatial types. SQL can also be used to insert, delete, update spatial types in addition to being able to query the spatial data with the help of spatial functions. For example to find out all the parks in city which overlap the rivers in the city can be found using the SQL query:

SELECT A.feature FROM parks A, rivers B
WHERE sdo_geom.relate(A.geometry, B.geometry, 'OVERLAP') = TRUE;

2.4   Spatial Indexing

The introduction of spatial indexing capabilities into the Oracle database engine through Oracle Spatial is a key feature. A spatial index acts much as any other index as a mechanism to limit searches within tables (or data spaces) based on spatial criteria. An index is required to be able to efficiently process queries like find objects within a data space that overlap a query area (usually defined by a query polygon) and find pairs of objects from within two data spaces that spatially interact with one another (spatial join).

A spatial index in spatial cartridge is a logical index. The entries in the index are dependent on the location of the geometries in a coordinate space, but the index values are in a different domain. Index entries take on values from a linearly ordered integer domain while coordinates for a geometries may be pairs of integer, floating, or double-precision numbers. Spatial cartridge uses a linear quadtree based indexing scheme, also known as z-ordering which maps geometric objects to a set of numbered tiles. Point data can be very well indexed by a recursive decomposition of space. Spatial object with extent, such as area or line features create a problem for this sort of index, because they are highly likely to cross index cell partition boundaries. Alternative indexing mechanism, such as R-trees, have been proposed based on overlapping index cells (a non-hierarchical decomposition). Oracle Spatial chooses to take another approach to the problem. Each item is allowed multiple entries in the index. This allows one to index features with extent by covering them with the decomposition tiles from a hierarchical decomposition.

Extensible Indexing in Oracle With Oracle's extensible indexing framework, applications can defines the structure and access methods for the application specific data. (This is called the domain index in Oracle.) The application can store the index data either inside the Oracle database (e.g. in the form tables) or outside the Oracle database (in the form of files). And the application can define routines that manage and manipulate the index to evaluate SQL queries. In effect, the application controls the structure and semantic content of the domain index. The database system interacts with the application to build, maintain, and employ the domain index. The main advantage of this extensible indexing framework is that the index is always in sync with the data. That is once the index is build, all the updates on the base table will automatically result in updates in the index data. Thus the users are not required to worry about the data integrity and correctness issues. Once the domain index is built, it is treated like a regular B-tree index. The database server knows the existence of this domain index and thus manages all the index related work using user defined functions. The extensible indexing framework also provides hooks into the optimizer to let the domain index creator educate the optimizer about the cost functions and selectivity functions associated with the domain index. The optimizer can then generate execution plans that make educated choices regarding domain indexes. Oracle Spatial built an indexing mechanism using this extensible indexing framework which is completely integrated with the database system. This also provides full concurrency control that is available to non spatial data and b-tree indexes in the database.

2.5   Query Processing

Queries and data manipulation statements can involve application-specific operators, like the Overlaps operator in the spatial domain. Oracle's extensible framework lets applications/users define operators and tie the operators to a domain index. This lets the optimizer choose a domain index in evaluating a user defined operator. Oracle Spatial defined operators which are very common for many of the spatial applications. The spatial queries are evaluated using the popular two-step method: a filter step and a refinement step. A spatial index is used during the filter step and the actual geometries are used in the refinement step. This two-step process is used in both the window-query case and the spatial join case.

For example, Oracle Spatial provides an SDO_RELATE operator which can be used compute if two geometries overlap with each other. If we want to find all the roads through a county where the road intersects the county boundary, the query will look like this:

```
SELECT a.id FROM roads A, counties B
WHERE B.name = 'MIDDLESEX'
AND SDO_RELATE(A.geometry, B.geometry, 'MASK=OVERLAP') = 'TRUE';
```

This query shows a simple example where a non spatial attribute and a spatial attribute is used in the same query. Assume that there is only one row in the counties table that satisfies the predicate on counties.name column. Then optimizer in this case will be able to choose a B-tree index on counties.name column and use the spatial index to evaluate the SDO_RELATE operator as a window query on the roads table.

3 Conclusions

In this paper, we described our experiences in implementing a spatial database on top of Oracle's extensible framework. We described how the query language, data modeling and query processing issues are addressed in this system. However, there is still more research required in areas like partitioning techniques to support parallel query processing and bulk loading, and spatial clustering. In addition, there is a growing need for a industry wide benchmark for measuring performance of different database systems supporting spatial databases.

References

1. S. Defazio et al. Integrating IR and RDBMS Using Cooperative Indexing. Proceedings of SIGIR., Seattle, Washington, 1995.
2. M.J. Egenhofer. What's Special About Spatial? Database requirements for Vehicle Navigation in Geographic space. Proceedings of ACM SIGMOD, 1993.
3. R.H. Guting. An Introduction to Spatial Database Systems. VLDB, 3:357:399, 1994.
4. W. Kim, J. Garza, and A. Keskin. Spatial data management in Database Systems. pp1-13, 3rd Intl. Symposium on Advances in Spatial Databases, 1993.
5. OGC. The Open GIS Consortium. http://www.opengis.com.
6. Spatial Databases- Accomplishments and Research Needs. S. Shekhar, S. Chawla, S. Ravada, A. Fetterer, X. Liu and C.-t. Lu. IEEE Transactions on Knowledge and Data Engineering. Vol 1, Number 1, January 1999.
7. Oracle8i Spatial User's Guide and Reference, Release 8i, 1999.
8. M. Stonebreaker and G. Kennitz. POSTGRES Next-Generation Database Management System. Communications of the ACM, 34(10):78-92, 1993.
9. M. Stonebreaker and D. Moore. Object Relational DBMSs: The next great wave. Morgan Kaufmann, 1997.
10. M. F. Worboys. GIS: A computing perspective. Taylor and Francis, 1995.

SICC: An Exchange System for Cadastral Information

(Extended Abstract)

Maurizio Talamo[1,2*], Franco Arcieri[3], Giancarlo Conia[4], and Enrico Nardelli[1,5]

[1] "Coordinamento dei Progetti Intersettoriali" of AIPA - "Autorità per l'Informatica nella Pubblica Amministrazione", Roma, Italia.
[2] Univ. of Roma "La Sapienza", Roma, Italia.
[3] Consultant to AIPA for the SICC project.
[4] SOGEI - "Societa' Generale di Informatica SPA", Roma, Italia.
[5] Univ. of L'Aquila, L'Aquila, Italia.

1 Introduction

We discuss in this paper the major issues we faced during the design, prototyping and implementation of the "Sistema di Interscambio Catasto-Comuni" (SICC), namely the system for italian cadastral data exchange among the main organizations involved in Italy with the treatment of cadastral information, that are Ministry of Finance, Municipalities, Notaries, and Certified Land Surveyors.

The definition and design phases, conducted with the direct involvement of all communities interested to cadastral data, allowed to identify a new and promising approach, namely the *access keys warehouse*, for the realization of large distributed spatial applications that have the absolute requirement of integrating legacy spatial databases.

2 Starting Point and Objectives

Cadaster, a Department of the italian Ministry of Finance, is the public registry of the real estates and land properties. It was established for fiscal purposes. The main key to access cadastral information concerning real estates and land properties is expressed in terms of a unique cadastral identification code, made up by Municipality code, map sheet number, parcel number and flat number.

A Municipality has the objective of planning and managing land use. For this purpose it mainly uses two keys. The cadastral identification code, as above defined, and the property location expressed in terms of street, civic number, and flat number.

---

* M.Talamo is the CEO of the Initiative for the Development of an IT Infrastructure for Inter-Organization Cooperation (namely, "Coordinamento dei Progetti Intersettoriali") of AIPA - "Autorità per l'Informatica nella Pubblica Amministrazione", http://www.aipa.it.

R.H. Güting, D. Papadias, F. Lochovsky (Eds.): SSD'99, LNCS 1651, pp. 360–364, 1999.
© Springer-Verlag Berlin Heidelberg 1999

Cadaster data is managed by the Land Department of the Ministry of Finance through its Land Offices ("Uffici del Territorio") that are present at the level of Provinces (one Office for each Province), which are a subdivision of the main administrative partition of Italy in Regions and an aggregation of Municipalities.

The Ministry of Finance and Municipalities use cadastral data for taxation of properties. According to italian law, taxes on real estates and land properties have to be based on their cadastral value ("rendita catastale"), computed on the basis of a number of objective parameters depending on the property characteristics. General parameters always used are the size and the location of the property. Once values for such parameters are know the cadastral value is automatically computed. The Ministry of Finance is currently evaluating a proposal to introduce also parameters related to the market value of the property.

Furthermore, through its Estate Public Registry Offices ("Conservatorie Immobiliari"), the Ministry of Finance also keeps record of and certify ownership rights and mortgage rights relative to properties.

Municipalities also have their databases about real estates and land properties. These are used, as set by the law, to support and manage actions in the sectors of Toponymy, Local Fiscality, Public Works, and Land Management.

Size of data bases managed by Municipalities is largely variable, considering that about 6.000 of the 8.102 italian municipalities have less then 5.000 citizens, but 8 of the 20 Region chief towns have more than one million inhabitants.

It is clear that there is a continuous exchange flow of cadastral data among Municipalities, Ministry of Finance, Notaries and Certified Land Surveyors. Currently the exchange of cadastral information takes place mostly using papers.

Note also that cadastral databases are not managed at a single central location but at the more than 100 Land Offices of the Ministry of the Finance. This means that there is not a single centralized system, but more than 100 systems, geographically distributed over the whole italian territory.

The cadastral databases contains about about 300.000 maps, approximately one third of it is in an electronic form, and about 1,5 millions of geodetic reference points. These maps ("Catasto Terreni") are the geodetic reference for land parcels and for the planimetry of the building possibly existing within land parcels. The planimetry of various flats inside the building is recorded, together with other descriptive data, in "Catasto Fabbricati" and has not a direct geodetic reference.

Typical inquiries on cadastral databases produce the certificates needed by notaries in all sale acts and buyers pay a fee to obtain them from the Cadaster. Usually the certificate is about location and cadastral value. Additionally, topographic and geodetic information (for land parcels) or planimetry (for real estates) may be asked.

Note that data of geometric type are often required during sale transactions to check if the current situation of the land property/ real estate is coherent with respect to to the situation recorded in the cadastral databases.

Every year in Italy there are about 1,5 millions requests of cadastral certificates and in one of the largest provinces there are about 100.000 yearly requests.

Updates to cadastral data bases are requests to change, for a given real estate/ land parcel, some piece of information of geometric nature or of descriptive nature. They can be related to the creation of a new building (with its flats) in a land parcel or to the variation of an existing building or flat or to the change of some descriptive data (e.g., the owner).

The number of yearly geometric updates to cadastral databases is about 250.000. These updates always triggers further updates, since a change affects one or more of the following aspects: property rights and mortgage rights, fiscal nature and parameter values, destination and allowable usage.

In 1995, the situation was the following :

- cadastral data recorded in Cadaster are not, in general, up to date with cadastral data recorded in Municipalities, and both are not, in general, exactly describing the situation in reality. It has been estimated through a sample of 5% of the overall data, that about 40% of the whole set of data held in cadastral data bases is not up to date with the reality. Please note that this refers to the *overall* set of data, including both data generated by the cadastral offices and data received from the outside. Data generated inside cadastral offices are usually up-to-date, hence the Cadaster is able to perform its main role. The greatest source of incoherence is in data received from the outside, and the consequence of this is a great difficulty to establish a reliable correlation between a situation represented in the cadastral databases and a situation represented in databases of other organizations.
- the way cadastral data change as consequence of actions taken by municipalities, on one side, and by the Ministry of Finance, on the other side, are usually different. This is the main reason for the lack of correlation between data held by Municipalities and data held by Ministry of Finance, notwithstanding the large efforts that are periodically taken to force the correlation. It has been estimated that about 10% of the overall data changes every year.

To deal with coherence maintenance issues in cadastral data exchange, as required by the law [1], AIPA started in 1995 the SICC project [2,3,4,5], with the participation of Ministry of Finance and ANCI, the association of italian municipalities.

The overall objective of the SICC project was to provide technical tools to overcome this situation without affecting the current relations and kind of interaction among interested entities and without changing their inner work organization.

At the same time, law required an organizational change towards a decentralization of activities. But this had anyway to be implemented in such a way to keep at the center the roles of validation and high-level control for cadastral information.

3   Our Solution

It is clearly not possible to proceed in such a situation with an approach based on building new spatial databases, even if the database is distributed and based on a "federation" concept. In fact, this would require a huge amount of resources and would not satisfy the need of keeping the current system working for the everyday needs of citizens.

Also, an approach based on the usual "data warehouse" concept would not be adequate, given the high dynamicity of data and the strong emphasys on the certification purposes of the overall system.

Hence we defined and used in the SICC project the concept of **Access Keys Warehouse**. This approach is conceptually simple but it has shown in the SICC project its high effectiveness.

With this approach a *data repository* containing all data items that can be found in various databases of a distributed systems *is set-up only from a virtual point of view*, while data items remains at their physical locations.

An Access Keys Warehouse is then made up by two main components (for more details see [6,7]:

- An *exchange identifier database*, that is physically built and contains *access keys* and *logical links* for data items in the various databases of the distributed system. The access keys are attribute names, selected from the existing attributes in the underlying databases: the main rule in order to select them is that their concatenation constitutes a unique identifier for the data item. Logical links provide the access paths to the physical (distributed) databases where further data elements about the identified data can be found.
  Note that attributes in the exchange identifiers database act towards legacy systems as access keys: their value is used to query legacy systems. Hence they are **not** physical pointers, and the legacy systems maintain their independence and transparency both with respect to location and to implementation.
  The exchange identifier database is populated using data existing in the various distributed locations.
- A *coherence manager*, that is triggered by updates occuring in the various databases of the distributed system. It activates data and control flows towards the distributed locations so that the various databases can be kept up to data as a consequence of a change happened to a data item in a specific location.

The use of the Access Keys Warehouse concept fully supports the SICC project targets, since it allows to progressively synchronize the various distributed databases. This increase in database correlation then means that data manipulation can be more and more de-centralized towards municipalities while keeping a central high-level control.

The first prototype of the SICC project was implemented in 1995 by AIPA and the italian National Research Council. This prototype proved the feasibility of the technical solution and of the organizational model proposed.

Then SOGEI, the italian company managing the computerized information system of the Ministry of Finance, developed a second prototype, with a better degree of integration among cadastral data and services. This prototype has been put into operations in peripheral offices of Neaples municipality in May 1997.

It was then subsequently validated, through the involvement of about 100 Municipalities ranging from Region chief towns to very small ones and a small sample of notaries and certified land surveyors, for about one year.

Finally, in September 1998 the engineereed system, named SISTER [4] and developed as well by SOGEI, has been put into nation-wide operation.

Access to the system is through a WEB-based interface and the effectiveness of its use is demonstrated by the sharp increase of requests managed by it during the first months. In the month of January 1999 there has been already more than 100.000 cadastral certification queries. Remember that such a query is usually paid by its final user.

The final phase of the whole project is running in 1999 and aims at extending the range of services provided to end users.

It is as well planned to use the *Access Keys Warehouse* approach in the implementation of other large distributed applications for the italian Public Administration, that are currently in the definition phase at AIPA.

References

1. Law n.133, 26 february 1994 and Law Decree n.557, 30 december 1993, art.9, (in italian).
2. M.Talamo, F.Arcieri, G.Conia, Il Sistema di Interscambio Catasto-Comuni (parte I), GEO Media, vol.2, Jul-Aug 1998, Maggioli Editore, (in italian).
3. M.Talamo, F.Arcieri, G.Conia, Il Sistema di Interscambio Catasto-Comuni (parte II), GEO Media, vol.2, Sep-Oct 1998, Maggioli Editore, (in italian).
4. Il Catasto Telematico, Notiziario Fiscale, n.11-12, pp.19–22, Nov-Dec 1998, Ministry of Finance, Roma (in italian).
5. C.Cannafoglia, A.De Luca, F.Molinari, G.F.Novelli, Catasto e Pubblicita' Immobiliare, pp 406-420, Nov. 1998, Maggioli Editore, (in italian).
6. F.Arcieri, E.Cappadozzi, P.Naggar, E.Nardelli, M.Talamo, Access Keys Warehouse: a new approach to the development of cooperative information systems, submitted to the 5th International Conference on Cooperative Information Systems (CoopIS'99), Edinbourgh, September 1999.
7. F.Arcieri, E.Cappadozzi, P.Naggar, E.Nardelli, M.Talamo, Specification and architecture of an Access Keys Warehouse, Technical Report n.7/99, Univ. of L'Aquila, March 1999.

Requirements of Traffic Telematics
to Spatial Databases

Thomas Brinkhoff

Fachhochschule Oldenburg (University of Applied Sciences), Fachbereich V
Ofener Str. 16/19, D-26121 Oldenburg, Germany
tbrinkhoff@acm.org
http://www.fh-oldenburg.de/iapg/personen/brinkhof/

Abstract. Traffic telematics comprises services like traffic information, navigation, and emergency call services. Providers of such services need large spatial databases especially for the representation of detailed street networks. However, the implementation of the services requires a functionality which cannot be offered by a state-of-the-art spatial database system. In this paper, some typical requirements concerning the determination of a vehicle position, the traffic-depending computation of routes, and the management of moving spatial objects are discussed in respect to spatial database systems. It will be shown, that these tasks demand for the capability to handle spatio-temporal data and queries.

1 Introduction

One of the most challenging and encouraging applications of state-of-the-art technology is the field of traffic telematics. In the last few years, several new companies or new divisions have been founded in order to establish services concerning the collection, processing and transmission of data and information in respect to the road traffic[1]. This process was set off (1) by the urgent need of the road users for information and assistance and (2) by the development of new technologies: traffic telematics would not be possible without the availability of mobile (voice and data) communication techniques (e.g. the GSM standard including the short message service SMS) [3] and without satellite location (especially GPS). Typical services around traffic telematics are the following [4]:

- traffic information services via (mobile) telephones or special car terminals,
- on-board and off-board navigation services,
- breakdown and emergency call services,

---

[1] In Germany, some of the companies working in the field of traffic telematics are the Mannesmann Autocom GmbH (http://www.passo.de), the Tegaron Telematics GmbH (http://www.tegaron.de), the DDG Gesellschaft für Verkehrsdaten GmbH, and the automobile club ADAC (http://www.adac.de). Also most of the car manufacturers have founded special divisions or have commissioned subsidiaries in order to establish such services.

R.H. Güting, D. Papadias, F. Lochovsky (Eds.): SSD'99, LNCS 1651, pp. 365-369, 1999.

- information and booking services, and
- fleet services.

In order to offer such services to customers, the service center of a provider needs the access to a large heterogeneous database storing alphanumerical data like customer records as well as spatial data. These spatial data consist of different vector and raster data sets: examples are administrative areas, postal code areas, and detailed topographical raster maps. Most important is the very detailed representation of the street network. Such street data are typically delivered according to the European GDF standard ("geographic data file" [2]) containing information like road names, house numbers, points of interest, and traffic restrictions.

Another typical aspect of the spatial database is the change of data: Some parts are relatively static (e.g. the street network) whereas other parts are permanently changing. For example, the information about traffic jams and the average speed on a road may change in intervals of only few minutes. A special motivation for using database systems in order to store changing or volatile data is the aspired robustness of the services: in a multi-computer and multi-process environment, the services can be synchronized best using a central database.

In the rest of the paper, some typical requirements are presented concerning spatial database systems in order to implement services around traffic telematics. Several of these requirements cannot be satisfied by state-of-the-art database systems.

2 Determining the Location of a Vehicle

In order to implement a breakdown and emergency call service, it is necessary to determine automatically the location of the customer's vehicle and to derive the road name and the house number interval (or the names of crossing roads). But also for other services it is essential to locate a vehicle. Examples are off-board navigation services (see section 3), where the actual position of the vehicle is automatically used as starting point of the route computation, and the *"floating car data" technique (FCD)* [5] where speed changes combined with the location of the vehicle are used for the detection of traffic jams.

The actual position of the vehicle which has been transmitted to the service center is used for locating the vehicle. That sounds easy. However, the position delivered by GPS is imprecise because of jamming or/and because of the interference originated by buildings. In some cases, the accuracy is only several hundreds of meters. In order to solve this problem, not only the actual position of the vehicle, but also additional positions are transmitted to the service center [5]. At this positions, the vehicle performed special maneuvers. In addition, the accuracy of the positions, the distance which the vehicle drove between the positions, and the direction of the car at the positions (incl. accuracy) are transmitted. These data are called *string of pearls*, an example is depicted in figure 1.

To locate a vehicle means to determine of the road segment(s) where the vehicle could be, including a probability of the location. For performing such a computation, it is necessary

- to handle imprecise positions,
- to handle (imprecise) conditions between imprecise positions, and

- to take the underlying network including (time-dependent) traffic restrictions into account.

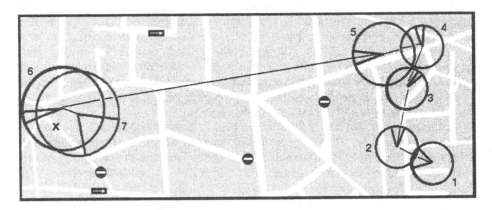

Fig. 1. String of pearls: seven pearls where the oldest pearl has number "1" and the most actual pearl is provided with number "7". The sectors of the pearls represent the direction of the vehicle including the possible error. The order of the pearls is illustrated by the line which connects the pearls. The small cross shows the real position of the vehicle. In the background, the street network is depicted; the symbols represent traffic restrictions.

The computation of the location can be done outside of the database system by a program using standard spatial queries. Because of performance and maintenance issues, it seems to be reasonable to locate the vehicle by a query which can be performed and optimized by a spatial database system. However, state-of-the-art database systems are not able to perform such queries.

3 Traffic-Depending Route Computation

On-board navigation terminals, which allow to compute the shortest route and to navigate the driver to the destination, are state of the art. The map is generally stored in the car on a compact disk. However, these terminals do not consider the actual traffic situation. A route computation which takes the traffic situation into consideration can be performed best in a service center where the (complete) traffic situation is known and where the current street network and the current traffic restrictions are maintained. Such an approach is called *off-board navigation*.

The computation of a route which takes the actual traffic situation into account is time-dependent. In order to compute the weight of a road segment used by the routing algorithm, (1) the time when this road segment is expected to be passed must be determined and (2) the expected speed at this road segment at the expected time must be computed. For performing such a spatio-temporal query, the database system must be able:

- to handle time,
- to perform updates efficiently,

- to compute routes efficiently, and
- to handle different times in one query depending on the route computed in the same query.

Especially, the last requirement is hard to fulfill.

4 Moving Objects

For fleet services, it is necessary to keep track on the positions of the vehicles. Also other (future) traffic telematics services such as toll charging via GPS/GSM, theft protection, and the integration of individual and public transport need the management of large sets moving spatial objects.

The support of motion requires that the spatial database system stores moving objects efficiently and supports spatial queries with time conditions in respect to the actual time, in respect to the past and, especially, in respect to the (near) future. The larger the number of moving objects, the more important performance issues will become.

The motion of vehicles (or other means of transport) is influenced and restricted by different factors:

- *the street network:* the street network channels the traffic. In consequence, no traffic can be observed outside of the network and most of the vehicles use mainly a small set of the network, i.e. they drive on major roads and motorways. This observation is also valid for public transport systems.
- *the traffic*: in the rush hour or in traffic jams, the average speed decreases. For example, this effect is used by the FCD technique in order to detect traffic jams.
- *time schedules:* the motion in public transport systems but also - on special parts of the network (e.g. ferry lines) - of individual vehicles is controlled by time schedules.
- *other conditions*: weather, day time, week day, holiday periods, etc. have influence on the average and the individual behavior of the vehicles.

For storing moving vehicles in a spatial database system, these factors have to be taken into consideration: First, the type of motion is correlated to performance issues: in order to support the storage of moving objects, the database system has to offer adequate data models and efficient access structures. For example, the structure of the network has to be taken into account in order to achieve high performance. Second, the motion influences the query processing process, especially if queries in respect to the (near) future are considered. This affects also the design of query languages. Furthermore, the integration of a knowledge base storing rules about the typical movement of vehicles could be useful in order to answer queries about moving vehicles.

These requirements illustrate that state-of-the-art (spatial) database systems need considerable improvements for an adequate support of moving objects.

5 Conclusions

Several requirements resulting from the field of traffic telematics were presented in this paper. Some of them could be solved by a state-of-the-art database (e.g. the storage of imprecise positions), others are topic of actual research activities (e.g. the CHOROCHRONOS project for spatio-temporal database systems [1]). For the efficient management of moving objects, for example, many questions are not solved [6]. Also the use of a spatio-temporal database system for the computation of traffic-dependent routes (see section 3) seems to be an unsolved problem.

A special problem is the integration of different techniques and solutions into one database system. Due to maintenance and operating reasons, it is an urgent need to use one standardized (spatial) database system to store the spatial data and implement the applications which are used by the different services. The use of several special-purpose systems and algorithms for fulfilling the requirements presented in this paper is extremely expensive (in respect to time and money). This observation will be strengthened by the expected transition of the static street network (actual update rate: quarterly) into a permanently updated network. In this case, an on-line update will be required which makes the maintenance of several databases and special-purpose systems more difficult and more expensive.

References

1. CHOROCHRONOS: A Research Network for Spatiotemporal Database Systems. http://www.dbnet.ece.ntua.gr/~choros/
2. ERTICO: Geographic Data Files (GDF). http://www.ertico.com/links/gdf/gdf.htm
3. Mehl, H.: Mobilfunk-Technologien in der Verkehrstelematik. Informatik-Spektrum 19 (1996) 183-190.
4. Mannesmann Autocom: PASSO Verkehrstelematikdienste. http://www.passo.de/
5. Vieweg, S.: Beitrag der Geo-Information zu Verkehrstelematik-Diensten im Individual-verkehr. Geoinformationssysteme GIS 5/98 (1998) 12-15.
6. Wolfson, O., Xu, B., Chamberlain, S., Jiang, L.: Moving Objects Databases: Issues and Solutions. International Conference on Scientific and Statistical Database Management, Capri, Italy (1998).

Author Index

Springer
and the
environment

At Springer we firmly believe that an international science publisher has a special obligation to the environment, and our corporate policies consistently reflect this conviction.
We also expect our business partners – paper mills, printers, packaging manufacturers, etc. – to commit themselves to using materials and production processes that do not harm the environment. The paper in this book is made from low- or no-chlorine pulp and is acid free, in conformance with international standards for paper permanency.

 Springer

Lecture Notes in Computer Science

For information about Vols. 1–1557
please contact your bookseller or Springer-Verlag